Kongelbeck

Low-Noise
Electronic Design

Low-Noise Electronic Design

C. D. MOTCHENBACHER

Honeywell, Inc.

F. C. FITCHEN

University of Bridgeport

A Wiley-Interscience Publication

JOHN WILEY & SONS, New York · London · Sydney · Toronto

Library of Congress Cataloging in Publication Data

Motchenbacher, C. D. 1931–
Low-noise electronic design.

"A Wiley-Interscience production."
Includes bibliographical references.
1. Electronic circuit design. 2. Electronic noise. 3. Electronic data processing—Electronic circuit design. I. Fitchen, Franklin C., joint author. II. Title.

TK7867.M69 621.3815'3 72–8713
ISBN 0–471–61950–7

Printed in the United States of America

10–9 8 7 6 5 4 3 2 1

TO GRETCHEN

PREFACE

This book is intended for use by practicing engineers and by students of electronics. It can be used for self-study or in an organized classroom situation as a quarter or semester course or short course. A knowledge of electric circuit analysis, the principles of electronic circuits, and mathematics through an introduction to the calculus is assumed.

The approach used in this text is practice or design oriented. The material is not a study of noise, but rather of methods to deal with the everpresent noise in electronic systems.

Electrical noise is a problem in industrial, military, and consumer equipment design. The designer must be cognizant not only of the sources of noise, but also of the methods of noise reduction that are available to him. He will strive toward an optimum design. In this quest he must use all available and applicable tools. One such tool is the digital computer. In Chapter 8, computer-aided design of low-noise systems is considered. Another powerful tool is the laboratory. Chapter 14 discusses noise measurement methods and techniques.

The book is divided into two parts. Part I, Chapters 1 through 6, is primarily concerned with noise mechanisms, noise models, and the analysis of noisy circuits and systems. Part II, starting with Chapter 7, focuses on the design of low-noise circuits and systems. Many design examples are given.

The treatment may be considered to be oriented toward low- and mid-frequency applications. The examples given are applicable to the instrumentation, audio, and control fields. Whereas there are few direct references to high-frequency communications, much of the material presented can be applied to that area of activity.

A summary of the most important points and topics is included at the end of each chapter. It is hoped that these summaries will be useful to the reader who does not have the time to start from the beginning. For the student who is encountering this material for the first time, the summaries serve to accentuate the more important aspects of the presentation. A controlled amount of repetition is included in the text, and many cross-references are made to other sections of the book for prerequisite or additional material. Problems are included at the end of each chapter and answers to the problems are given following Appendix IV.

This book is an outgrowth of our research, development, design, and teaching experiences. Special recognition is given to Honeywell, Incorporated, for cooperation and support.

C. D. MOTCHENBACHER
F. C. FITCHEN

Hopkins, Minnesota
Bridgeport, Connecticut
July 1972

CONTENTS

SYMBOLS

Symbol	*Definition*	*First used or Defined in Equation*
A_i	Current gain	10–1
A_v	Voltage gain	1–5
A_{vo}	Midband voltage gain	1–5
BW	Bandwidth	2–27
C	Capacitance	1–16
C	Correlation coefficient	1–8
C	Coulombs, electric charge	1–12
$C_{b'c}$, $C_{b'e}$	Hybrid-π transistor model capacitances	4–4
C_p	Shunt capacitance	3–10
CMV	Common-mode voltage	Ch. 11
C_π	Base-emitter capacitance ($C_{b'e}$)	12–66
E, E_1, E_2 etc.	rms noise voltages	1–7
$E_A, E_B, \ldots, E_F$	rms noise voltage of resistors $R_A, R_B, \ldots, R_F$	12–3
E_b	Thermal noise of $r_{bb'}$	4–10
E_f	rms noise voltage with $1/f$ *power* spectrum	1–10
E_n	rms equivalent noise voltage generator	2–7
E_{nb}	rms narrowband noise voltage	2–24
E_{ni}	Equivalent input noise	2–1
E_{no}	rms noise at output of network	1–16
E_{ns}	rms noise of source resistance	3–3
E_{nT}	Total equivalent input noise voltage	12–73
E_{n1}, E_{n2}, E_{n3}	Noise voltage generators of stages 1, 2, 3	12–11
E_s	rms noise of source resistance R_s	4–10
E_{sh}	Shot noise voltage of a diode	14–19

Symbol	*Definition*	*First used or Defined in Equation*
E_t	rms thermal noise voltage of a resistance or conductance	1–3a
E_{wb}	rms wide band noise voltage	2–24
F	Noise factor	2–9
f	Frequency	1–10
1/f	(read "one-over-f") noise source whose power level is inversely proportional to frequency	1–9
f_{hfe}	Beta-cutoff frequency$=f_\beta$	4–5
f_h	High frequency 3 dB point	1–9
f_l	Lower frequency 3 dB point	1–9
f_L	Lower noise-corner frequency Hybrid-π transistor model	4–8
F_{opt}	Optimum noise factor	2–14
f_T	Gain-bandwidth product	4–5
G_a	Available power gain	2–17
g_m	Transconductance	4–2
G_o	Peak power gain	1–4
I_B	dc base current	4–14
I_b	Shot noise of base current	4–10
I_{bb}	Burst noise current generator	5–1
I_C	dc collector current	4–14
I_c	Shot noise of collector current	4–10
I_{CBO}	Collector-base leakage current with emitter open	8–3
I_{DC}	Direct current	1–12
I_E	dc emitter current	1–13
I_{ex}	Excess 1/f noise current	3–11
I_f	1/f noise current in Hybrid-π model	4–8
I_n	rms equivalent noise current generator	2–7
I_{no}	rms noise current out of a network	4–10
I_{nT}	Total equivalent input noise current	12–74
I_{n1}, I_{n2}, I_{n3}	Noise current generators of stages 1, 2, 3	12–11
I_0	Saturated value of reverse diode current	1–13
I_{sh}	rms shot noise current	1–12
I_t	rms thermal noise current of a resistance	1–7
k	Boltzmann's constant 1.38×10^{-23} W-sec/°K	1–1
K, K_1, K_2 etc.	Arbitrary constants	1–9
K_t	System transfer voltage gain	2–3
K_{tc}	Transfer voltage gain of cascaded stages	12–16

Symbol	*Definition*	*First used or Defined in Equation*
N_f	Noise power of signal source with 1/f frequency distribution	1–9
N_i	Input noise power	2–10
N_o	Output noise power	2–10
N_o	Available noise power	2–17
N_t	Available noise power	1–1
NEP	Noise equivalent power	3–14
NF	Noise figure	2–11
NI	Noise index in μV/V/Decade	9–2
Q	Quality factor of a tuned circuit	8–9
q	Electronic charge 6.02×10^{-19} coulomb	1–12
R	Resistance, real part of Z	1–2
r_b	Base resistance for 1/f noise	4–9
r_b	Base resistance in T-equivalent transistor model	4–30
$r_{bb'}$, $r_{b'e}$, r_{ce}, $r_{b'c}$	Hybrid-π transistor model resistances	4–3
R_C	Collector resistor	10–4
R_E	Emitter resistor	10–10
R_e	Absolute value of emitter impedance	10–3
r_e	Schockley emitter resistance, emitter resistance in common-base T model	1–15
R_i	Input resistance	10–6
R_L	Load resistance	10–1
R_n	Equivalent noise resistance	2–15
R_o	Optimum source resistance	2–13
R_p	Parallel resistor	3–5
R_s	Resistance of signal source	2–2
r_x	Base resistance ($r_{bb'}$)	12–1
r_0	Output resistance ($\simeq r_{ce}$)	12–1
r_π	Base-emitter resistance ($r_{b'e}$)	12–1
S/N	Signal-to-noise ratio	3–3
S_i	Input signal power	2–10
$S(f)$	Noise spectral density	1–6
S_o	Output signal power	2–10
T	Temperature in degrees Kelvin (°K)	1–1
T	Transformer turns ratio	7–3
T_s	Equivalent noise temperature	2–16
V_{BE}	Potential applied between base and emitter	1–13
V_{CC}	dc collector supply voltage	10–10

Symbol	*Definition*	*First used or Defined in Equation*
V_m	Peak value of ac signal	11–1
V_o	Signal voltage at the output of a network	2–3
V_s	Signal voltage at the source	2–3
V_1, V_2, etc.	Signal voltages	1–7
X_C	Reactance of a capacitance	1–16
Z_i	Input impedance	2–2
Z_s	Source impedance	2–5
α	Transistor current gain, T-equivalent circuit	4–32
β_o	Short-circuit current gain	4–1
Δf	Noise bandwidth	1–4
γ	Exponent in 1/f noise model	4–8
Λ	q/kT	4–2
ω	$2\pi f$	1–16

Low-Noise
Electronic Design

INTRODUCTION

Sensors, detectors, and transducers are basic to the instrumentation and control fields. They are the "fingers" and "eyes" that reach out and measure. They must translate the characteristics of the physical world into electrical signals. We process and measure these signals and interpret them to be the reactions taking place in a chemical plant, the environment of an orbiting satellite, or the odor of an onion. An engineering problem often associated with sensing systems is the level of electrical noise generated in the sensor and in the electronic system.

In recent years, new high-resolution sensors and high-performance systems have been developed. All sensors have a basic or limiting noise level. The system designer must interface the sensor with electronic circuitry that contributes a minimum of additional noise. To raise the signal level an amplifier must be designed to complement the sensor. Achieving optimum system performance is a primary consideration in this book.

The following chapters answer several important questions. When we are given a sensor with specific impedance, signal and noise characteristics, how can we design or select an amplifier for minimum noise contribution, and concurrently maximize the signal-to-noise ratio of the system? If we have a sensor-amplifier system complete with known signal and noise, we must determine the major noise source; is it sensor, amplifier, or pickup noise? Are we maximizing the signal? Is improvement possible?

This book is a study of low-noise electronic design, not a treatise on noise as a physical phenomenon. It is divided into two parts. The first six chapters are concerned with *Noise Mechanisms and Models*; Part II, Chapters 7 through 13, deals with *Design Techniques and Examples*. The final chapter is a treatment of noise measurement methods.

In Part I we are concerned with the sources of noise; these sources include sensors and other devices, amplifiers, and associated circuitry. Noise models are developed for circuit components and for subsystems, and the relations between noise sources, biasing circuits, and operating point selection are examined. Analysis uses the noise models of system components to predict the expected minimum noise performance.

To realize the noise performance predicted from a system analysis many design decisions are required. Part II examines all the areas of possible noise problems, including component noise, noise in power supplies, shielding, biasing circuit noise, feedback, multistage configurations, as well as the associated gain, stability, and frequency response behavior. Design examples are given of noiseless biasing, low-noise power supplies, amplifier pairs, and complete low-noise systems.

SYMBOLS

To designate electric circuit quantities the system of symbols used in this book conforms to standard practice in the semiconductor field.*

1. *Dc values of quantities are indicated by capital letters with capital subscripts* (I_B, V_{CE}). *Direct supply voltages have repeated subscripts* (V_{BB}, V_{CC}).
2. *Rms values of quantities are indicated by capital letters with lowercase subscripts* (I_c, V_{ce}).
3. *Maximum or peak values are designated as rms values but bear an additional final subscript* m (V_{cm}, V_{bem}).
4. *The time-varying components of voltages and currents are designated by lowercase letters with lowercase subscripts* (i_e, v_{ce}).
5. *Instantaneous total values are represented by lowercase letters with capital subscripts* (i_C, v_{CE}).

An example may be helpful. A time-varying waveform.

$$i_b = I_{bm} \sin \omega t$$

when added to a dc wave of average value I_B, is represented by its total value

$$i_B = I_B + i_b$$

So a complete expression is

$$i_B = I_B + I_{bm} \sin \omega t$$

The rms value of the time-varying portion of this composite wave is I_b.

The symbols for rms, peak, and instantaneous quantities (2, 3, and 4) apply only to the time-varying portion of a waveform. No symbols are proposed for the rms and peak values of a wave containing both time-varying and dc components.

PREFIXES

The following prefixes are used to indicate decimal multiples or submultiples of units:

* "IEEE Standard Letter Symbols for Semiconductor Devices," *IEEE Trans. Electron Devices,* ED-11, No. 8, August 1964.

Multiple	Prefix	Symbol
10^{12}	tera	T
10^{9}	giga	G
10^{6}	mega	M
10^{3}	kilo	k
10^{-3}	milli	m
10^{-6}	micro	μ
10^{-9}	nano	n
10^{-12}	pico	p
10^{-15}	fcmto	f

Thus 1.7 GΩ is an impedance equal to 1.7×10^9 Ω, and 6.0 fA is a current of 6.0×10^{-15} A.

GENERAL REFERENCES

Specific references are given at the end of each chapter. Information concerning many of the topics covered in this book is available only in the periodical literature. The topics of electrical noise and noise in communications systems, however, are well treated in available books.

Below, a short list is presented of available books devoted to allied material.

A. R. Bennett, *Electrical Noise*, McGraw-Hill, New York, 1960.

W. R. Davenport, Jr., and W. L. Root, *Random Signals and Noise*, McGraw-Hill, New York, 1958.

J. J. Freeman, *Principles of Noise*, Wiley, New York, 1958.

S. Goldman, *Frequency Analysis, Modulation and Noise*, McGraw-Hill, New York, 1948.

R. A. King, *Electrical Noise*, Chapman and Hall, London, 1966.

W. A. Rheinfelder, *Design of Low-Noise Input Circuits*, Hayden Book, New York, 1964.

A. van der Ziel, *Noise*, Prentice-Hall, Englewood Cliffs, N.J., 1954.

A. van der Ziel, *Fluctuation Phenomena in Semiconductors*, Butterworth, London, 1959.

A. van der Ziel, *Noise: Sources, Characterization, Measurement*, Prentice-Hall, Englewood Cliffs, N.J., 1970.

R. F. Shea, *Amplifier Handbook*, McGraw-Hill, New York, 1966, (Chapter 7, "Amplifier Noise," by E. G. Nielsen).

PART I

Noise Mechanisms and Models

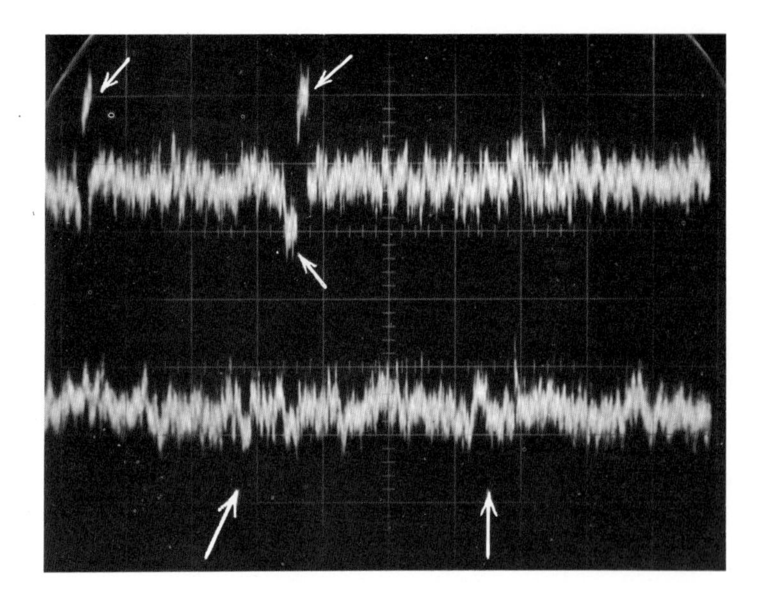

"Popcorn noise," discussed in Chapter 5, is shown in the traces. The top trace is considered to represent a moderate level of this noise. The bottom trace is a low level. Some devices exhibit popcorn noise with five times the amplitude shown in the top trace. Horizontal sensitivity is 2 msec/cm.

Chapter 1

NOISE MECHANISMS

The problems caused by electrical noise are apparent in the output device of an electrical system, but the causes of noise are unique to the low signal level portions of the system. The "snow" that may be observed on a TV receiver display is the result of internally generated noise in the first stages of signal amplification.

This chapter defines the major types of noise present in electrical systems, and discusses methods of representing these sources for the purpose of noise circuit analysis. In addition, concepts such as noise bandwidth and spectral density are introduced.

1-1 NOISE DEFINITION

Noise, in the broadest sense, can be defined as *any unwanted disturbance that obscures or interferes with a desired signal.* Disturbances often come from sources external to the system being studied and may result from electrostatic or electromagnetic coupling between the circuit and the 60Hz power lines, a radio transmitter, or fluorescent lights. Cross talk between adjacent circuits, hum from dc power supplies, or microphonics caused by the mechanical vibration of components are all sources of disturbances. With the exception of pickup from electrical storms and galactic radiation, most of these types of disturbances are "man-made"; they can often be minimized or eliminated by adequate shielding, filtering, or the layout of circuit components.

We use the word "noise" to represent basic random-noise generators or spontaneous fluctuations that result from the physics of the devices and materials that make up the electrical system. Thus the thermal noise apparent in all electrical conductors at temperatures above absolute zero is an example of noise as discussed in this book. This fundamental or true noise cannot be predicted exactly, nor can it be totally eliminated, but it can be manipulated.

Noise is important. The limit of resolution of a sensor is often noise. The dynamic range of a system is determined by noise. The highest signal level that can be processed is limited by the characteristics of the circuit, but the smallest detectable level is set by noise.

In addition to the familiar effects of noise in communications systems, noise is a problem in control and computing systems. For example, the presence of spikes of random noise makes it difficult to design a circuit that triggers (switches) at a specific signal amplitude. When noise of varying amplitude is mixed with the signal, noise peaks can cause a level detector to trigger falsely. To reduce the probability of false triggering, noise reduction is recommended.

Suppose that we have a system that is too noisy, but are uncertain whether the noisiness is caused by disturbances or by fundamental noise. We add shielding. A general rule for frequencies above 1000 Hz or impedance levels over 1000 Ω is to use conductive shielding (aluminum or copper). For low frequencies and lower impedances, we can use magnetic shielding (supermalloy, mu-metal), and twist wire pairs. (Inter-8 weave wire from Perfection Mica Company may help.) We can also put the preamp on a separate battery supply. If these efforts help, we can try more shielding. The work may be moved to another location, or measurements can be made during the quieter evening hours. If these techniques do not reduce the disturbance, look to fundamental noise mechanisms. Fundamental or true noise is the type considered almost exclusively in this book.

1-2 NOISE PROPERTIES

Noise is a totally random signal. It consists of frequency components that are random in both amplitude and phase. Although the long-term rms value can be measured, the exact amplitude at any instant of time cannot be predicted. If the instantaneous amplitude of noise could be predicted, noise would not be a problem.

It is possible to predict the randomness of noise. Much noise has a Gaussian or normal distribution of instantaneous amplitudes with time [1]. The common Gaussian curve is depicted in Fig. 1-1 along with a photograph of electrical noise as obtained from an oscilloscope. The area under the Gaussian curve represents the probability that a particular event will occur. Since probability can take on values from 0 to 1, the total area represents unity. The waveform is centered about the zero voltage level, corresponding to probability of .5 that the instantaneous value of the noise waveform is either

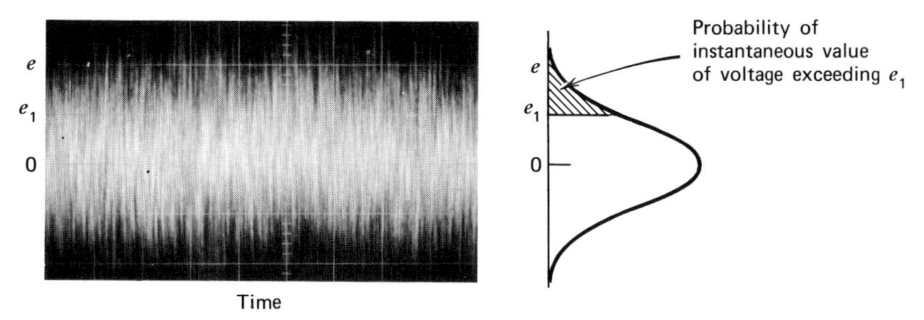

Fig. 1-1. Noise waveform and Gaussian distribution of amplitudes.

above or below that level. If we consider a value such as e_1, the probability of exceeding that level at any instant in time is shown by the cross-hatched area. To a good engineering approximation, common electrical noise lies within plus or minus three times the root-mean-square (rms) value of the noise wave. The peak-to-peak voltage is less than six times the rms for 99.7% of the time.

The preceding paragraph mentioned the rms value of a noise waveform. The rms definition is based on equivalent heating effect. The most common type of electronic voltmeter rectifies the wave to be measured, measures the average value of the rectified wave, and indicates the rms value on a scale calibrated by multiplying the average value by 1.11 to simulate rms. This type of meter correctly indicates the rms value of a sine wave, *but noise is not sinusoidal*, and the reading of a noise waveform will be 11.5% low. Correction can be made by multiplying the reading by 1.13. This is further discussed in Section 14-9.

1-3 THERMAL NOISE

The three main types of noise mechanisms are referred to as *thermal noise*, *low-frequency* $(1/f)$ *noise*, and *shot noise*. Thermal noise is the most often encountered and is considered first.

Thermal noise is caused by the random thermally excited vibration of the charge carriers in a conductor. This carrier motion is similar to the Brownian motion of particles. From studies of Brownian motion, thermal noise was predicted. It was first observed by J. B. Johnson of Bell Telephone Laboratories in 1927, and a theoretical analysis was provided by H. Nyquist in 1928. Because of their work thermal noise is called Johnson noise or Nyquist noise.

In every conductor at a temperature above absolute zero the electrons are in random motion, and this vibration is dependent on temperature. Since each electron carries a charge of 1.59×10^{-19} C, there are many little current surges as electrons randomly move about in the material. Although the average current in the conductor resulting from these movements is zero, instantaneously there is current fluctuation that gives rise to a voltage across the terminals of the conductor.

The *available noise power* (N_t) in a conductor is found to be proportional to the absolute temperature and to the bandwidth of the measuring system. In equation form this is

$$N_t = kT\,\Delta f \tag{1-1}$$

where k = Boltzmann's constant = 1.38×10^{-23} W-sec/°K

T = temperature of the conductor in degrees Kelvin (°K)

Δf = *noise bandwidth* of the measuring system in hertz.

At room temperature (290°K), for a 1-Hz bandwidth, evaluation of Eq. 1-1 gives $N_t = 4 \times 10^{-21}$ W. This is -204 dB with reference to 1 W.

The noise power predicted by Eq. 1-1 is that caused by thermal agitation of the carriers. Other noise mechanisms can exist in a conductor, but they are excluded from consideration here. Thus the thermal noise represents a minimum level of noise in a resistive element.

In Eq. 1-1 the noise power is proportional to the bandwidth. There is equal noise power in each hertz of bandwidth; the power in the band from 1 to 2 Hz is equal to that from 1000 to 1001 Hz. This results in thermal noise being called "white noise." White implies that the noise is made up of many frequency components just as white light is made up of many colors. A Fourier analysis gives a flat plot of noise versus frequency. The comparison to white light is not exact, for white light consists of equal energy per wavelength, not per hertz.

Thermal noise ultimately limits the resolution of any measurement system. Even if an amplifier could be built perfectly noise-free, the resistance of the signal source would still contribute noise.

Equation 1-1 can be changed into a much more useful form. *Available power is the power that can be supplied by a source when it is feeding a resistance load equal to its internal resistance.* Since available power can be expressed as $E^2/4R$, Eq. 1-1 can be rewritten:

$$N_t = kT\,\Delta f = \frac{E_t^2}{4R} \tag{1-2}$$

Therefore the *rms noise voltage* (E_t) of a resistance R is

$$E_t = \sqrt{4kTR\,\Delta f} \tag{1-3a}$$

where R = resistance or real part of the conductor's impedance

$4kT = 1.61 \times 10^{-20}$ at room temperature (290°K)

When working with noise voltages it is often necessary to use mean square values of quantities. A symbol commonly used for the mean square value of thermal noise is $\overline{e_t^2}$. The overbar represents a mean or average value. Rms is simply the square root of a mean square value, therefore $E_t = \sqrt{\overline{e_t^2}}$. *Since they are equivalent, both* $\overline{e^2}$ *and* E^2 *can be used to represent mean square.* An alternate expression for thermal noise voltage is

$$E_t^2 \quad \text{or} \quad \overline{e_t^2} = 4kTR\,\Delta f \tag{1-3b}$$

Equation 1-3 is very important in noise work. It provides the limit that must be kept in mind. We shall see that the measure of an amplifier's performance, noise figure, is only a measure of the noise the amplifier adds to the thermal noise of the source resistance.*

Several important observations can be made from Eq. 1-3. Noise voltage is proportional to the square root of bandwidth, no matter where the frequency band is centered. *Reactive components do not generate thermal noise.* The resistance used in the equation is not simply the dc resistance of the device or component, but is more exactly defined as the real part of the complex impedance. In the case of an inductance, it may include eddy current losses. In the case of a capacitor, it can be caused by dielectric losses. It is obvious that cooling a conductor decreases its thermal noise.

A chart for rapidly determining thermal noise voltage is given in Fig. 1-2. To determine the noise voltage from a resistance, select the resistance on the horizontal axis, move vertically to the diagonal line corresponding to the system *noise bandwidth*, follow the horizontal line to the left- or right-hand axis, and read the rms noise voltage. It can be seen that the numerical values repeat themselves every other decade, thus the graph can be extended for resistances or bandwidths not shown.

As an example we can determine the rms noise voltage of a 1000-Ω resistor with an amplifier noise bandwidth of 1 Hz to be 4 nV. This is a good number to memorize as a benchmark. From this we can scale up or down by the square root of resistance or bandwidth.

The effect of broadband thermal noise must be minimized. Equation 1-3 implies that there are several practical ways. The sensor resistance must be kept as low as possible, and additional series resistance elements must be

* A more complete expression for thermal noise is $E_t^2 = 4kTRp(f)\,df$, where $p(f)$ is referred to as the Planck factor: $p(f) = (hf/kT)(e^{hf/kT} - 1)^{-1}$. $h = 6.62 \times 10^{-34}$ J-sec is Planck's constant. The term $p(f)$ is usually ignored since $hf/kT \ll 1$ at room temperature for frequencies into the microwave band. Therefore $p(f) = 1$ for all practical purposes [2].

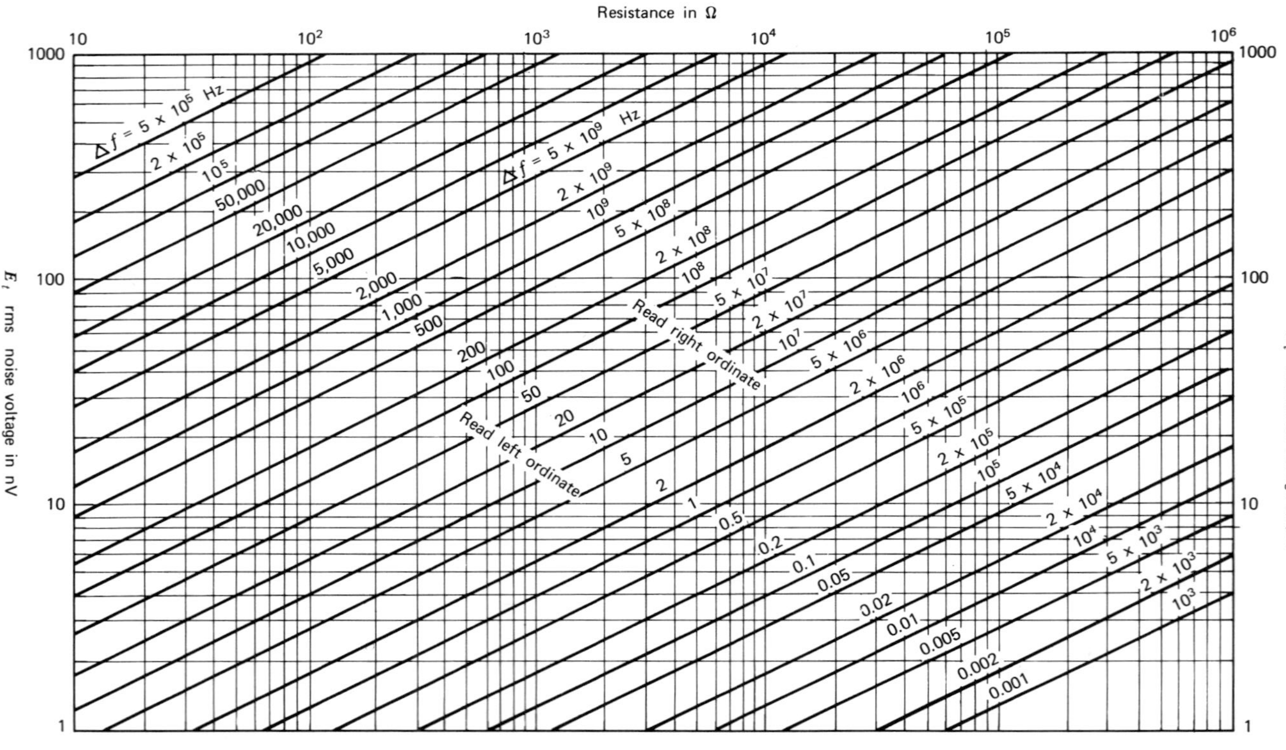

$E_t = 1.27 \times 10^{-4}\sqrt{R\,\Delta f}\ \mu V$

Fig. 1-2. Plot of the equation for Johnson noise in a resistance at 24°C: $E_t = 1.27 \times 10^{-4}\sqrt{R\Delta f}\,\mu V$.

avoided. Also, it is desirable to keep bandwidth as narrow as possible, while maintaining the bandwidth necessary to pass the signal signature. When designing a system, frequency limiting can be incorporated in one of the later stages. For a laboratory application, frequency limiting can be obtained from wave analyzers, tuned voltmeters, and filters. It is usually undesirable to do the frequency limiting with the sensor or the input coupling network. This tends to decrease signal and sensor noise, but it does not attenuate the amplifier noise that is generated following the coupling network.

Even though we have shown that there is a time-varying current and available power in every conductor, this is not a new power source! You cannot put a diode in series with a noisy resistor and use it to power a transistor radio. If the conductor were connected to a load (another conductor), the noise power of each would merely be transferred to the other. If a resistor at 300°K, room temperature, were connected in parallel with a resistor at 0°K, there would indeed be a power transfer from the higher temperature resistor to the lower. The warmer resistor would try to cool down and the other would try to warm up until they came into thermal equilibrium. At that point there would be no further power transfer.

Thermal noise has been extensively studied. Expressions are available for predicting the number of maxima per second present in thermal noise, and also the number of zero crossings expected per second in the noise waveform. These quantities are dependent on the width of the passband. Formulas are given in Problem 3 at the end of this chapter.

1-4 NOISE BANDWIDTH

Noise bandwidth is not the same as the commonly used 3-dB bandwidth. There is one definition of bandwidth for signal and another for noise.

The bandwidth of an amplifier or a tuned circuit is classically defined as the frequency span between half-power points. The half-power points are values on the frequency axis where the signal transmission has been reduced by 3 dB from the central or midrange reference value. A 3-dB reduction represents a loss of 50% in power level and corresponds to a voltage level equal to 70.7% of the voltage at the center frequency reference.

The noise bandwidth, Δf, is the frequency span of a rectangularly shaped power gain curve equal in area to the area of the actual power gain versus frequency curve. Noise bandwidth is the area under the power curve, the integral of power gain versus frequency, divided by the peak amplitude of the curve. This can be stated in equation form:

$$\Delta f = \frac{1}{G_o} \int_0^\infty G(f)\, df \tag{1-4}$$

where Δf = noise bandwidth in hertz

$G(f)$ = power gain as a function of frequency

G_o = peak power gain

Since power gain is proportional to the network voltage gain squared, the equivalent noise bandwidth can also be written as

$$\Delta f = \frac{1}{A_{vo}^2} \int_0^\infty [A_v^2(f)]\, df \tag{1-5}$$

where $A_v(f)$ = voltage gain as a function of frequency

A_{vo} = midband voltage gain

The equation is more useful in the second form.

The plot shown in Fig. 1-3 is typical of a broadband amplifier. The shape of the curve may appear strange because it has a linear frequency scale instead of the more-common logarithmic scale. The area of the dotted rectangle is equal to the area under the $A_v^2(f)$ curve. Thus the bandwidth from 0 to Δf is the noise bandwidth.

If the high-frequency response, as shown in Fig. 1-3, is determined by a

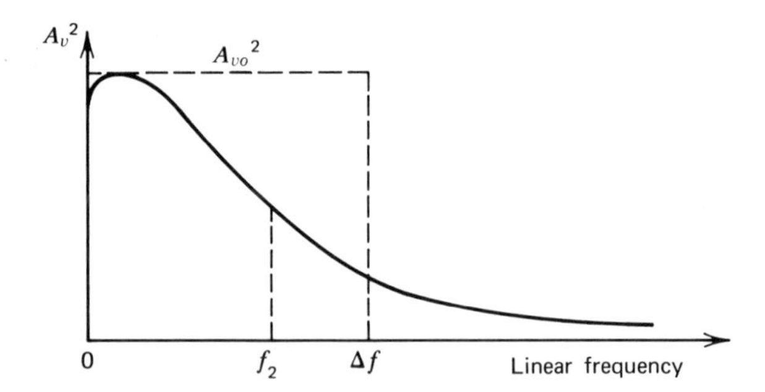

Fig. 1-3. Definition of noise bandwidth.

single *R-C* time constant, the voltage gain falls off at 6 dB/octave, and Δf is equal to $\pi/2$ or $1.571 \times f_2$, the upper cutoff frequency of the voltage gain curve. Similarly, if the frequency response of the amplifier is determined by two *R-C* stages with the same time-constant, the high-frequency voltage gain falls off at 12 dB/octave, and the noise bandwidth Δf equals 1.222 times the common break frequency f_2. As the roll-off becomes sharper with a large number of cascaded stages, Δf approaches the 3-dB bandwidth.

When an equation for the gain curve is not available, graphical methods can be used for determining noise bandwidth. A plot of network voltage gain *squared* versus frequency is made on linear graph paper. The squares under the curve are counted. A rectangle is then formed with the ordinate equal to the midband voltage-gain-squared value, having the same number of squares as previously determined. The horizontal dimension of this rectangle is Δf.

Care must be exercised in measurements pertaining to noise. We must not allow the measuring equipment to change the bandwidth of the system. Also, we must bear in mind that to arrive at useful results, the network response must continue to fall off as we reach higher and higher frequencies.

The term spectral density is used to describe the noise content in a unit of bandwidth. We know that noise consists of many frequency components; to indicate how these components are distributed we could plot mean square noise per unit bandwidth against frequency. For a thermal noise source the spectral density $S(f)$ is

$$S(f) = \frac{E_t^2}{\Delta f} = 4kTR \qquad \text{V}^2/\text{Hz} \tag{1-6}$$

It is characteristic of white noise sources that the plot of $S(f)$ versus f be simply a horizontal line.

When measuring noise we often work with the rms value of a noise quantity. Thus we might obtain a kind of spectral density by dividing the rms value of a noise voltage by the square root of the noise bandwidth, and obtain

$$\frac{E}{\sqrt{\Delta f}} \qquad \text{in units of V/Hz}^{1/2}$$

The result of this mathematical operation can be interpreted as simply the rms noise voltage in 1 Hz of bandwidth. Note that $E/\sqrt{\Delta f}$ is a symbol for a quantity that can be measured; the units are V/Hz$^{1/2}$. Often this density function is symbolized by $E/\sqrt{\sim}$, or in the case of a current, $I/\sqrt{\sim}$. Since a bandwidth of 1 Hz is almost always used, the units for these functions are referred to as "volts per hertz" and "amps per hertz."

1-5 THERMAL NOISE EQUIVALENT CIRCUITS

In order to perform a noise analysis of an electronic system, every element that generates thermal noise is represented by an equivalent circuit composed of a noise voltage generator in series with a noiseless resistance. Suppose, then, that we have a noisy resistance R connected between terminals a and b. For analysis, we substitute the equivalent shown in Fig. 1-4a, a noiseless resistance of the same ohmic value, and a series noise generator with rms value E_t equal to $(4kTR\,\Delta f)^{1/2}$. This generator is supplying the circuit with multi-frequency noise; it is specified by the rms value of its total output.

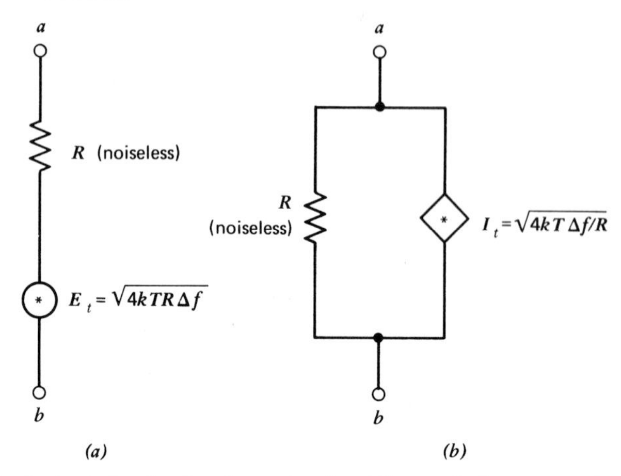

Fig. 1-4. Equivalent circuits for thermal noise. The symbols representing the voltage and current generators are used for noise sources exclusively.

According to Norton's theorem, the series arrangement shown in Fig. 1-4*a* can be replaced by an equivalent constant-current generator in parallel with a resistance. The noise current generator I_t will have a rms value equal to E/R, or in this instance,

$$I_t = \left(\frac{4kT\,\Delta f}{R}\right)^{1/2} \tag{1-7}$$

A graph of I_t vs. R is shown in Fig. 1-7.

If a voltmeter with infinite input impedance and zero self-noise were connected between *a* and *b*, the thermal noise voltage could be measured. However, because real voltmeters also contribute noise, such a direct reading is not accurate.

The system of symbols that we employ in noise analysis uses the letters E and I to represent noise quantities. *The letter* V *is reserved for signal voltage.* Because noise generators do not have an instantaneous phase characteristic as is attributed to sine waves in the phasor method of representation, no specific polarity indication is included in the noise source symbols in Fig. 1-4. Polarity of noise sources is discussed in Chapter 2.

1-6 ADDITION OF NOISE VOLTAGES

When sinusoidal signal voltage sources of the *same frequency and amplitude* are connected in series, the resultant voltage has twice the common amplitude if they are in phase, and combined they can deliver four times the power of one source. If, on the other hand, they differ in phase by 180°, the net voltage and power from the pair is zero. For other phase conditions they may be combined using the familiar rules of phasor algebra.

If two sinusoidal signal voltage sources of *different frequencies* with rms amplitudes V_1 and V_2 are connected in series, the resultant voltage has rms amplitude equal to $(V_1^2 + V_2^2)^{1/2}$. The mean square value of the resultant wave V_r^2 is the sum of the mean square values of the components ($V_r^2 = V_1^2 + V_2^2$).

Equivalent noise generators represent a very large number of component frequencies with a random distribution of amplitudes and phases. When independent noise generators are series connected, the separate sources neither help nor hinder one another. *The output power is the sum of the separate output powers, and consequently it is valid to combine such sources so that the resultant mean square voltage is the sum of the mean square voltages of the individual generators.* This statement can be extended to noise-current sources in parallel.

The generators E_1 and E_2 shown in Fig. 1-5 represent uncorrelated noise

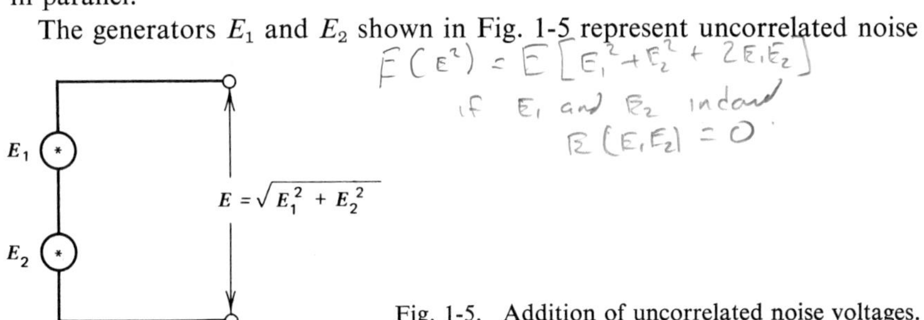

Fig. 1-5. Addition of uncorrelated noise voltages.

sources. We form the sum of these voltages by adding mean square values. Thus the mean square of the sum, E^2, is given by

$$E^2 = E_1^2 + E_2^2$$

Taking the square root of a quantity such as E^2 represents rms. It is not valid to sum the rms voltages of series noise sources. However, one can often neglect the smaller of the two noise signals when their rms values are in a 10:1 ratio. The smaller signal only adds 1% to the overall voltage. A 3:1 ratio has only a 10% effect on the total.

If two resistors are connected in parallel, the total thermal noise voltage is that of the equivalent resistance. Similarly, with two resistors in series, the total noise voltage is determined by the arithmetic sum of the resistances.

1-7 CORRELATION

When noise voltages are produced independently and there is no relationship between the instantaneous values of the voltages, they are uncorrelated.

Uncorrelated voltages are treated according to the discussion of the preceding section.

Two waveforms that are of identical shape are said to be 100% *correlated* even if their amplitudes differ. An example of correlated signals would be two sine waves of the same frequency and phase. The instantaneous or rms values of fully correlated waveforms can be added arithmetically.

A problem arises when we have noise voltages that are partially correlated. This can happen when each contains some noise that arises from a common phenomenon, as well as some independently generated noise. In order to sum partially correlated waves, the general expression is

$$E^2 = E_1^2 + E_2^2 + 2CE_1E_2 \tag{1-8}$$

The term C is called the correlation coefficient and can have any value between -1 and $+1$, including 0. When $C = 0$, the voltages are uncorrelated, and the equation is the same as given in Fig. 1-5. When $C = 1$, the signals are totally correlated. Then rms values E_1 and E_2 can be added linearly. A -1 value for C implies subtraction of correlated signals, for the waveforms are then 180° out of phase.

Very often one can assume the correlation to be zero with little error. If the two voltages are equal and fully correlated, summing gives twice their separate rms values, whereas the uncorrelated summing is 1.4 times their separate rms values. Thus the maximum error caused by the assumption of statistical independence is 30%. If the signals are partially correlated or one is much larger than the other, the error is smaller.

1-8 EXCESS NOISE

The thermal noise voltage of a resistor was previously discussed. Many resistors also exhibit *excess noise* when direct current is present. This noise contribution is greatest in composition carbon resistors and is usually not important in wirewound resistors. Excess noise is so named because it is present in addition to the fundamental thermal noise of the resistor.

Excess noise usually occurs whenever current flows in a discontinuous medium such as a composition carbon resistor. A carbon resistor is made up of carbon granules squeezed together, and current tends to flow unevenly through the resistor. There are something like microarcs between the carbon granules. This behavior gives rise to a 1/f noise spectrum. A 1/f spectrum means that the noise *power* varies inversely with frequency. Thus the noise voltage increases as the square root of the decreasing frequency. By decreasing frequency 1:10, the noise voltage increases by 3:1.

Low-frequency noise results in equal power in each decade of frequency.

In other words, the noise power in the band from 10 to 100 Hz is equal to that of the band from 0.01 to 0.1 Hz. Since the noise in each of these intervals is uncorrelated, the mean square values can be added. Total noise power increases as the *square root* of the number of frequency decades.

Excess noise in a resistor can be measured in terms of a noise index expressed in dB. *The noise index is the number of microvolts of noise in the resistor per volt of dc drop across the resistor in each decade of frequency.* Thus, even though the noise is caused by current flow, it can be expressed in terms of the direct voltage drop rather than resistance or current. The noise index of some brands of resistors may be as high as 10 dB which corresponds to 3 μV/(dc volt)/(decade). This can be a significant contribution. Resistor noise is further discussed in Chapter 9.

1-9 LOW-FREQUENCY NOISE

Low-frequency or *1/f noise* has several unique properties. If it were not such a problem it would be very interesting. The spectral density of this noise increases without limit as frequency decreases. Firle and Winston have measured 1/f noise as low as 6×10^{-5} Hz [3]. This frequency is but a few cycles per day.

When first observed in vacuum tubes, this noise was called "flicker effect," probably because of the flickering observed in the plate current. Many different names are used, some of them uncomplimentary. In the literature, names like excess noise, pink noise, semiconductor noise, low-frequency noise, and contact noise will be seen. These all refer to the same thing. The term "red noise" is applied to a noise power spectrum that varies as $1/f^2$.

The noise power typically follows a $1/f^{\alpha}$ characteristic with α usually unity, but α has been observed to take on values from 0.8 to 1.3 in various devices. The major cause of 1/f noise in semiconductor devices is traceable to properties of the surface of the material. The generation and recombination of carriers in surface energy states and the density of surface states are important factors. Improved surface treatment in manufacturing has decreased 1/f noise, but even the interface between silicon surfaces and grown oxide passivation are centers of noise generation.

As pointed out by Halford, 1/f noise is quite common [4]. Not only is it observed in tubes, transistors, diodes, and resistors, but it is also present in thermistors, carbon microphones, thin films, and light sources. The fluctuations of a membrane potential in a biological system have been reported to have flicker noise. No electronic amplifier has been found to be free of flicker noise at lowest frequencies. Halford points out that $\alpha = 1$ is the most common value, but there are other mechanisms with different alphas. For example, fluctuations of the frequency of rotation of the earth have an α of 2 and the power spectral density of galactic radiation noise has $\alpha = 2.7$.

Since 1/f noise power is inversely proportional (K_1) to frequency, it is possible to determine the noise content in a band by integration of $K_1 f^{-1}$ over the range of frequencies in which our interest lies. The result is

$$N_f = K_1 \ln \left(\frac{f_h}{f_l} \right) \tag{1-9}$$

Symbols f_h and f_l are the upper and lower frequency limits of the band being considered.

The mean square value of the corresponding noise voltage is

$$E_f^2 = K \ln \left(\frac{f_h}{f_l} \right) = K \ln \left(1 + \frac{\Delta f}{f_l} \right) \simeq K \frac{\Delta f}{f} \tag{1-10}$$

When the band is 1-Hz wide, $f_h = f_l + 1$. Then Eq. 1-10 can be written

$$S(f) = \frac{K}{f} \qquad \text{V}^2/\text{Hz} \tag{1-11}$$

This is the spectral density of 1/f noise.

Because 1/f noise power continues to increase as the frequency may be decreased, we might ask the question, why is the noise not infinite at dc? Although the noise voltage in a 1-Hz band may theoretically be infinite at dc or 0 frequency, there are practical considerations that keep the total noise manageable for most applications. The noise power per decade of bandwidth is constant, but a decade such as that from 0.1 to 1 Hz is narrower than the decade from 1 to 10 Hz. But, when considering the 1/f noise in a dc amplifier, there is a lower limit to the frequency response set by the length of time the amplifier has been turned on. This low-frequency cutoff attenuates frequency components with periods longer than the "on" time of the equipment.

A numerical example may be of assistance. Consider a dc amplifier with upper cutoff frequency of 1000 Hz. It has been on for 1 day. Since 1 cycle/day corresponds to about 10^{-5} Hz, its bandwidth can be stated as 8 decades. If it is on for 100 days we may add 2 more decades or $\sqrt{2}$ times its 1-day noise. The noise per hertz approaches infinity, but the total noise does not.

A fact to remember concerning a 1/f noise-limited dc amplifier is that measurement accuracy cannot be improved by increasing the length of measuring time. In contrast, when measuring white noise, the accuracy increases as the square root of the measuring time.

1-10 SHOT NOISE

In tubes, transistors, and diodes there is a noise current mechanism called *shot noise*. Current flowing in these devices is not smooth and continuous, but rather it is the sum of pulses of current caused by the flow of carriers, each carrying one electronic charge.

Consider the case of a vacuum tube diode with a high plate voltage so that all the electrons emitted from the cathode are collected. We assume no space-charge region near the cathode to smooth out the current flow. Each electron carries a charge q, and when it arrives at the plate it results in an impulse of current. This pulsing flow is a granule effect, and the variations are referred to as shot noise. The rms value of the shot noise current is observed to be given by

$$I_{sh} = \sqrt{2qI_{DC}\,\Delta f} \tag{1-12}$$

where q = electronic charge, 1.59×10^{-19} C

I_{DC} = direct current in A

Δf = noise bandwidth in Hz

We note that the shot-noise current is proportional to the square root of the noise bandwidth. This means that it is white noise containing constant noise power per hertz of bandwidth. A convenient graph of shot noise current vs. diode current is shown in Fig. 1-7.

Shot noise is associated with current flow across a potential barrier. Such a barrier exists at every *pn* junction in semiconductor devices and at the cathode surface in a vacuum tube. No barrier is present in a simple conductor, therefore no shot noise is present. The most important barrier is the emitter-base junction in a bipolar transistor. The *V-I* behavior of that junction is described by the familiar diode equation

$$I_E = I_0(\epsilon^{qV_{BE}/kT} - 1) \tag{1-13}$$

where I_E = emitter current in A

I_0 = saturated value of reverse current

V_{BE} = potential applied between base and emitter

Suppose that we consider separately the two currents that make up I_E in Eq. 1-13:

$$I_E = I_1 + I_2$$

where $I_1 = -I_0$, $\quad I_2 = I_0\epsilon^{qV_{BE}/kT}$

Current I_1 is caused by thermally generated minority carriers, and I_2

represents the diffusion of majority carriers across the barrier. Each of these currents has full shot noise, and even though the direct currents they represent flow oppositely, their mean-square noise values can be added.

Under reverse biasing, $I_2 \rightarrow 0$ and the shot noise current of I_1 dominates. On the other hand, when the diode is strongly forward biased, the shot noise current of I_2 dominates. For zero bias, there is no external direct current, and I_1 and I_2 are equal and opposite. The mean square value of shot noise is twice the reverse bias noise current:

$$I_{sh}^2 = 4qI_0\,\Delta f \tag{1-14}$$

The equivalent circuit representation for a shot noise source is a current generator as previously noted in Eq. 1-12. For the case of the forward-biased *pn* junction, a noiseless resistance parallels this noise current generator. By differentiating Eq. 1-13 with respect to V_{BE}, we obtain a conductance. The reciprocal of that conductance is referred to as the Schockley emitter resistance r_e and is given by

$$r_e = \frac{kT}{qI_E} \tag{1-15}$$

At room temperature $r_e = 0.026/I_E$. The element r_e is not a thermally noisy component, for it is a dynamic effect and not a bulk or material characteristic.

An equivalent circuit representing shot noise at a forward-biased *pn* junction is shown in Fig. 1-6. Mathematically the mean square value of shot

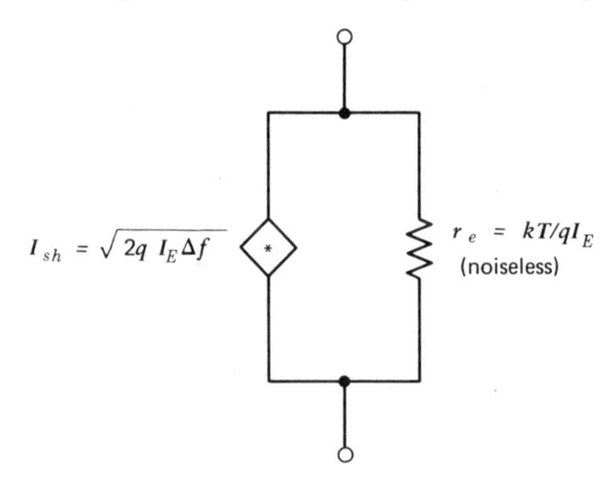

Fig. 1-6. Shot noise equivalent circuit for forward-biased *pn* junction.

noise is equal to thermal noise for an unbiased junction, and equal to one-half of the resistive thermal noise voltage at a forward-biased junction. Further discussion is available in the literature [5,6].

The noise voltage of a forward-biased junction is the product of shot noise current I_{sh} and diode resistance r_e.

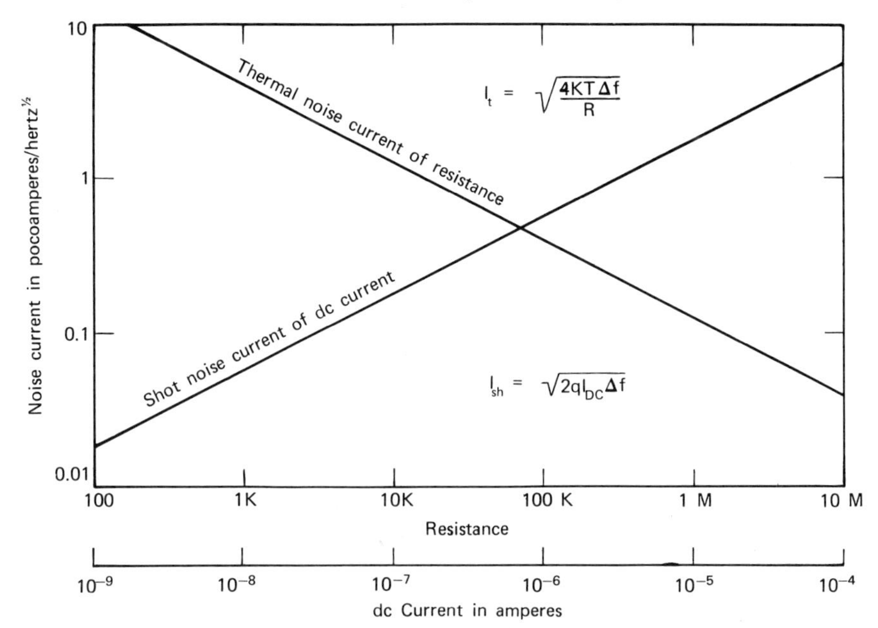

Fig. 1-7. Plot of noise current of shot noise and thermal noise.

1-11 CAPACITIVE SHUNTING OF THERMAL NOISE

The thermal noise expression $E_t = (4kTR\,\Delta f)^{1/2}$ predicts that an open circuit (infinite resistance) generates an infinite noise voltage. This is not observed because in a practical situation there is always some shunt capacitance that limits the voltage. Thus an actual noisy resistance is shown in Fig. 1-8.

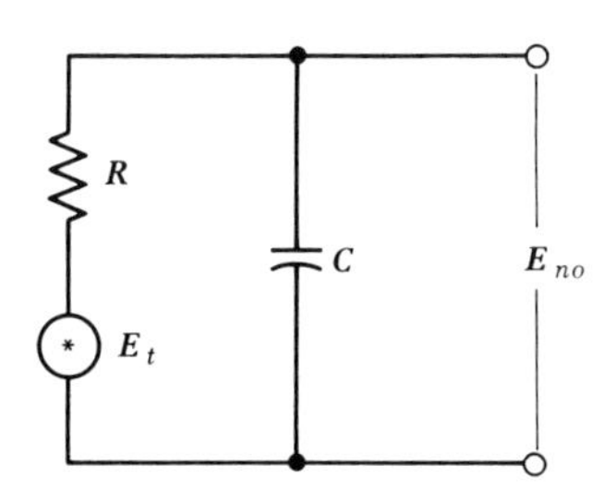

Fig. 1-8. Noise shunted by capacitor.

The thermal noise voltage E_t increases as the square root of resistance. Low-frequency noise from E_t directly affects the output noise E_{no}. Higher frequency components from E_t are more effectively shunted by the capacitor C. Increasing R increases the noise voltage, but decreases the cutoff frequency and consequently the noise bandwidth. A plot of the resulting noise voltage versus frequency is shown in Fig. 1-9 for values of resistance of R, $4R$, and $9R$.

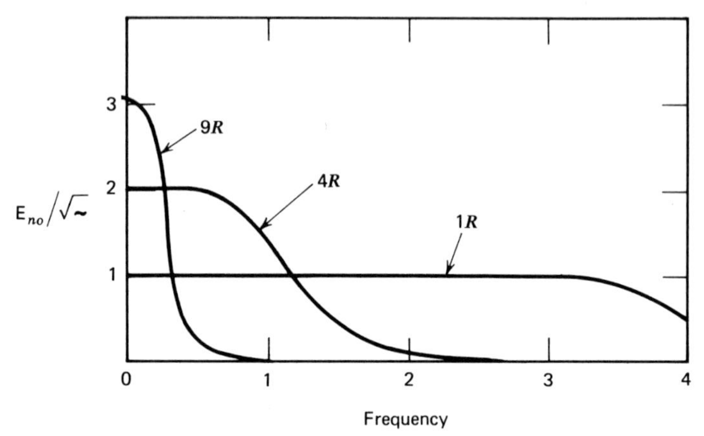

Fig. 1-9. Noise spectral density versus frequency for the circuit of Fig. 1-8.

The value of E_{no} can be calculated using the voltage division principle. Since E_{no} is not a phasor, interest is in its magnitude only. Thus

$$E_{no} = E_t \left| \frac{-jX_c}{R - jX_c} \right| = \frac{E_t}{\sqrt{1 + \omega^2 C^2 R^2}}$$

Therefore

$$E_{no}^2(f) = \frac{4kTR\,\Delta f}{1 + \omega^2 C^2 R^2} \tag{1-16}$$

Let us now integrate the noise voltage $E_{no}^2(f)$ as given by Eq. 1-16 over the entire frequency spectrum. This operation provides a measure of the total noise energy in a frequency band that is solely limited by R and C. We obtain

$$\int_0^\infty [E_{no}^2(f)]\, df = \frac{kT}{C} \tag{1-17}$$

Equation 1-17 is independent of the resistance, as we would expect from the earlier conclusion that available noise power is generated in R, but does not depend on the value of R. Clearly the total noise energy is limited by temperature and capacitance.

1-12 NOISE CALCULATOR

Quan-Tech Laboratories, Incorporated, has designed a special-purpose slide rule to assist in performing noise calculations.* A summary of the ten scales of the noise calculator is given here.

* Available free from Quan-Tech Laboratories, Inc., 43 South Jefferson Road, Whippany, N.J. 07981. Write to Brian Whitlock.

1. *Thermal noise voltage developed by a resistance for* $\Delta f = 1$. Scales are E_t and R. Refer to Eq. 1-3*a*.

2. *Thermal noise current developed by a resistance for* $\Delta f = 1$. Scales are I_t and R. Refer to Eq. 1-9.

3. *Shot noise current for* $\Delta f = 1$. Scales are I_{sh} and I_{DC}. Refer to Eq. 1-12.

4. *Shot noise voltage for* $\Delta f = 1$ *according to* $E_{sh} = (2qI_{DC})^{1/2}R$. Scales are E_{sh} and I_{DC}.

5. *Noise figure according to* NF = 20 log (total noise/thermal noise of R_s) in dB. Scales are R_s, total noise spectral density, and NF in dB. Refer to Eq. 2-12.

6. *Addition of uncorrelated noise voltages and currents.* Scales are E_1 or I_1, E_2 or I_2, and resultant E or I. Refer to Eq. 1-8.

7. *Conversion between noise index and noise voltage for resistor excess noise.* Scales are noise index in dB and noise voltage in μV/V. Refer to Eq. 9-2.

8. *Wideband to narrowband thermal noise voltage conversion.* Scales are Δf, S, and E_t. Refer to Eq. 1-6.

9. *Conversion between voltage ratio and dB* according to dB = 20 log V, where V is voltage ratio.

10. *Ohm's law* according to $E = IR$. Scales are R, I, and E.

SUMMARY

a. Noise is any unwanted disturbance that obscures or interferes with a signal.

b. Thermal noise is present in every electrical conductor:

$$E_t = (4kTR\,\Delta f)^{1/2}$$

When evaluated this yields 4 nV for 1000 Ω and $\Delta f = 1$ Hz.

c. Noise bandwidth Δf is the area under the $A_v^2(f)$ curve divided by A_{vo}^2, the reference value of gain squared.

d. For circuit analysis a noisy resistance can be replaced by a noise voltage generator in series with a noiseless resistance, or a noise current generator in parallel with a noiseless resistance.

e. Noise quantities can be added according to

$$E^2 = E_1^2 + E_2^2 + 2CE_1E_2$$

C is the correlation coefficient, $-1 \le C \le +1$. Usually $C = 0$.

f. Excess noise is generated in most components when direct current is present.

g. 1/f noise is especially troublesome at low audio frequencies.

h. Shot noise is present when direct current flows across a potential barrier:

$$I_{sh} = (2qI_{DC}\,\Delta f)^{1/2}$$

i. The total thermal noise energy in a resistance is finite and is limited by the effective capacitance across its terminals and the absolute temperature.

PROBLEMS

1. Determine the rms thermal noise voltage of resistances of 1 kΩ, 50 kΩ, and 1 MΩ for each of the following bands: 50 kHz, 1 MHz, and 20 MHz. Use Fig. 1-2 and check using Eq. 1-3*a*. Consider $T = 290°\text{K}$.

2. Calculate the rms thermal noise voltage of a 50 mH inductance in a 1-Hz band. Consider that the inductor has series resistance of 1000 Ω and that the band is centered around 10 kHz.

3. (a) The statistically expected number of maxima per second in white noise with upper and lower frequency limits f_2 and f_1 is [7]

$$\left[\frac{3(f_2^{\,5} - f_1^{\,5})}{5(f_2^{\,3} - f_1^{\,3})}\right]^{1/2}$$

Find the maxima for $f_1 = 0$ and $f_2 = 10^6$ Hz.

(b) The expected total number of zero crossings per second is given by

$$\left[\frac{4(f_2^{\,2} + f_1 f_2 + f_1^{\,2})}{3}\right]^{1/2}$$

If $f_1 = 0$ and $f_2 = 10$ MHz, evaluate the number of zero crossings. Show that for narrowband noise the assumption that $f_1 \simeq f_2$ yields

$$(f_1 + f_2)$$

4. Determine the noise bandwidth of a circuit with $A_v^{\,2}$ response described as follows:

$$0 \le f \le 1\text{ kHz} \qquad A_v^{\,2} = f$$

$$1\text{ kHz} \le f \le 20\text{ kHz} \qquad A_v^{\,2} = 1000$$

$$20\text{ kHz} \le f \le 100\text{ kHz} \qquad A_v^{\,2} = 1000 - 0.0125(f - 20{,}000)$$

$$100\text{ kHz} < f \qquad A_v^{\,2} = 0$$

5. Calculate E_t for a noisy resistance of 500,000 Ω. Transform this into the noise current generator form and determine I_t. $T = 290°\text{K}$ and $\Delta f = 10^5$ Hz.

The resistor is to be connected in parallel with another noisy resistor of 250,000 Ω. Determine the mean square and rms values of the noise voltage present at the terminals of the pair.

6. Find the resistance of a *pn* junction that exhibits 200 nV rms of shot noise. Assume $\Delta f = 10^6$ and $I_{DC} = 10$ mA. Compare your answer with the value of r_e predicted by Eq. 1-13. What conclusions can you reach?

REFERENCES

1. Bennett, A. R., *Electrical Noise*, McGraw-Hill, New York, 1960, p. 42.
2. Van der Ziel, A., *Noise*, Prentice-Hall, Englewood Cliffs, N.J., 1954, pp. 8–9.
3. Firle, J. E., and H. Winston, *Bull. Am. Phys. Soc.*, **30**, 2 (1955).
4. Halford, D., "A General Model for f^{α} Spectral Density Random Noise with Special Reference to Flicker Noise 1/f," *Proc. IEEE*, **56**, 3 (March 1968), 251.
5. Thornton, R. D., D. DeWitt, E. R. Chenette, and P. E. Gray, *Characteristics and Limitations of Transistors*, Wiley, New York, 1966, pp. 138–145.
6. Baxandall, P. J., "Noise in Transistor Circuits," *Wireless World*, **74**, 1397–1398 (November 1968), 388–392—(December 1968), 454–459.
7. Rice, S. O., "Mathematical Analysis of Random Noise," *Bell System Technical Journal*, **23**, 4 (July 1944).

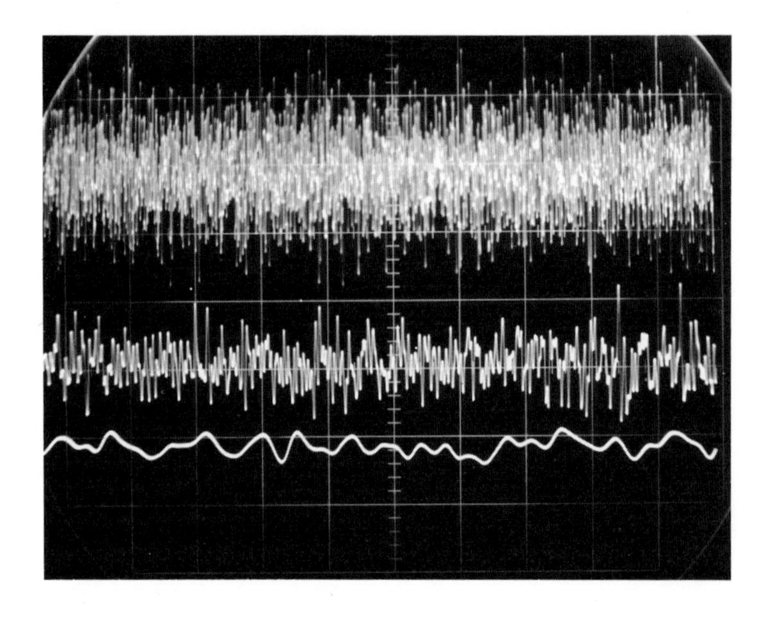

White noise is shown as it appears on oscilloscope display with horizontal sweep of 1 msec/cm. Top waveform, bandwidth is dc to 200 kHz; center waveform, bandwidth is dc to 20 kHz; bottom waveform, bandwidth is dc to 2 kHz. Note that the bandwidth reduction affects both the peak amplitude and the rms value.

Chapter 2

AMPLIFIER NOISE

Since every electrical component is a potential source of noise, a network such as an amplifier that contains many components could be difficult to analyze from a noise standpoint. Therefore, a noise model is helpful to simplify noise analysis. The E_n-I_n amplifier model discussed in this chapter contains only two noise parameters. The parameters are not difficult to measure.

The concept of noise figure is introduced, and it is shown that optimization of the noise figure is possible. From a study of the noise contributions of the stages of a cascaded network, it can be concluded that the major noise source is the first signal processing stage. If a high level of power amplification is available from that stage, noise contributions from other portions of the electronics can be considered to be negligible.

2-1 THE NOISE VOLTAGE AND CURRENT MODEL

There are universal noise models for any two-port network [1]. The network is considered as a noise-free black box, and the internal sources of noise are represented by a pair of noise generators located at one port, usually the input. This noise model, shown in Fig. 2-1, is used to represent an amplifier; it can also apply to passive circuits, tubes, transistors, tunnel diodes, integrated circuit (IC) amplifiers, and so on. The figure also includes the signal source V_s and noisy source resistance R_s.

Amplifier noise is represented completely by a zero impedance voltage generator E_n in series with the input port, an infinite impedance current generator I_n in parallel with the input, and by a complex correlation coefficient C (not shown). Each of these terms may be frequency dependent. The thermal noise of the signal source is represented by noise generator E_t.

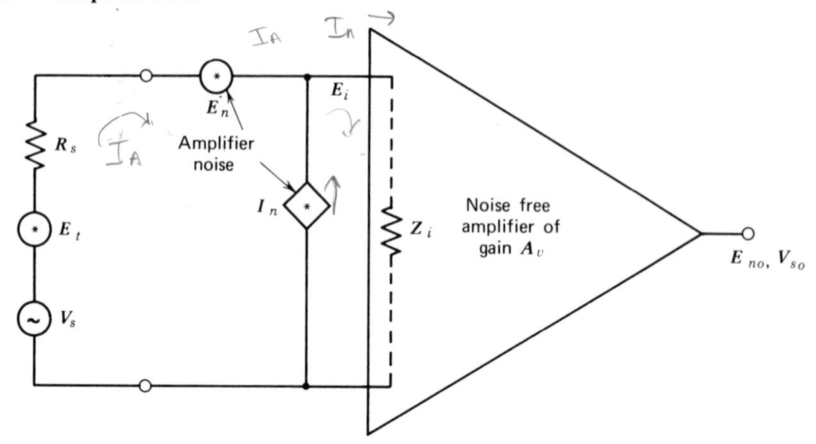

Fig. 2-1. Amplifier noise model and signal source.

In a practical design, we are usually concerned with the signal-to-noise ratio at the output of the system. That is where we are using the signal for indication, display, level detecting, or relay driving. Because we are considering signal and noise in an electronic system that has stages of amplification, response shaping, and so forth, it is usually quite difficult to evaluate the results of even minor circuit modifications on the signal-to-noise ratio. By referring all noise to the input port, and considering the amplifier to be noise free, it is easier to appreciate the effects of such changes on both signal and noise.

Both E_n *and* I_n *parameters are required to represent adequately an amplifier.*

2-2 EQUIVALENT INPUT NOISE

Although we have reduced the number of noise sources to three in the system shown in Fig. 2-1 by using the E_n-I_n model for the electronic circuitry, additional simplifications are welcomed. *Equivalent input noise,* E_{ni}, *will be used to represent all three noise sources.* This parameter refers all noise sources to the signal source location. Since both the signal and the noise equivalent are then present at that point in the system, the S/N can be easily evaluated. We proceed to determine E_{ni}.

The levels of signal voltage and noise voltage that reach Z_i in the circuit are multiplied by the noiseless voltage gain A_v. We will determine those levels. Voltage and current division principles can be applied, and since we wish to sum noise contributions, noise must be expressed in mean square values. The total noise at the output port is

$$E_{no}^2 = A_v^2 E_i^2 \tag{2-1}$$

Therefore

$$E_{no}^2 = A_v^2\left[\frac{(E_n^2 + E_t^2)Z_i^2}{(R_s + Z_i)^2} + \frac{I_n^2 Z_i^2 R_s^2}{(R_s + Z_i)^2}\right] \tag{2-2}$$

The transfer function from input signal source to output port we call system gain K_t. By definition,

$$K_t = \frac{V_o}{V_s} \tag{2-3}$$

Note that K_t is different from the voltage gain A_v; it is dependent on both amplifier input impedance and generator impedance and varies with frequency. The rms output signal can be expressed by

$$V_o = \frac{A_v V_s Z_i}{R_s + Z_i} \tag{2-4}$$

Substituting Eq. 2-4 into Eq. 2-3 gives an expression for the system gain K_t in terms of network parameters:

$$K_t = \frac{A_v Z_i}{R_s + Z_i} \tag{2-5}$$

The total output noise given in Eq. 2-2 divided by the system gain given in Eq. 2-5 yields an expression for equivalent input noise:

$$\frac{E_{no}^2}{K_t^2} = E_{ni}^2 \tag{2-6}$$

The expression for *equivalent input noise* is

$$E_{ni}^2 = E_t^2 + E_n^2 + I_n^2 R_s^2 \tag{2-7}$$

This equation is important for the analysis of many noise problems; it can be applied to systems using any type of active device. Note the simplicity of its terms. The mean square equivalent input noise is the sum of the mean square values of the three noise voltage generators. *This single noise source located at* V_s *can be substituted for all sources of system noise.*

Amplifier input resistance and capacitance are not present in the equivalent input noise expression. A noise analysis is somewhat simplified when we can omit from consideration the loading caused by the amplifier input impedance.

As previously mentioned, the noise voltage and current generators may not be completely independent. To be absolutely correct, we must introduce the correlation coefficient C as discussed in Section 1-7. A modified form of Eq. 2-7 results:

$$E_{ni}^2 = E_t^2 + E_n^2 + I_n^2 R_s^2 + 2CE_n I_n R_s \tag{2-8}$$

The correlation term can be approached, if desired, from the standpoint that it is another noise generator in the system of Fig. 2-1. It could be represented as a voltage generator with rms value $(2CE_nI_nR_s)^{1/2}$ in series with E_n or a current generator $(2CE_nI_n/R_s)^{1/2}$ in parallel with I_n.

2-3 MEASUREMENT OF E_n AND I_n

Another reason for wide acceptance of the E_n-I_n model is the ease of measurement of its parameters. The thermal noise of the source resistance, E_t, can be obtained from the chart of Fig. 1-3.

It can be observed that if R_s *is purposely made to equal zero, two terms in Eq. 2-7 drop out, and the resulting equivalent input noise is simply the noise generator* E_n. A measurement of total output noise under the $R_s = 0$ condition therefore equals A_vE_n. Division of total output noise by A_v gives a value for E_n.

We now have determined two of the three components of the equivalent input noise expression. The third component, I_nR_s, is most easily determined by making the source resistance very large. The source thermal noise contribution is proportional to the square root of source resistance, whereas the I_nR_s term is proportional to the first power of resistance and dominates at a sufficiently large value of source resistance. *To determine* I_n, *then, we measure the total output noise with a large source resistance and divide by the system gain* as measured with this source resistance in series with the input. This gives E_{ni}, which is mostly the I_nR_s term. Dividing by R_s yields the I_n component. If there is still a contribution from the thermal noise of the source, it can be subtracted from E_{ni}.

The values of E_n and I_n vary with frequency, operating point, and the type of input device. Values have been measured and are compiled in appendices I, II and III.

Equivalent input noise, E_{ni}, applies to a specific frequency band. If it is necessary to operate over a wide bandwidth, the noise at each frequency can be calculated or measured and a mean square summing made. In effect, this process is integrating the equivalent input noise versus frequency expression over the bandwidth being considered.

2-4 INPUT NOISE EXAMPLES

Plots of equivalent input noise voltage versus source resistance are given in Fig. 2-2. Also shown in these examples are values of the components E_n and I_nR_s for two amplifiers and the behavior of source resistance thermal noise E_t.

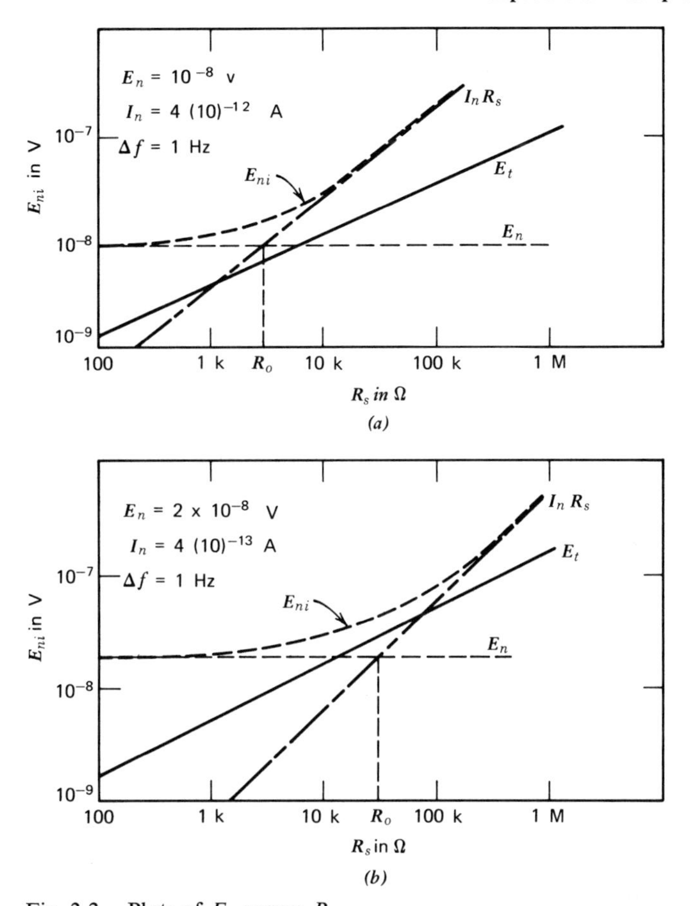

Fig. 2-2. Plots of E_{ni} versus R_s.

In each graph, the equivalent input noise E_{ni} is bounded by three separate lines. Each line corresponds to a term in Eq. 2-7. At low values of source resistance, E_n alone is important. As the source resistance increases, the thermal noise of the source becomes significant. At sufficiently high values of R_s, the total equivalent input noise is equal to the $I_n R_s$ term.

In Fig. 2-2*a*, the E_n-I_n noise dominates the thermal noise throughout the range of values of R_s. In Fig. 2-2*b*, the noise current, I_n, is an order of magnitude smaller. The total noise is limited therefore by thermal noise through a part of the range. Lowering the values of E_n and I_n widens the thermal noise limited region. System noise limited by sensor thermal noise is an ideal case.

Curves of the Fig. 2-2 type apply to all kinds of active devices. The levels, however, differ. With a vacuum-tube amplifier we expect E_n to be larger than shown, whereas I_n may be only 1/100 of the value given in the figure.

2-5 NOISE FIGURE

Noise factor, F, also called *noise figure*, is a figure-of-merit for a device or a circuit with respect to noise. According to IEEE standards, *the noise factor of a two-port device is the ratio of the available output noise power per unit bandwidth to the portion of that noise caused by the actual source connected to the input terminals of the device, measured at the standard temperature of 290°K* [2]. This definition of noise factor in equation form is

$$F = \frac{\text{total available output noise power}}{\text{portion of output noise power caused by } E_t \text{ of source resistance}} \tag{2-9}$$

An equivalent definition of noise factor is

$$F = \frac{\text{input signal-to-noise ratio}}{\text{output signal-to-noise ratio}} = \frac{S_i/N_i}{S_o/N_o} \tag{2-10}$$

Noise factor, since it is a power ratio, can be expressed in dB. The logarithmic expression for noise figure NF is

$$\text{NF} = 10 \log F \tag{2-11}$$

As in much of the literature, the symbol F refers to noise factor and NF is used for the logarithmic noise figure.

Noise figure is a measure of the signal-to-noise degradation attributed to the amplifier. For a perfect amplifier, one that adds no noise to the thermal noise of the source, noise factor $F = 1$ and the noise figure $\text{NF} = 0$. It is sometimes possible to approximate a perfect system.

The noise figure NF can be defined in terms of E_n and I_n. Thus

$$\text{NF} = 10 \log \frac{E_{ni}^2}{E_t^2} = 10 \log \frac{E_t^2 + E_n^2 + I_n^2 R_s^2}{E_t^2} \tag{2-12}$$

This equation shows that the noise figure can also be expressed as the ratio of the total mean square equivalent input noise to the mean square thermal noise of the source.

An illustration of noise factor can be obtained from Fig. 2-2*b*. Noise factor is proportional to the square of the ratio of total equivalent input noise (dotted curve) to thermal noise (solid curve). At low resistances, the ratio of total noise to thermal noise is very large and the noise figure is large, representing poor performance. As source resistance increases, thermal noise increases, but total input noise does not. The noise factor therefore decreases. The plot of total input noise E_{ni} is closest to thermal noise when $E_n = I_n R_s$. This is the point of minimum noise factor. As we go to higher source resistances, total noise follows the $I_n R_s$ curve, and the noise factor again becomes large.

The definition of NF used here is based on a reference temperature of 290°K. When this definition is applied to sensors that are cooled, apparent negative values of NF can result.

The term *spot noise factor* F_o, when used to describe system noise, is simply a narrowband F. Often F_o is defined for $\Delta f = 1$ Hz, and a test frequency such as 1000 Hz can be used. Spot noise factor is clearly a function of frequency and is sometimes simply written as $F(f)$.

The principal value of F is to compare amplifiers or amplifying devices; it is not necessarily the appropriate indicator for optimizing noise performance. Because of the definition, F can be reduced by an increase in the thermal noise of the source resistance. Since a change of this type has little bearing on amplifier design, F is not as useful for system otimization as quantities such as E_{ni} or S_o/N_o.

2-6 OPTIMUM SOURCE RESISTANCE

The point at which total equivalent input noise approaches closest to the thermal noise curve in Fig. 2-2*b* is significant. At this point, the amplifier adds minimum noise to the thermal noise of the source. The noise figure reaches a minimum value. This optimum source resistance is called R_{opt} or R_o and may be obtained from

$$R_o = \left.\frac{E_n}{I_n}\right|_{E_n = I_n R_s} \tag{2-13}$$

The value of noise factor at this point can be called F_{opt}. A rearrangement of Eq. 2-12 can yield

$$F_{\text{opt}} = 1 + \frac{E_n I_n}{2kT\,\Delta f} \tag{2-14}$$

Noise figure variations are illustrated in Fig. 2-3. The minimum value of noise figure occurs at $R_s = R_o$. As the product of E_n and I_n increases, not only does F_{opt} increase, but NF is more highly sensitive to source resistance variations. The lowest curve gives a good noise figure over a wide range of source resistance, whereas the upper curve represents poor operation when the source is not equal to R_o. From an engineering standpoint, if the noise figure is less than 3 dB, there is little advantage to be gained from continued noise reduction effort because half the noise is from the source.

Optimum source resistance, R_o, *is not the resistance for maximum power transfer*. There is no direct relation between R_o and the amplifier input impedance Z_i. R_o is determined by the amplifier noise mechanisms and has a bearing on the maximum signal-to-noise ratio. Optimum power transfer is

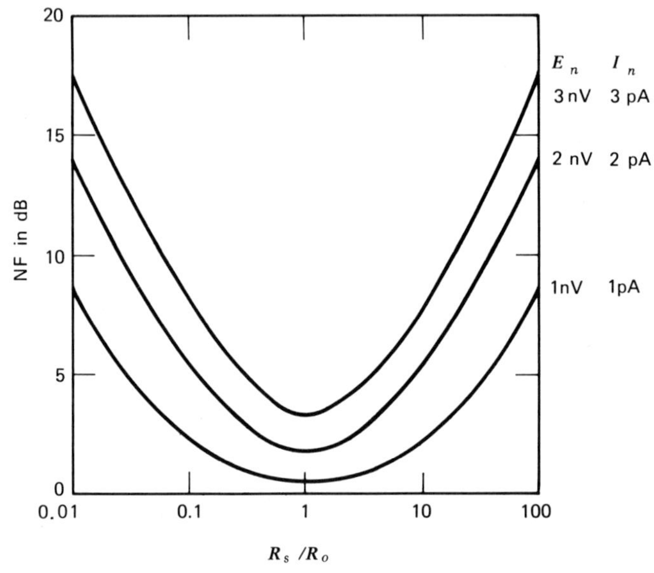

Fig. 2-3. Noise figure versus source resistance.

based on maximizing signal only. The input impedance of an amplifier is strongly affected by circuit conditions such as feedback, but the noise of the amplifier is unaffected by feedback, except insofar as the feedback resistors generate noise.

2-7 NOISE RESISTANCE AND NOISE TEMPERATURE

Noise resistance $\mathrm{R_n}$ *is the value of a resistance that would generate thermal noise of value equal to the amplifier noise.* An expression for R_n is desired. Equating thermal noise to amplifier noise gives

$$4kTR_n\,\Delta f = E_n^2 + I_n^2R_s^2$$

Therefore

$$R_n = \frac{E_n^2 + I_n^2R_s^2}{4kT\,\Delta f} \tag{2-15}$$

This equivalent noise resistance is *not* related to amplifier input resistance, nor does it bear any relation to source resistance.

Noise temperature $\mathrm{T_s}$ *is the value of the temperature of the source resistance that generates thermal noise equal to the amplifier noise.* Proceeding as before, we equate terms:

$$4kT_sR_s\,\Delta f = E_n^2 + I_n^2R_s^2$$

Therefore

$$T_s = \frac{E_n^2 + I_n^2 R_s^2}{4kR_s \, \Delta f} \tag{2-16}$$

Equation 2-16 yields T_s in °K. R_n and T_s are most useful for vacuum tubes where $E_n \gg I_n R_s$.

2-8 NOISE IN CASCADED NETWORKS

We now consider the problem of locating the important noise sources within a system. If we derive a usable expression for the noise factor of cascaded networks in terms of the characteristics of each network, we will then be able to predict for design purposes ways of minimizing system noise.

The system to be analyzed is shown in Fig. 2-4. It consists of a signal

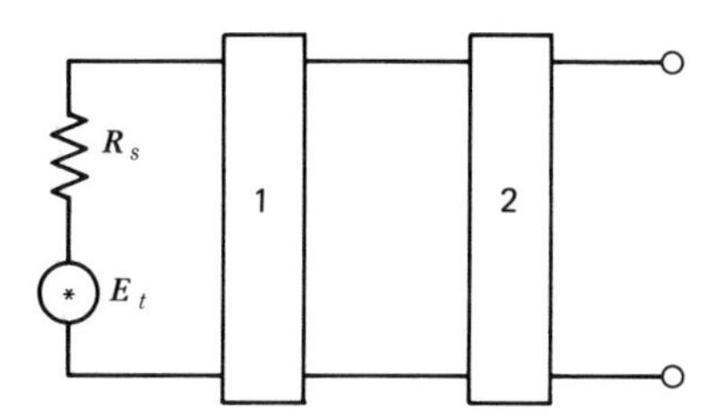

Fig. 2-4. Cascaded networks.

source with internal thermal noise and two cascaded networks. Equation 2-10 gives noise factor as the quotient of input S/N to output S/N. The available power gain of the system is represented by G_a, and the available thermal noise power is $N_i = E_t^2/4R_s$. Therefore an alternate expression for F is

$$F = \frac{N_o}{G_a kT \, \Delta f} \tag{2-17}$$

where N_o is the available noise power at the load terminals.

Equation 2-17 is not a useful design equation in its present form because we do not know N_o and G_a at this point.

The available noise power at the input to network 2, N_{i2}, is

$$N_{i2} = N_{o1} = F_1 G_1 kT \, \Delta f \tag{2-18}$$

Equation 2-18 is simply a rearrangement of Eq. 2-17. The second stage, *considered separately*, behaves according to

$$F_2 = \frac{N_{o2}}{G_2 kT \, \Delta f} \tag{2-19}$$

The noise originating in the second stage is $N_{o2} - G_2kT\,\Delta f$, or from Eq. 2-19 it is

$$F_2G_2kT\,\Delta f - G_2kT\,\Delta f = (F_2 - 1)G_2kT\,\Delta f \tag{2-20}$$

The $G_2kT\,\Delta f$ term represents thermal noise power in the hypothetical source resistance for network 2.

The total output noise N_{oT} is given by the sum of terms from Eqs. 2-18 and 2-20:

$$N_{oT} = G_2(F_1G_1kT\,\Delta f) + (F_2 - 1)G_2kT\,\Delta f = (F_1G_1G_2 + F_2G_2 - G_2)kT\,\Delta f \tag{2-21}$$

The noise factor of the cascaded pair is

$$F_{12} = \frac{N_{oT}}{G_1G_2kT\,\Delta f} \tag{2-22}$$

By substitution of Eq. 2-21 into Eq. 2-22 we obtain

$$F_{12} = F_1 + \frac{F_2 - 1}{G_1}$$

If the analysis is extended to three stages, we obtain the classical relation developed by Friis [3]:

$$F_{123} = F_1 + \frac{F_2 - 1}{G_1} + \frac{F_3 - 1}{G_1G_2} \tag{2-23}$$

One concludes then that *the noise factor of a cascaded network is primarily influenced by first-stage noise, provided that the gain of that stage is large.*

When network 1 is a combination of passive circuit elements, for example, a coupling or equalizing network, its available power gain is less than unity, and the overall system noise is severely influenced by noise contributions represented by F_2.

2-9 NARROWBAND AND WIDEBAND NOISE

Noise quantities can be specified on either a narrowband or a wideband basis. Noise spectral density, as introduced in Section 1-4, is obtained in the laboratory using a filter with a very narrow passband. Wideband noise, on the other hand, is limited by the system itself or by the noise-measuring device.

In this section we examine the relations between narrowband and wideband noise quantities, with particular emphasis on the bandwidth limiting imposed on the electronic system by a nonideal amplifier.

Suppose that we have a source of white noise and wish to relate wideband noise values to spot or narrowband noise values. Consider the noise bandwidth of the system to be Δf, the narrowband mean square noise voltage to be E_{nb}^2, and the wideband mean-square noise voltage to be E_{wb}^2. System response is shown in Fig. 2-5*a*. Then it must be true that

$$E_{wb}^2 = \int_{f_l}^{f_h} A_v^2(f) E_{nb}^2 \, df \qquad (2\text{-}24)$$

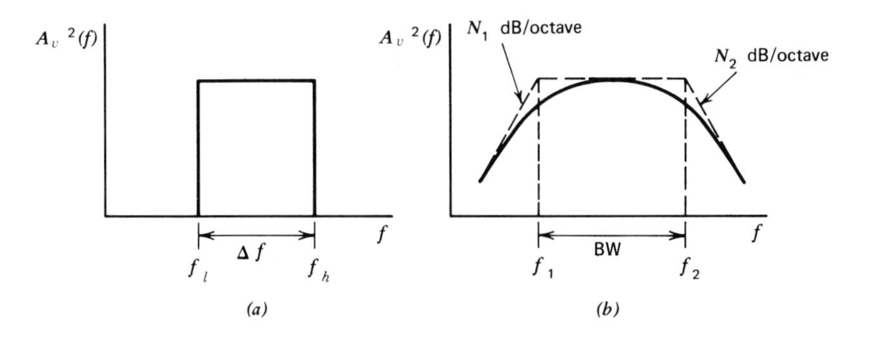

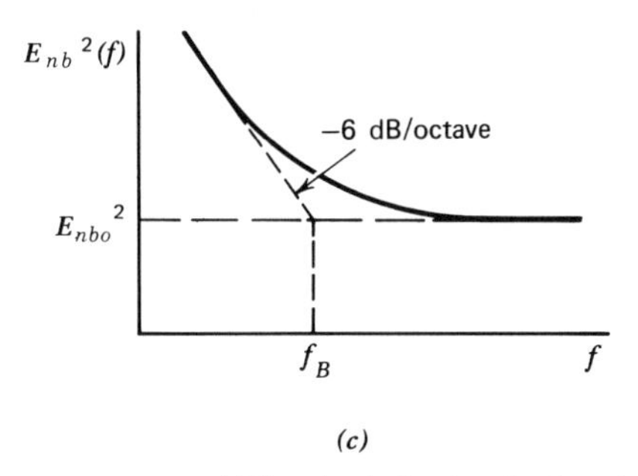

Fig. 2-5. Response curves and 1/f noise behavior for bandwidth conversions.

The limits of the noise bandwidth are f_l and f_h and $\Delta f = f_h - f_l$. For $A_v^2(f)$, a constant equal to unity for convenience, and E_{nb}^2 constant with frequency, it follows that

$$E_{wb}^2 = E_{nb}^2 \, \Delta f \qquad (2\text{-}25)$$

This case results in the simplest relation between wideband and narrowband noise; rms noise voltage increases as the square root of the bandwidth.

As a second case, *we assume white noise with mean square voltage* E_{nb}^2, *and system response as given in Fig. 2-5b.* For the $A_v^2(f)$ function, the break frequencies are f_1 and f_2 and the roll-offs are N_1 and N_2 dB/octave. A mathematical approximation to the magnitude behavior shown in this curve is

$$A_v^2(f) = \left(\frac{f}{f_1}\right)^{N_1/6} \quad \text{for} \quad 0 < f < f_1$$

$$A_v^2(f) = 1 \quad \text{for} \quad f_1 < f < f_2$$

$$A_v^2(f) = \left(\frac{f_2}{f}\right)^{N_2/6} \quad \text{for} \quad f_2 < f < \infty$$

Both N_1 and N_2 are considered to be positive in these equations. In order to determine the wideband noise we again integrate the product of response and white noise over the spectrum. The result is

$$E_{wb}^2 = E_{nb}^2\left[\text{BW} + \frac{f_1}{N_1/6 + 1} + \frac{f_2}{N_2/6 - 1}\right] \tag{2-26}$$

where BW $= f_2 - f_1$. The first term in the brackets is the noise contribution at midfrequencies, the second term is the contribution below f_1, and the third term is the contribution above f_2.

A set of typical numbers is used for an example. For $N_1 = N_2 = 24$ dB/octave, $f_1 = 50$ Hz, $f_2 = 20$ kHz, we determine BW to be just about 20 kHz. The high-frequency term contributes BW/3 and the contribution of the low-frequency term is negligible. Therefore, the wideband mean square noise voltage is approximately 4BW/3 times larger than the narrowband noise voltage. Here we have made approximations to the actual response: exact analysis yields 1.22 BW. If the $A_v^2(f)$ response falls off at 12 dB/octave and 36 dB/octave, exact analysis yields 1.57 BW and 1.12 BW, respectively.

A third example is the practical case where the noise is not white but varies with frequency as shown in Fig. 2-5c and the response is as shown in Fig. 2-5b. The break frequency f_B is assumed to lie between f_1 and f_2. This noise behavior is represented by

$$E_{nb}^2(f) = E_{nbo}^2 \quad \text{for} \quad f_B < f < \infty$$

$$E_{nb}^2(f) = E_{nbo}^2\left(\frac{f_B}{f}\right) \quad \text{for} \quad 0 < f < f_B$$

where E_{nbo}^2 is the reference or midfrequency value of E_{nb}^2.

To integrate, the spectrum is broken into four portions:

$$E_1^2 = \int_0^{f_1} E_{nbo}^2\left(\frac{f_B}{f}\right)\left(\frac{f}{f_1}\right)^{N_1/6} df = E_{nbo}^2\frac{6f_B}{N_1} \quad \text{for} \quad 0 < f < f_1 \tag{2-27a}$$

$$E_2^2 = \int_{f_1}^{f_B} E_{nbo}^2\left(\frac{f_B}{f}\right) df = E_{nbo}^2 \ln\frac{f_B}{f_1} \quad \text{for} \quad f_1 < f < f_B \tag{2-27b}$$

$$E_3{}^2 = \int_{f_B}^{f_2} E_{nbo}{}^2\, df = E_{nbo}{}^2(f_2 - f_B) \qquad \text{for } f_B < f < f_2 \qquad (2\text{-}27c)$$

$$E_4{}^2 = \int_{f_2}^{\infty} E_{nbo}{}^2 \left(\frac{f_2}{f}\right)^{N_2/6} df = E_{nbo}{}^2 \frac{f_2}{N_2/6 - 1} \qquad \text{for } f_2 < f < \infty \qquad (2\text{-}27d)$$

The total wideband noise is the sum of Eqs. 2-27*a* through 2-27*d*:

$$E_{wb}{}^2 = E_1{}^2 + E_2{}^2 + E_3{}^2 + E_4{}^2$$

Therefore

$$E_{wb}{}^2 = E_{nbo}{}^2\left[f_2\left(\frac{N_2}{N_2 - 6}\right) + f_B\left(\frac{6 - N_1}{N_1}\right) + \ln\frac{f_B}{f_1}\right] \qquad (2\text{-}28)$$

As a numerical example we use the following values: $f_1 = 50$ Hz, $f_2 = 20$ kHz and $N_1 = N_2 = 24$ dB/octave. Consider that $f_B = 1$ kHz. Then the multiplier of $E_{nbo}{}^2$ becomes

$$\frac{4f_2}{3} - \frac{3f_B}{4} + \ln 20$$

When evaluated the multiplier is 25,920. The final term involving f_B is found to be negligible in this example.

2-10 NOISE CIRCUIT ANALYSIS

An introduction to the circuit analysis of noisy networks was given in Section 1-7. Earlier in this chapter, in the derivation of Eq. 2-2, noise circuit analysis techniques were used. Here we expand on the preceding discussions in order to clarify the theory and extend it to further applications.

Refer to Fig. 2-6*a*. A sinusoidal voltage source is feeding two noiseless

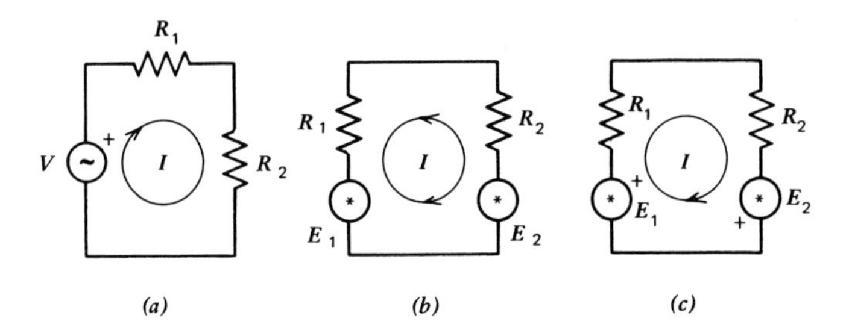

Fig. 2-6. Circuits for analysis examples.

resistances. Kirchhoff's voltage law allows us to write

$$V = IR_1 + IR_2 \tag{2-29}$$

Now suppose that we wanted to equate mean square values of the three terms in Eq. 2-29. Let us square each term:

$$V^2 = (IR_1^2) + (IR_2)^2$$

This operation is not valid! Why? There is 100% correlation between IR_1 and IR_2, for they contain the same current I. Therefore a correlation term must be present, and the correct expression is

$$V^2 = I^2R_1^2 + I^2R_2^2 + 2CIR_1IR_2 \tag{2-30a}$$

Since C must equal unity because only one current exists, the equation becomes

$$V^2 = I^2(R_1 + R_2)^2 \tag{2-30b}$$

The rule for series circuit analysis is simply that *when resistances or impedances are series connected, they should be summed first, and then, when dealing with mean square quantities, the sum should be squared.*

If V had been a noise source E, the same rule applies, for there is only one current in the circuit.

Now consider the circuit shown in Fig. 2-6*b*. Two uncorrelated noise voltage sources (or sinusoidal sources of different frequencies) are in series with two noiseless resistances. The current in this circuit must be expressed in mean square terms:

$$I^2 = \frac{E_1^2 + E_2^2}{(R_1 + R_2)^2} \tag{2-31}$$

No correlation term is present.

A convenient method for noise circuit analysis, when more than one source is present, employs superposition. The superposition principle states: *in a linear network the response for two or more sources acting simultaneously is the sum of the responses for each source acting alone with the other voltage sources short-circuited and the other current sources open-circuited.* Let us use superposition on the circuit of Fig. 2-6*b*. The loop currents caused by E_1 and E_2, each acting independently, are

$$I_1 = \frac{E_1}{R_1 + R_2}; \qquad I_2 = \frac{E_2}{R_1 + R_2}$$

And, for uncorrelated quantities,

$$I^2 = I_1^2 + I_2^2$$

Therefore

$$I^2 = \frac{E_1^{\,2}}{(R_1 + R_2)^2} + \frac{E_2^{\,2}}{(R_1 + R_2)^2} = \frac{E_1^{\,2} + E_2^{\,2}}{(R_1 + R_2)^2} \tag{2-32}$$

This agrees with Eq. 2-31.

In Fig. 2-6*c* sources E_1 and E_2 are correlated. Polarity symbols have been added to show that the generators are aiding. Then

$$I^2 = \frac{E_1^{\,2} + E_2^{\,2} + 2CE_1E_2}{(R_1 + R_2)^2} \tag{2-33}$$

where $0 < C \leq +1$ for aiding generators. If the polarity symbol on either generator were at its opposite terminal, C would take on values between 0 and -1. Note that when full correlation exists, it is valid to equate rms quantities ($E = IR_1 + IR_2$).

Suppose that we have a circuit such as shown in Fig. 2-7 in which there are several uncorrelated noise currents. We wish to determine the total

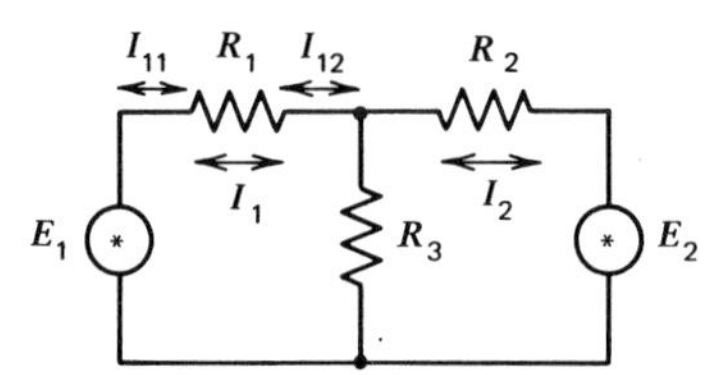

Fig. 2-7. Two-loop circuit.

current I_1 through R_1. For this example superposition is used; the contribution of E_1 to I_1 is termed I_{11}, and the contribution of E_2 to I_1 is I_{12}. It follows that

$$E_1^{\,2} = I_{11}^{\,2}\left(R_1 + \frac{R_2R_3}{R_2 + R_3}\right)^2 \quad \text{and} \quad E_2^{\,2} = I_2^{\,2}\left(R_2 + \frac{R_1R_3}{R_1 + R_3}\right)^2$$

observe that $I_{12} = I_2R_3/(R_1 + R_3)$. Therefore

$$I_{11}^{\,2} = \frac{E_1^{\,2}(R_2 + R_3)^2}{(R_1R_2 + R_1R_3 + R_2R_3)^2} \quad \text{and} \quad I_{12}^{\,2} = \frac{E_2R_3^{\,2}}{(R_1R_2 + R_1R_3 + R_2R_3)^2}$$

Hence

$$I_1^{\,2} = \frac{E_1^{\,2}(R_2 + R_3)^2 + E_2^{\,2}R_3^{\,2}}{(R_1R_2 + R_1R_3 + R_2R_3)^2} \tag{2-34}$$

When finding the total current resulting from several uncorrelated noise sources, the contributions from each source must be added in such a way so that the magnitude of the total current is increased by each contribution. Therefore neither the E_1 nor the E_2 terms in Eq. 2-34 could accept negative signs. An argument based on the heating effect of the currents, or one based on combining currents of different frequencies, can be used to justify this statement.

When performing a noise analysis of multisource networks, it is convenient to ascribe polarity symbols to uncorrelated sources in order that the proper addition (and no subtraction) of effects take place. This is shown in the analysis example given in Section 4-10.

SUMMARY

a. A universal noise model for amplifiers uses generators E_n and I_n at the input port.

b. All noise sources in a sensor-amplifier system can be represented by equivalent input noise E_{ni} (or I_{ni}), a voltage (or current) generator located in series (in parallel) with the signal source.

c. Equivalent input noise for a simple sensor-amplifier system is

$$E_{ni}^2 = E_t^2 + E_n^2 + I_n^2 R_s^2$$

d. From data on E_{ni} versus R_s, both E_n and I_n can be determined.

e. Noise factor is useful to compare amplifiers and devices:

$$F = \frac{S_i/N_i}{S_o/N_o}$$

It is often expressed in dB as noise figure.

f. The optimum source resistance ($R_o = E_n/I_n$) is defined for $E_n = I_n R_s$. This results in optimum noise factor.

g. For source resistance thermal noise equal to amplifier noise, noise resistance and noise temperature are defined.

h. For cascaded networks,

$$F = F_1 + \frac{F_2 - 1}{G_1} + \frac{F_3 - 1}{G_1 G_2}$$

System noise is predominantly first-stage noise when the gain of that stage is high.

i. Conversion of wideband to narrowband noise voltages follows

$$E_{wb}^2 = E_{nb}^2 \, \Delta f$$

when noise is white and the passband is well-defined.

j. Noise analysis of multisource networks can use the superposition principle.

PROBLEMS

1. Derive Eq. 2-14 from Eq. 2-12.

2. Determine the noise figure and noise factor for the following values of $(E_n^2 + I_n^2R_s^2)/E_t^2$: (a) 0, (b) 1, and (c) 2.

3. Find the noise factor in dB for a system with noise temperature equal to ambient (290°K).

4. Derive Eq. 2-17.

5. A system is composed of two noisy resistances in series. Resistance R_1 is the signal source, and R_2 is the load or output. Determine the noise figure for this network. Can you conclude that the broadband noise figure and the spot noise figure are identical for this system?

REFERENCES

1. "Representation of Noise in Linear Twoports," *Proc. IRE,* **48**, 1 (January 1960) 69–74.
2. "IRE Standards on Methods of Measuring Noise in Linear Twoports, 1959," *Proc. IRE,* **48**, 1 (January 1960), 60–68.
3. Friis, H. F., "Noise Figures of Radio Receivers," *Proc. IRE,* **32**, 7 (July 1944), 419–422.

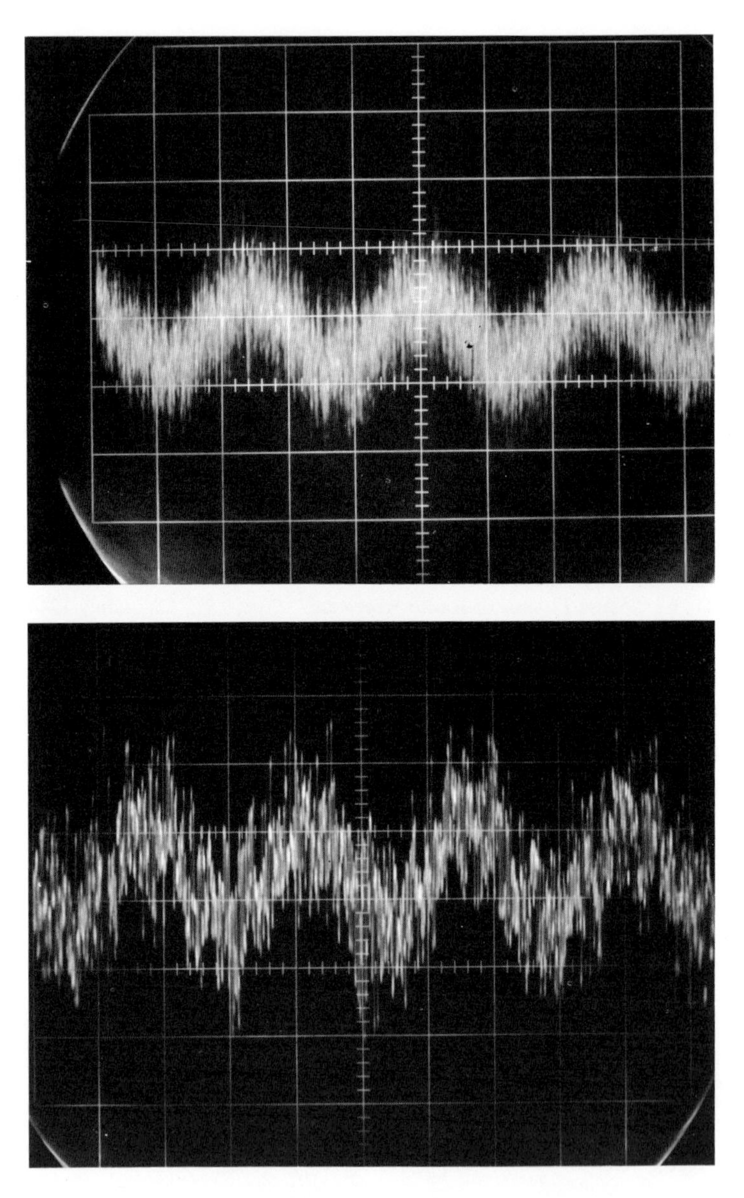

The signal-to-noise ratio is about unity in each photo. Top waveform is sinusoidal signal in white noise; bottom waveform is sinusoidal signal in 1/f noise. White noise appears "furry" or "grassy" on an oscilloscope; 1/f noise appears "rough" and "jumpy."

Chapter 3

NOISE IN SENSORS

In instrumentation, measurement, and control systems it is necessary to monitor some physical process or mechanism. The objective may be to measure a process variable, such as fluid flow, without disturbing it. The usual method is to insert a sensor or transducer that converts a small amount of the flow energy into an electrical signal. Since we do not want to affect the flow, very little power can be removed from the process, and, therefore, the resulting electrical signal output from the sensor is often weak.

In the design of new sensing devices, and in the application of existing devices, many characteristics are important. The sensitivity or gain is an important parameter, as is the frequency or transient response. The linearity of the transfer characteristic must be considered. Output impedance, input and output ranges of the variable, and power supply drain are important in certain applications. And certainly the noise properties of sensors have a great effect on the success of an application.

The engineer's task is to amplify the weak sensor signal without masking it by noise. All sensors have internal noise generators, and can be characterized by their basic signal-to-noise ratio.* Methods of analysis of sensor-amplifier systems are introduced in this chapter.

3-1 NOISE MODELS

To develop the noise model of a sensor we can start with its circuit diagram. From this we draw an ac equivalent circuit that includes all impedances and generators. To each resistance and current generator we add the appropriate noise generators to develop a noise equivalent circuit. The resistances

* The words *sensor*, *source*, *transducer*, and *detector* are used interchangeably in this book to represent the device that generates the electrical signal to be measured.

have thermal noise and possibly excess noise. The current generators may have shot noise, l/f noise, and/or excess noise. These mechanisms are described in Chapter 1. Using this noise equivalent circuit, an expression for equivalent input noise can be derived.

It is advantageous, however, to study the entire sensor-electronic system. A typical system may include a coupling device or network, as well as an amplifier. The noise equivalent circuit of the coupling network is easily obtained, and the E_n-I_n representation is valid for the amplifier. When we combine these three parts, we obtain a noise equivalent for the system.

The derivation of the equivalent input noise for the system follows three steps:

1. Determine the total output noise.
2. Calculate the system gain.
3. Divide the total output noise by system gain to get equivalent input noise.

In Chapter 8, noise models and equivalent input noise expressions are derived for six classes of sensors. These sensors are

1. Resistive sensor
2. Biased resistive source.
3. *RLC* source.
4. Biased diode sensor.
5. Transformer model.
6. Piezoelectric sensor.

In this chapter, we develop methods for the analysis of general sensor-amplifier combinations and determine S/N ratios. The effects of shunt resistance and capacitance on equivalent input noise are determined.

3-2 A GENERAL NOISE MODEL

The diagram shown in Fig. 3-1 contains the E_n-I_n representation for the system electronics. The sensor is described by its signal voltage V_s, its internal impedance Z_s, and a noise generator E_{ns} that represents all sources of sensor noise. To generalize the diagram a coupling network section represented by impedance Z_c and a noise source E_{nc} is included in shunt with the sensor.

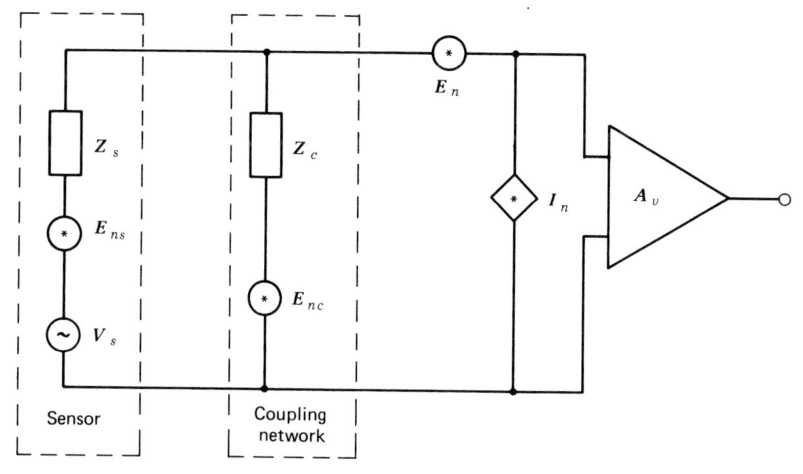

Fig. 3-1. General noise model.

A general form for the equivalent input noise voltage is

$$E_{ni}^2 = A^2 E_{ns}^2 + B^2 E_n^2 + C^2 I_n^2 R_s^2 + \cdots \tag{3-1}$$

A general form for an equivalent input noise current is

$$I_{ni}^2 = J^2 I_{ns}^2 + \frac{K^2 E_n^2}{R_s^2} + L^2 I_n^2 + \cdots \tag{3-2}$$

To evaluate these noise equivalents, the A^2, B^2, and so on, coefficients must be known.

Several cases are discussed to show the source of these constant terms. Generally speaking, the A term is dependent on the sensor impedance Z_s and the coupling network Z_c. The B term is caused by shunt impedance in the sensor and coupling network. The C term is determined by series impedances in the sensor and coupling network.

Equivalent input noise generators E_{ni} and I_{ni} are described in Eqs. 3-1 and 3-2. Either form can be used. The choice depends on the type of sensor being employed. If the signal source is a current generator, the equivalent noise current expression is more convenient. If the signal is a voltage generator, the equivalent voltage form may be more convenient.

The method of calculating E_{ni} or I_{ni} is the same for any transducer. Starting from a noise equivalent diagram, the total output noise E_{no} is calculated using Kirchhoff's laws. The equivalent input noise is the output noise E_{no} divided by system gain K_t. The system gain may be either a voltage gain, a current gain, or a transfer gain as needed.

3-3 EFFECT OF PARALLEL LOAD RESISTANCE

The simplest type of sensor is represented by a resistance in series with a signal voltage generator as shown in Fig. 3-2. Also shown is a shunt network consisting of R_p and E_{np}. The practical purpose of this network may be to supply the sensor with bias power. For example, the sensor may be a precision potentiometer with its input being the mechanical displacement of its shaft. Unless the potentiometer is supplied with electrical bias power, V_s would always be zero in the diagram of Fig. 3-2.

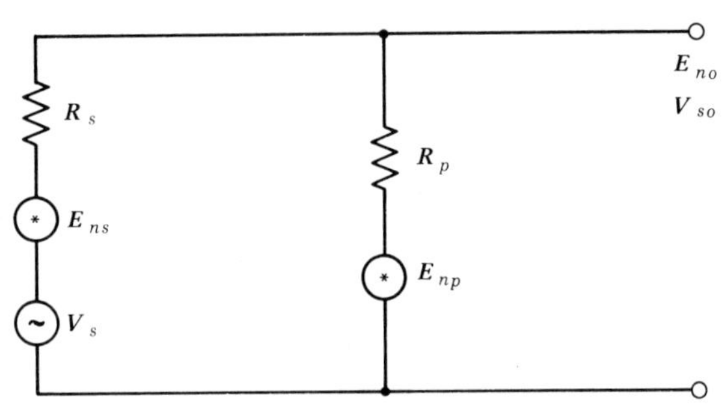

Fig. 3-2. Sensor model shunted by resistance.

The signal V_s and noise E_{ns} of the sensor are in series with the source resistance. The signal-to-noise power ratio is simply the ratio of V_s^2 to E_{ns}^2. When a load resistor such as R_p or other coupling network elements are added, the output signal-to-noise ratio is degraded.

Let us calculate the effect of shunt resistance on signal-to-noise power ratio or equivalent input nosie. This is done in two ways. First we directly calculate the output noise and output signal. Next we repeat the results by deriving an expression for the equivalent input noise E_{ni}.

As previously noted for $R_p \to \infty$,

$$S/N = \frac{V_{so}^2}{E_{no}^2} = \frac{V_s^2}{E_{ns}^2} \tag{3-3}$$

When R_p is not infinite, it must be included in the expression. Assume, for example, that $R_p = R_s$. The output noise, output signal, and signal-to-noise are determined for the case of $E_{ns} = E_{np}$:

$$E_{no}^2 = \left(\frac{E_{np}}{2}\right)^2 + \left(\frac{E_{ns}}{2}\right)^2 = \frac{E_{ns}^2}{2}$$

The output signal is

$$V_{so} = \frac{V_s}{2}$$

Therefore

$$S/N = \frac{V_{so}^2}{E_{no}^2} = \frac{(V_s/2)^2}{E_{ns}^2/2} = 0.5\left(\frac{V_s^2}{E_{ns}^2}\right) \tag{3-4}$$

From Eq. 3-4 we conclude that a shunt resistor decreases the signal more than the noise and the result is a decrease in signal-to-noise ratio. For the matched condition, source resistance equal to load resistance, the S/N power ratio is reduced by 50% from the unloaded value.

A more complete analysis is now made of the circuit shown in Fig. 3-3.

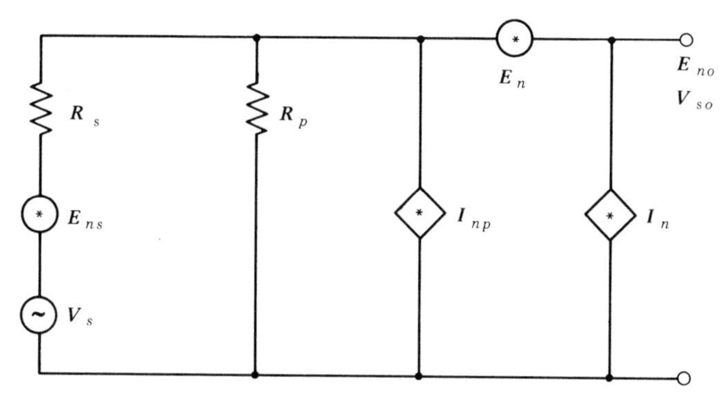

Fig. 3-3. Amplifier and sensor models with shunt resistance.

Again, a noisy shunt resistance R_p is present. For convenience we consider its noise to be represented by the current generator $I_{np} = (4kT\,\Delta f/R_p)^{1/2}$. Amplifier noise generators E_n and I_n are included.

To examine the effect of all noise sources we calculate the equivalent input noise. Three steps to be followed are the following:

1. From the equivalent circuit determine the output noise E_{no}.
2. Calculate the system gain K_t (the transfer function from sensor to output).
3. Divide output noise by system gain to obtain the equivalent input noise.

Following this procedure, we obtain

1. $$E_{no}^2 = E_{ns}^2\left(\frac{R_p}{R_s + R_p}\right)^2 + E_n^2 + (I_n^2 + I_{np}^2)\left(\frac{R_pR_s}{R_p + R_s}\right)^2 \tag{3-5}$$

2. $$K_t = \frac{R_p}{R_s + R_p}$$

3. $$E_{ni}^2 = \frac{E_{no}^2}{K_t^2} = E_{ns}^2 + \left(\frac{R_s + R_p}{R_p}\right)^2 E_n^2 + (I_n^2 + I_{np}^2)R_s^2 \tag{3-6}$$

There are two differences between Eq. 3-6 and the basic equation for equivalent input noise, Eq. 2-7. As predicted in Eq. 3-1, the E_n term has a coefficient; it is dependent on the shunt resistance R_p. If R_p were made very large, the coefficient of E_n would approach unity. The second difference is the additional thermal noise generator I_{np} representing shunt resistance. Amplifier input impedance is not important, as was discussed in Section 2-2.

In practice, the shunt resistor should be as large as possible to reduce its noise contribution. If the thermal noise of this component can be made zero, an improvement in performance results. An inductance has no thermal noise and could be used in certain applications in place of R_p.

The input signal-to-noise ratio for the entire system of Fig. 3-3 is considered to be V_s^2/E_{ns}^2. The output signal-to-noise ratio is V_{so}^2/E_{no}^2. Both V_{so} and E_{no} are equally amplified by a noiseless amplifier, so we can omit amplifier voltage gain from calculations. Therefore $V_{so} = V_s R_p/(R_p + R_s)$. Noise output E_{no} is given in Eq. 3-5. The output ratio is, therefore,

$$\frac{S_o}{N_o} = \frac{V_s^2}{E_{ns}^2 + E_n^2/K_t^2 + (I_n^2 + I_{np}^2)R_s^2}$$

The noise factor is $(S_i/N_i)/(S_o/N_o)$. From the preceding discussion we obtain

$$F = 1 + \frac{E_n^2}{E_{ns}^2 K_t^2} + \frac{(I_n^2 + I_{np}^2)R_s^2}{E_{ns}^2} \tag{3-7}$$

For the special case of no amplifier noise contribution and no contribution from the shunt resistor, F reverts to unity, as expected.

3-4 EFFECT OF SHUNT CAPACITANCE

Although a capacitance is virtually noise-free, it can be the cause of an increase in the equivalent input noise. A shunt capacitance does not affect the *sensor* signal-to-noise ratio. It decreases the signal and noise, but not the ratio.

Consider the noise equivalent circuit, including shunt capacitance, shown in Fig. 3-4. We shall study this circuit using the method of the preceding section.

1. $$E_{no}^2 = E_{ns}^2\left(\frac{1}{1 + \omega^2 R_s^2 C^2}\right) + E_n^2 + I_n^2\left(\frac{R_s^2}{1 + \omega^2 R_s^2 C^2}\right) \tag{3-8}$$

2. $$K_t^2 = \frac{1}{1 + \omega^2 R_s^2 C^2}$$

3. $$E_{ni}^2 = E_{ns}^2 + (1 + \omega^2 R_s^2 C^2)E_n^2 + I_n^2 R_s^2 \tag{3-9}$$

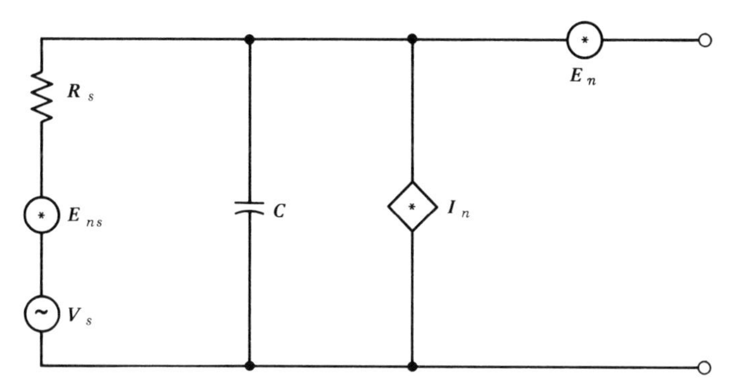

Fig. 3-4. Shunt capacitance along with amplifier and sensor models. Because E_n has no series impedance, I_n can be located on either side of E_n.

Although the E_n^2 term increases with shunt capacity, as evident from the coefficient of that term in Eq. 3-9, there are no other noise changes. Therefore the S/N for the sensor alone is not changed. Only the effective amplifier noise contribution is increased. This shunt capacitance *is not* the input capacitance of the amplifier. The amplifier input capacitance drops out of the expression for equivalent input noise, but its effects may be implied in the values of E_n and I_n.

3-5 NOISE OF A RESONANT CIRCUIT

Another interesting model is the resonant inductor type of sensor. The noise equivalent circuit for such a sensor is shown in Fig. 3-5. Following the method of this chapter the expression for input noise is

$$E_{ni}^2 = E_{ns}^2 + [(1 - \omega^2 C_p L_s)^2 + (\omega R_s C_p)^2]E_n^2 + (R_s^2 + \omega^2 L_s^2)I_n^2 \qquad (3\text{-}10)$$

Both the E_n and I_n terms are increased by the sensor impedance. The coefficient of the I_n term is primarily determined by the series elements R_s and L_s. The coefficient of the E_n term is primarily determined by the shunt capacity, although there is also a resonance term present.

Let us examine the resonance multiplier on the E_n noise mechanism. When $\omega^2 C_p L_s = 1$, the coefficient has a minimum value of $\omega R_s C_p$. If R_s is small, the noise voltage term E_n may be negligible. The thermal noise of the source E_{ns} and the equivalent noise current mechanisms are unaffected by this resonance.

Resonance offers a way to reduce the noise of an amplifier for a low-resistance sensor. A purposely added series inductance and shunt capacitor

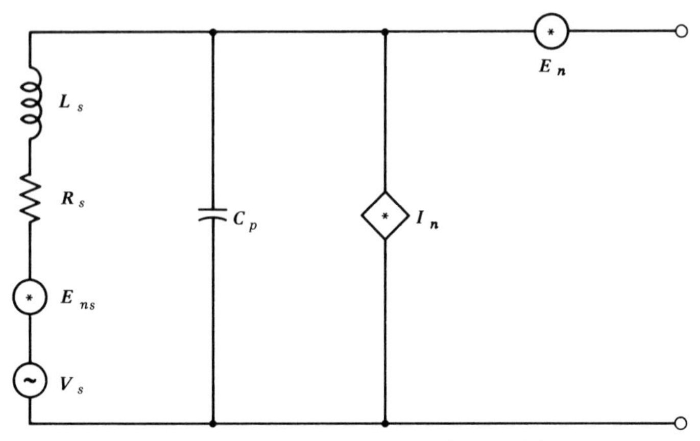

Fig. 3-5. Amplifier and resonant sensor noise models.

could resonate the circuit. At low-source resistances, the E_n term may dominate. The resonance can remove most of the contribution of the E_n term, giving a very low equivalent input noise dominated by the thermal noise of the sensor. Although this is a narrowband technique, it may have advantages over a matching transformer.

3-6 BIASED RESISTIVE DETECTOR EXAMPLE

As an example of the material discussed in this chapter, let us consider a system using a photoconductive infrared cell as its sensor. This device can be classified as a biased resistive type of detector. The system diagram is shown in Fig. 3-6*a*. From this diagram the ac equivalent circuit of Fig. 3-6*b* is drawn. The everpresent shunt capacity is included, but the reactance of coupling capacitor C_c is assumed to be much lower than the source resistance at the lowest frequency of interest.

Noise generators have been added to the ac equivalent circuit in Fig. 3-7. The noise sources are represented by their current generator equivalents, for in this instance it is convenient to add parallel current generators. The additional noise generator I_{ex} represents the excess 1/f noise generated by the bias direct current flowing through the infrared cell. A solution for equivalent input noise yields

$$E_{ni}^2 = (I_{ns}^2 + I_{nL}^2 + I_n^2 + I_{ex}^2)Z_p^2 + E_n^2 \tag{3-11}$$

where

$$Z_p = R_s \| R_L \| (-jX_c)$$

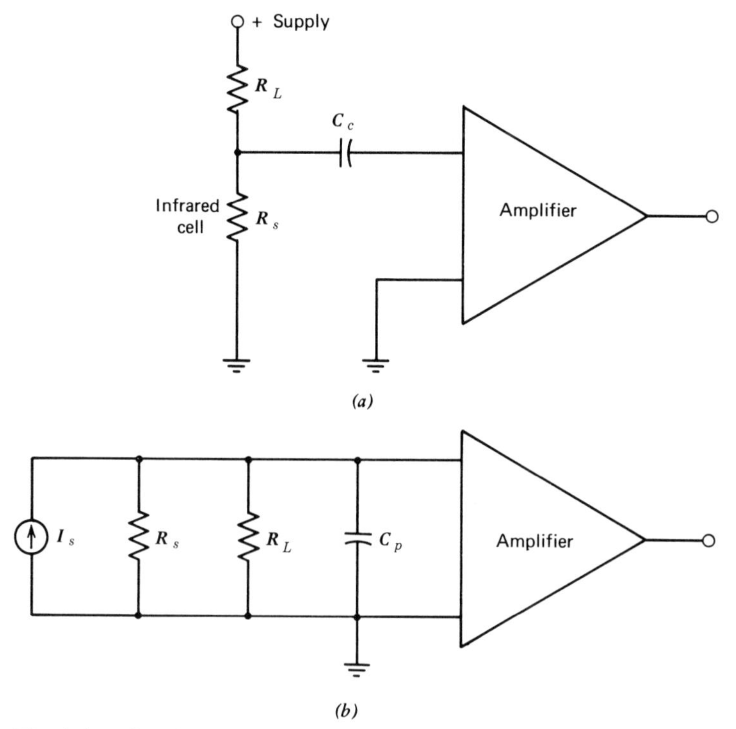

Fig. 3-6. Biased resistive detector.

As a current generator, equivalent input noise is

$$I_{ni}^2 = I_{ns}^2 + I_{nL}^2 + I_n^2 + I_{ex}^2 + \frac{E_n^2}{Z_p^2} \tag{3-12}$$

Two significant noise current generators are caused by the thermal noise of the source resistance and the load resistance. Since the value of I_{nL} is inversely

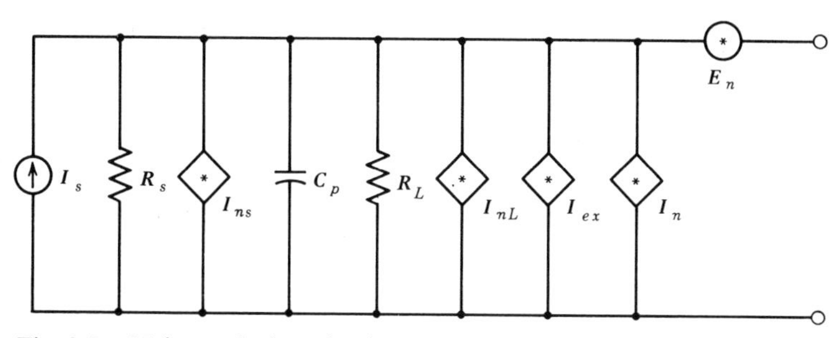

Fig. 3-7. Noise equivalent circuit.

related to the resistance, we again desire R_L to be large. R_L can be replaced by a noise-free inductance in some applications.

3-7 PIN PHOTODIODE SENSOR SYSTEM

We continue the discussion of the preceding section with reference to a specific sensor, the PIN photodiode. In this section several of the engineering considerations relevant to the application of this device are discussed. The device is a detector for visible and near infrared radiation. It is contained in a TO-18 can with a glass-window cap to admit radiation [1]. The diode is a biased detector, with the biasing voltage between 5 and 20 V dc for best performance.

The PIN photodiode is a semiconductor "sandwich" composed of an "I" or intrinsic layer of silicon with *p*-type silicon diffused into its upper face and *n*-type silicon forming the lower face. The external surfaces of the sandwich are covered with gold to prevent optical radiation from reaching the semiconductor material.

The I-layer is purposely thick; the *p*-layer is very thin. A dc bias of less than 50 V is applied to the structure with its positive connected to the *n*-material and negative to the *p*-material. No connection is made to the *I*-layer.

Light energy can reach the diode through an aperture in the gold coating above the *p*-layer. When a photon is absorbed by the silicon, a hole and an electron are liberated from a broken covalent bond. Because the *p*-layer is so thin, the bond that is broken is in the I-layer. The applied potential has establishes an electric field in that layer that accelerates the hole toward the *p*-layer and the electron toward the *n*-layer. Naturally, this charge movement constitutes a current, the value of which is dependent on the intensity of light impinging on the device.

Electrically the PIN diode's behavior is similar to a conventional junction diode. The reverse bias controls the capacitance of the structure; increasing the value of the bias to 20 V reduces that capacitance to less than 1 pF in some types. The thin *p*-layer can be represented by a low-valued resistance; the current leakage through the device under reverse bias can be modeled by a large resistance.

In the small-signal and noise model of Fig. 3-8, R_p is the series resistance of the semiconductor, less than 50 Ω. Elements C_d and R_d are the diode's capacitance and leakage resistance, typically 5 pF and 10 GΩ, respectively. I_s is the signal current resulting from illumination. Thermal noise generated in R_p and R_d has no effect on operation because of the values of those resistances. The major noise source is shot noise:

$$I_d^2 = 2qI_{DC}\,\Delta f \tag{3-13}$$

where I_{DC} is the dark or leakage current that can be as low as 100 pA.

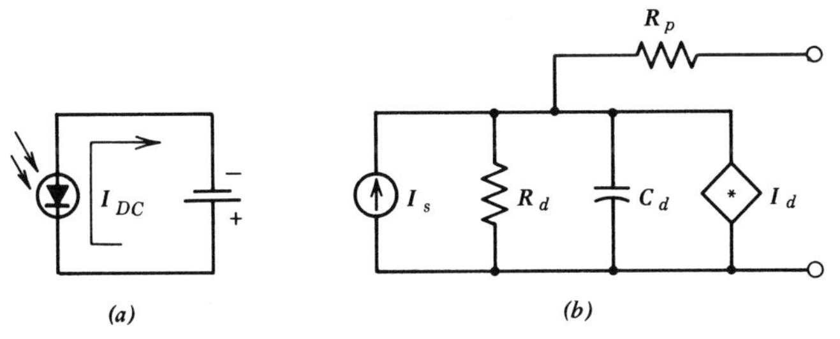

Fig. 3-8. (*a*) Biasing for photodiode; (*b*) small-signal and noise equivalent circuit.

Some excess 1/f noise is generated in the photodiode. It can be included in the I_d generator of the figure. It has a noise corner of about 20 to 30 Hz and rises at about 10 dB/decade for frequencies below that break.

The signal source, I_s, in Fig. 3-8, is capable of high-current responsivity (CR); it can provide an output of some 0.5 μA/μW in the band of light wavelengths from 0.5 to 0.8 μm. This band includes part of the visible spectrum and a portion of the infrared spectrum. The quoted responsivity corresponds to 0.75 electrons/photon or a quantum efficiency of 75%.

Noise equivalent power, NEP, is a figure of merit used for comparison of sensors. *NEP is the value of the input signal* (*in this case light power*) *that produces an output electrical signal equal to the noise output present when no input is applied.* Thus in equation form we have

$$\text{NEP} = \frac{\text{noise current (A/Hz}^{1/2})}{\text{current responsivity (A/W)}} \qquad \text{W/Hz}^{1/2}$$

or

$$\text{NEP} = \frac{I_n}{\text{CR}} \tag{3-14}$$

A low value for NEP corresponds to a highly sensitive detector. Using numbers given in a preceding paragraph, the NEP for a PIN sensor can be as low as -110 dBm (dBm refers to dB below a 1 mW reference). The corresponding numerical value of NEP for this case is 1.1×10^{-14} W/Hz$^{1/2}$. For an input light power level of 1 pW, the S/N ratio for the device, found by dividing signal current by noise current, is nearly 40 dB.

The threshold sensitivity of a PIN photodiode system is adversely affected if the diode is forced to feed a low-resistance load. A low value of shunting resistance results in a disproportionate amount of thermal noise at the input terminals, relative to the signal and noise from the photodiode sensor. In the simplified noise equivalent circuit shown in Fig. 3-9, sensor shot noise is I_d, shunt resistance is R_L, and the noise current associated with R_L is shown as I_t. Applying the definition of NEP to this circuit, we may obtain

$$\text{NEP} = \frac{\sqrt{I_d^2 + I_t^2}}{\text{CR}} \tag{3-15}$$

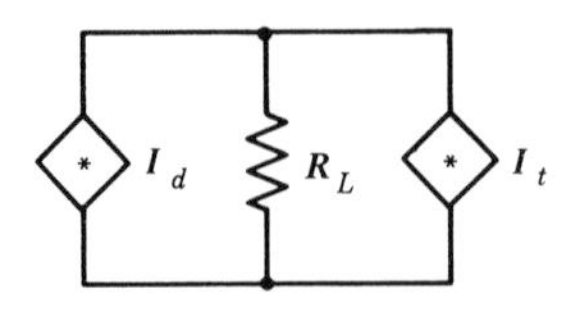

Fig. 3-9. Simplified noise equivalent circuit for PIN photodiode and amplifier.

It is clear that I_t^2 must be somewhat less than I_d^2 to reach an NEP level comparable with the unloaded figure given earlier. For $I_{DC} = 100$ pA, $I_d^2 = 3.2 \times 10^{-29}$ A^2. For I_t^2 to equal I_d^2, R_L must be 500 MΩ. Whereas it is usually not necessary for R_L to be that large in a practical application, it may be recognized that $I_t^2 = 4kT/R_L$, and consequently noise current increases for decreasing values of R_L.

The amplifier requirement can be satisfied in several ways. A bipolar transistor, selected for low noise, can be connected as an emitter-follower input stage; its operating point and load resistance are selected for low-noise and high-input impedance. A field-effect transistor (FET) may be ideal for this application. If the FET is connected in the common-drain configuration, the Miller effect is not present and thus amplifier input capacitance is low.

The high-resistance levels in the sensor-amplifier network result in a low value of upper cutoff frequency for the system. That frequency is approximately $1/2\pi(C_d + C_I)R_P$, with C_I the capacitance of the amplifier and $R_P = R_L \| R_I$. We see that for $C_d = C_I = 5$ pF, $R_P = 10$ MΩ, the cutoff frequency is a few thousand hertz. To broadband the system, two approaches can be followed. A "lead" network can be connected near the amplifier output to compensate for the high-frequency loss, or negative feedback can be employed around the amplifier.

The PIN photodiode sensor is a biased detector of the type discussed in the preceding section. Equation 3-12 can be used to study the effects of the components on equivalent input noise. The biased diode sensor is again studied in Section 8-9, and equations are presented for transfer voltage gain and equivalent input noise in terms of the values of the circuit elements.

SUMMARY

a. The three major noise contributors in an electronic system are sensor, amplifier, and coupling network.
b. Each contributor can be replaced by its noise equivalent circuit for noise analysis.
c. To determine equivalent input noise of a system, total output noise is divided by the system transfer gain.

d. Total noise at a location in an electronic system can be treated as the sum of the mean square values of the contributions of all sources at that location, each source acting independently.
e. E_{ni}, in its simplest form, is dependent on three noise sources: thermal sensor noise E_{ns} and amplifier parameters E_n and I_n. In general, the coupling network between sensor and amplifier increases the effective E_n and I_n portions of E_{ni}.
f. When the sensor impedance is reactive, resonance can sometimes be employed to reach acceptable noise performance.
g. Noise equivalent power (NEP) for a sensor is the ratio of noise to responsivity.

PROBLEMS

1. How would the addition of a shunt inductor affect the result given in Eq. 3-9 for input noise?
2. Derive Eq. 3-10.
3. Let E_{ns} in Eq. 3-7 include two sources of noise: a thermal noise contribution E_t and a shot noise contribution E_{sh}. How would you modify that equation taking into account that the basic definition of F considers available thermal noise power from the source resistance as its denominator?
4. Derive an expression for the noise factor F of the circuit of Fig. 3-5.
5. Repeat Problem 4 for the circuit of Fig. 3-7.

REFERENCE

1. *Solid State Devices*, Hewlett-Packard, Palo Alto, Cal., 1968, pp. 195–213.

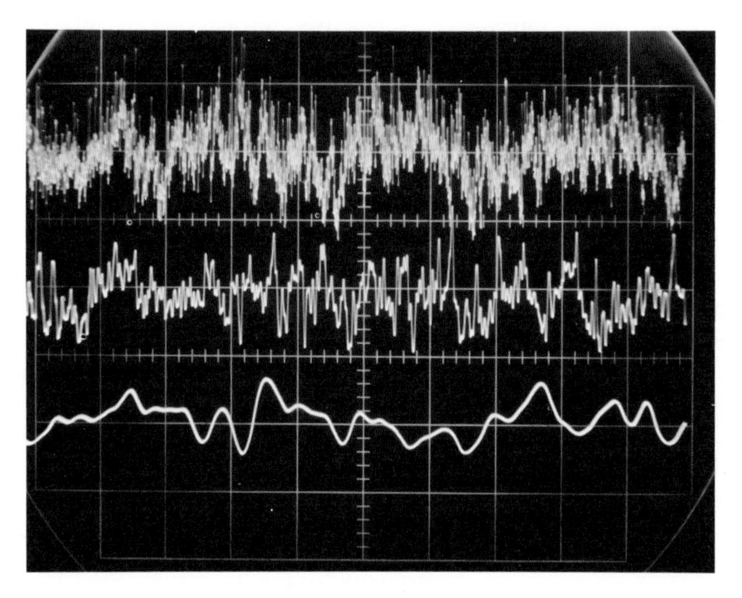

These traces show the effect of bandwidth limiting on l/f noise. Top waveform, bandwidth is 2 kHz; middle waveform, bandwidth is 200 Hz; bottom waveform, bandwidth is 20 Hz. Horizontal sensitivity is 50 msec/cm. Note that the peak amplitude is not proportionately reduced by bandwidth limiting.

Chapter 4

BIPOLAR TRANSISTOR NOISE MECHANISMS

The common bipolar transistor contains sources of thermal noise, 1/f noise, and shot noise, which are discussed in this chapter. The widely used hybrid-π small-signal equivalent circuit is modified to include noise sources in order to represent transistor noise behavior.

From the noise circuit model, we determine the equivalent input noise parameter E_{ni} for the bipolar transistor. The resulting expression is used to predict noise versus frequency behavior and forms the basis for derivation of expressions for E_n and I_n.

The conditions necessary for minimizing noise figure are considered. Because noise relations are expressed in terms of operating point currents, transistor parameters, temperature, and frequency, the circuit designer can apply the results to create low-noise electronic circuits.

Comparisons are made of various transistor types and results are presented that suggest the correct type for a given application.

4-1 THE HYBRID-π MODEL

The *hybrid-π* transistor model has been widely accepted as a representation of the small-signal behavior of the junction transistor. The parameters of the hybrid-π are generally not frequency dependent, so the model can be used over a wide band of frequencies. The basic model contains seven parameters as shown in Fig. 4-1. Between external base terminal b and internal terminal b' is the base-spreading resistance $r_{bb'}$. Elements $r_{b'e}$ and $C_{b'e}$ represent the significant portion of the input impedance of the transistor.

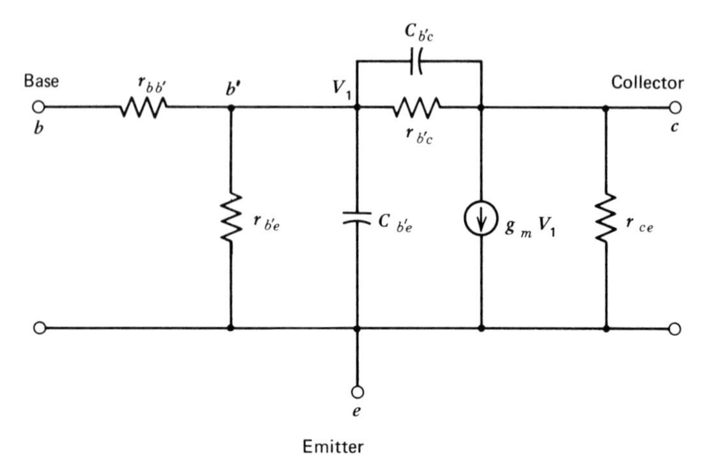

Fig. 4-1. Hybrid-π transistor small-signal equivalent circuit.

The amplification property of the device is represented by the dependent current generator $g_m V_1$, with V_1 the signal potential between b' and emitter terminal e. Elements r'_{bc} and $C_{b'c}$ are caused by the base-width modulation effect and by depletion-layer capacitance. These internal feedback mechanisms within the transistor often can be omitted in developing a simplified model. Element r_{ce} is representative of most of the output resistance of the transistor.

The more important parameters of the hybrid-π model can be expressed in terms of easily measured quantities. *The short-circuit current gain, called* h_{fe} *or* β, *is not a parameter of the hybrid-*π. However, if we assume a short circuit between c and e in Fig. 4-1, and consider low-frequency operation where the capacitances may be neglected, the low-frequency current gain β_o can be expressed in terms of the parameters of the model:

$$\beta_o \simeq \frac{g_m V_1}{V_1/r_{b'e}} = g_m r_{b'e} \tag{4-1}$$

Equation 4-1 assumes that $r_{b'c}$ is very large.

The parameter g_m can be derived from the diode equation relating I_C and V_{BE} [1]:

$$g_m = \frac{qI_C}{kT} = \Lambda I_C \tag{4-2}$$

At room temperature $\Lambda = q/kT = 40$. *Note that Eq. 4-2 relates a small-signal parameter to the dc collector current.*

Base-emitter resistance can also be expressed in terms of I_C. Substituting Eq. 4-2 into 4-1 yields

$$r_{b'e} = \frac{\beta_o}{g_m} = \frac{\beta_o}{\Lambda I_C} \tag{4-3}$$

The *gain-bandwidth product* f_T is the frequency at which the short-circuit current gain equals unity. This parameter is related to the hybrid-π parameters in the following manner:

$$C_{b'e} = \frac{g_m}{2\pi f_T} - C_{b'c} \tag{4-4}$$

The beta-cutoff frequency, f_{hfe} *or* f_β*, is the frequency at which beta has declined to 0.707 of its low-frequency reference value,* β_o. It can be shown that

$$f_{hfe} \simeq \frac{f_T}{g_m} \tag{4-5}$$

Another parameter useful to noise studies is the so-called Shockley emitter resistance, r_e (sometimes r_e'). This parameter, previously mentioned in Chapter 2, is simply the reciprocal of g_m:

$$r_e = \frac{1}{g_m} = \frac{1}{\Lambda I_C} \tag{4-6}$$

At room temperature, $r_e = 0.026/I_C$.

A typical set of hybrid-π parameters is

$$r_{b'e} = 97\ \text{k}\Omega \qquad r_{ce} = 1.6\ \text{M}\Omega$$
$$r_{bb'} = 278\ \Omega \qquad r_{b'c} = 15\ \text{M}\Omega$$
$$g_m = 0.0036\ \text{mho} \qquad C_{b'c} = 4\ \text{pF}$$
$$C_{b'e} = 25\ \text{pF}$$

These values apply to a type 2N4250 transistor at operating-point coordinates of $I_C = 0.1$ mA and $V_{CE} = 5$ V.

The basic hybrid-π model is slightly modified when considering monolithic integrated circuit transistors. Because of the longer distance for collector current to travel to the external contact, a small resistance is often added in series with the collector. An *npn* IC transistor is fabricated on a substrate of *p*-type semiconductor. The collector-to-substrate junction is reverse-biased. The capacitance associated with that junction can be added in the equivalent circuit between collector and a separate substrate terminal. The substrate terminal is almost always connected to ac ground.

4-2 THE NOISE MODEL

The noise mechanisms of bipolar transistors have been widely investigated and the results are well accepted [2]. Noise theory indicates that a good low-noise transistor must have a high short-circuit current gain β_o and low level of leakage current. There are many low-cost, low-noise transistors

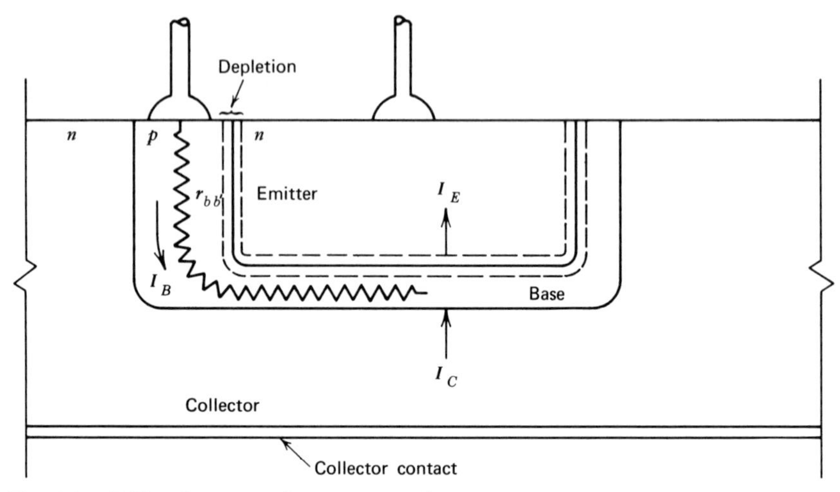

Fig. 4-2. Diffused-*npn* transistor cross-section.

available that meet these requirements. In the following derivations it is assumed that we have a high-gain, low-leakage transistor.

The cross-section of a diffused *npn* transistor is shown in Fig. 4-2. The *base-spreading resistance* r_{bb}' is the resistance of the lightly doped base region between the external base contact and the active base region. This is a true resistance, and therefore exhibits thermal noise. Fluctuations in base current I_B and collector current I_C are responsible for shot noise at the respective junctions. The flow of base current I_B through the base-emitter depletion region gives rise to 1/f noise [3]. Therefore, four noise generators are shown in the hybrid-π noise model of Fig. 4-3.

The thermal noise generators representing the source and load resistors are also shown. Feedback elements $C_{b'c}$ and $r_{b'c}$ have been removed for simplicity. This limits the application of this equivalent circuit to frequencies less than $f_T/\sqrt{\beta_o}$. At frequencies above f_{hfe}, the noise mechanisms are partially correlated and the total noise is larger than predicted by this simple model.

The E_b noise voltage generator is caused by the thermal noise of the base-spreading resistance. The noise current generator I_b is the shot noise of the total base current, and I_c is the shot noise of the collector current. These generators may be predicted theoretically from the discussions given in Chapter 1:

$$E_b^2 = 4kTr_{bb'}\,\Delta f$$

$$I_b^2 = 2qI_B\,\Delta f$$

$$I_c^2 = 2qI_C\,\Delta f$$

The source and load resistance thermal noise generators are

$$E_s^2 = 4kTR_s\,\Delta f$$

$$E_L^2 = 4kTR_L\,\Delta f$$

The 1/f noise generator warrants additional study.

The 1/f noise contribution is represented by a single noise current generator I_f connected as shown in Fig. 4-3. It has been observed experimentally that the 1/f noise current flows through the entire base resistance in alloy-junction germanium transistors, but a reduced value of $r_{bb'}$ is necessary to model correctly silicon planar transistors.

Following the discussion given in Section 1-9, the spectral density of 1/f noise current can be given by

$$I_f^2 = \frac{KI_B^{\gamma}}{f^{\alpha}} \tag{4-7}$$

Exponent γ ranges between 1 and 2, but often can be taken as unity. The experimental constant K takes on values from 1.2×10^{-15} to 2.2×10^{-12}. It has been found that this constant can be replaced by $2qf_L$ where q is the electronic charge, 1.6×10^{-19} C, and f_L is a constant having values from 3.7 kHz to 7 MHz. The value for f_L is a representation of the noise corner frequency. Its value does not match the corner frequency, but there is gross correlation. Exponent α is usually unity. Thus the form for 1/f noise used in this chapter is

$$I_f^2 = \frac{2qf_LI_B^{\gamma}}{f} \tag{4-8}$$

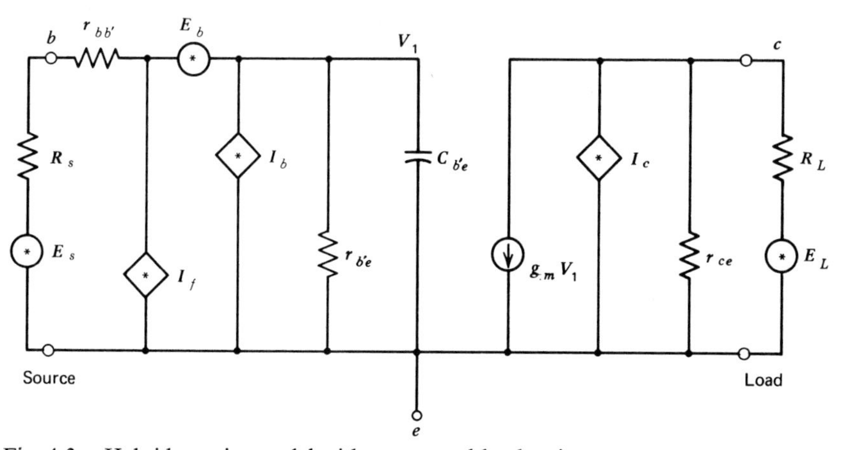

Fig. 4-3. Hybrid-π noise model with source and load resistances.

The 1/f noise voltage generator is the product of noise current given by Eq. 4-8 and the net resistance shunting I_f^2. Because the $r_{bb'}$ contribution to 1/f noise is less than theory predicts for planar transistors, it is necessary to define a new r_b, smaller than the value of $r_{bb'}$, to match experimental data. Unless this change is made, calculations predict excessive 1/f noise. *The effective* r_b *is approximately one-half of* $r_{bb'}$. An expression for the 1/f noise voltage generator is

$$E_f^2 = \frac{2qf_L I_B^\gamma r_b^2}{f} \tag{4-9}$$

The 1/f noise results from the trapping and detrapping of carriers in surface energy states. This is a process-dependent noise mechanism. It can be caused by defects such as impurities and dislocations. Transistors that have a high beta at very low collector currents seem to have little 1/f noise. The 1/f noise is strongly affected by surface properties. Planar transistors are passivated so their surfaces are fairly well protected. There are epoxy-encapsulated transistors with a low level of 1/f noise, but for the *most* critical applications it would still seem desirable to use a selected hermetically sealed transistor.

4-3 TRANSISTOR EQUIVALENT INPUT NOISE

To determine an overall signal-to-noise ratio, all transistor noise mechanisms are referred to the input port. To derive equivalent input noise E_{ni}, our method is to calculate the *total noise at the transistor output*, the *gain from source to output*, and finally *divide the output noise by the gain.*

If the output is shorted in Fig. 4-3, then the output noise current is

$$I_{no}^2 = I_c^2 + (g_m V_1)^2$$

or

$$I_{no}^2 = I_c^2 + g_m^2\left(\frac{(E_b^2 + E_s^2)Z_{b'e}^2}{(r_{bb'} + R_s + Z_{b'e})^2} + \frac{(I_b^2 + I_f^2)Z_{b'e}^2(r_{bb'} + R_s)^2}{(r_{bb'} + R_s + Z_{b'e})^2}\right) \tag{4-10}$$

For an input signal V_s, *the output short-circuit signal current* is

$$I_o = g_m V_1 = \frac{g_m V_s Z_{b'e}}{r_{bb'} + R_s + Z_{b'e}} \tag{4-11}$$

We define a transfer gain:

$$K_t = \frac{I_o}{V_s} = \frac{g_m Z_{b'e}}{r_{bb'} + R_s + Z_{b'e}} \tag{4-12}$$

We can now calculate the equivalent input noise E_{ni} as the ratio of Eqs. 4-10 to 4-12 according to

$$E_{ni}^2 = \frac{I_{no}^2}{K_t^2}$$

Therefore, *in terms of impedances and noise generators,* E_{ni} *is*

$$E_{ni}^2 = E_b^2 + E_s^2 + (I_b^2 + I_f^2)(r_{bb}' + R_s)^2 + \frac{I_c^2(r_{bb'} + R_s + Z_{b'e})^2}{g_m^2 Z_{b'e}^2} \tag{4-13}$$

Substituting the values for the noise generators ($\Delta f = 1$ Hz) into Eq. 4-13 gives the equivalent input noise:

$$E_{ni}^2 = 4kT(r_{bb'} + R_s) + 2qI_B(r_{bb'} + R_s)^2 + \frac{2qf_L I_B^{\gamma}(r_b + R_s)^2}{f} + \frac{2qI_C(r_{bb'} + R_s + Z_{b'e})^2}{g_m^2 Z_{b'e}^2} \tag{4-14}$$

The effective value of base resistance is used in the l/f term. Note that the last term contains $Z_{b'e}$ and is therefore frequency dependent. At low frequencies this term can be simplified to

$$\frac{2qI_C(r_{bb'} + R_s + r_{b'e})^2}{\beta_o^2}$$

At high frequencies, up to about $f_T/\sqrt{\beta_o}$, this term can be simplified to

$$\frac{2qI_C(r_{bb'} + R_s + 1/\omega C_{b'e})^2}{g_m^2/\omega^2 C_{b'e}^2} \simeq 2qI_C(r_{bb'} + R_s)^2\left(\frac{f}{f_T}\right)^2 \tag{4-15}$$

The final expression for equivalent input noise in terms of transistor parameters, temperature, operating-point currents, frequency, and source resistance is

$$E_{ni}^2 = 4kT(r_{bb'} + R_s) + 2qI_B(r_{bb'} + R_s)^2 + \frac{2qI_C}{\beta_o^2}(r_{bb'} + R_s + r_{b'e})^2 + \frac{2qf_L I_B^{\gamma}(R_s + r_b)^2}{f} + 2qI_C(r_{bb'} + R_s)^2\left(\frac{f}{f_T}\right)^2 \tag{4-16}$$

This form for E_{ni}^2 is an approximation. In the model of Fig. 4-3 the feedback capacitance $C_{b'c}$ has been omitted. This assumption results in actual noise greater than predicted by the equation. Nevertheless, Eq. 4-16 is a good engineering approximation to actual behavior. The errors are primarily at high frequencies, beyond transistor cutoff.

The first three terms in Eq. 4-16 are not frequency dependent and therefore form the *limiting noise* of any transistor. The first term $4kTr_{bb'}$ is the thermal noise voltage of the base resistance and the term $2qI_B(r_{bb'})^2$ is the shot noise

Table 4-1. Transistor Parameters and Noise Constants

Transistor	β_O @ I_C =					f_T in MHz @ I_C =										I_{CBO} (max)
Type	10 mA	1 mA	100 μA	10 μA	1 μA	10 mA	1 mA	100 μA	10 μA	1 μA	$C_{b'c}$ pF	$r_{bb'}$ Ω	r_b Ω	f_L MHz	γ	nA
2N930	355	292	200	125	77	129	107	32.5	5.4	0.495	8.0	750	350	1.7	1.3	1
2N2484	400	353	322	277	180	132	101	32.2	4.88	0.475	6.0	380	200	1.6	1.5	1
2N3117	714	662	535	417	130	163	141	37.5	5.64	0.566	4.5	1000	130	7.0	1.5	1
2N3242A	208	127	71	38	12	82.3	35.9	6.34	0.838	0.053	10.0	380	100	1.3	1.4	1
2N3391A	704	510	345	196	111	122	67.4	13.2	1.61	0.162	7.0	800	200	0.2	1.2	10
2N3800	416	425	422	400	250	310	157	34.3	4.85	0.521	4.0	330	130	0.36	1.2	1
2N4044	—	418	385	277	150	—	209	88.0	14.7	1.55	0.8	750	220	0.33	1.3	0.1
2N4058	201	250	235	192	120	106	106	43.2	7.02	0.759	5.0	280	100	0.18	1.4	10
2N4104	781	746	625	455	300	143	132	36.2	5.61	0.556	4.5	900	500	0.16	1.4	1
2N4124	418	299	200	114	60	360	172	39.7	4.9	0.495	4.0	110	40	0.011	1.1	5
2N4125	170	144	118	87	45	255	101	22.0	3.63	0.302	4.5	50	48	0.0037	1.2	5
2N3964 / 2N4250	389	373	347	290	150	160	79.2	16.4	2.64	0.194	4.0	150	65	0.0048	1.2	1
2N4403	278	200	136	80	48	182	68.6	11.9	1.34	0.133	8.5	40	8	0.0095	1.1	10
2N4935	122	120	69	30	15	302	492	103	12.2	0.99	0.25	90	40	0.72	1.3	1
2N5086	357	279	183	105	45	224	133	27.3	3.81	0.348	4.0	180	30	0.016	1.1	1
2N5118	—	135	222	227	192	—	307	71.9	15.2	1.6	—	2500	10,000	3.2	1.7	0.1
2N5138	215	154	77	55	29	131	64.2	14.2	2.14	0.169	5.0	100	20	1.25	1.3	1
TIS94	397	312	244	172	100	213	93	18.1	2.39	0.229	—	240	240	0.04	1.2	1
MM4048	232	227	215	200	166	256	137	34.7	5.04	0.492	4.0	130	100	0.20	1.5	1

voltage associated with the base current. The term $2qI_C(r_{b'e})^2/\beta_o^2$ is the collector current shot noise; it can be rearranged to the form $2qI_C r_e^2$. Another form for this voltage is $2kTr_e$.

It was mentioned that signal-to-noise ratio is a means of circuit and system evaluation. From determination of E_{ni}^2 we have a measure of output S/N. Both the signal V_s and the E_{ni} are at the same location in the system; they are amplified by the same gain when referred to the output port. Therefore

$$\frac{S_o}{N_o} = \frac{V_s^2}{E_{ni}^2} \tag{4-17}$$

Hence S_o/N_o is inversely related to E_{ni}^2. Should we wish to compare noise factors,

$$F = \frac{E_{ni}^2}{E_s^2} \tag{4-18}$$

This follows from the definition given in Section 2-5.

Transistor parameters measured on 19 types are given in Table 4-1. These values result from independent testing and do not necessarily agree with those available from manufacturers' specification sheets. Of specific interest is the variation of f_T with I_C.

4-4 NOISE VOLTAGE AND NOISE CURRENT MODEL

We now modify Eq. 4-16 to write *the total expression for* E_{ni}^2 *of the transistor in the absence of source noise*. By definition, this yields E_n^2 *of the transistor noise model*:

$$E_n^2 = 4kTr_{bb'} + 2qI_C r_e^2 + 2qI_B r_{bb'}^2 + \frac{2qf_L I_B^{\gamma} r_b^2}{f} + 2qI_C r_{bb'}^2 \left(\frac{f}{f_T}\right)^2 \tag{4-19}$$

The other parameter of the model is obtained from Eq. 4-16 by assuming that R_s is very large valued, specifically $R_s \gg r_{b'e}$. After dividing each term by R_s^2, *we obtain the* I_n^2 *parameter*:

$$I_n^2 = 2qI_B + \frac{2qf_L I_B^{\gamma}}{f} + 2qI_C\left(\frac{f}{f_T}\right)^2 \tag{4-20}$$

Derivation of the parameters of the E_n-I_n model is now complete.

4-5 LIMITING CASE FOR MIDBAND NOISE

Removing all frequency-dependent terms and the 1/f contribution from the expressions for E_n and I_n gives the *limiting noise condition for the midband region*. This is sometimes termed the *shot noise region* because the shot noise

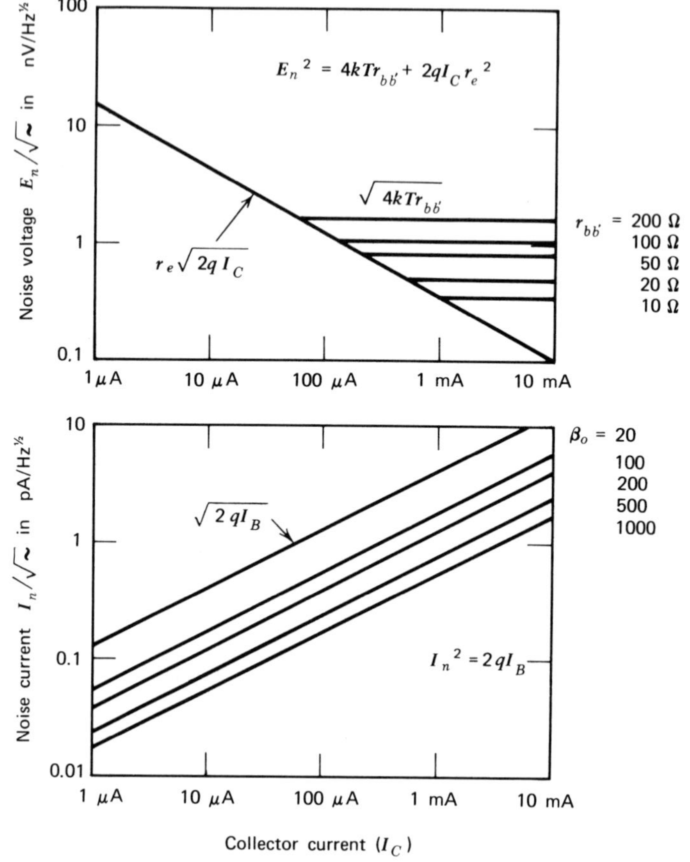

Fig. 4-4. Graphs of Eqs. 4-21 and 4-22.

current generators dominate in a well-designed circuit. Equations 4-19 and 4-20 become

$$E_n^2 = 4kTr_{bb'} + 2qI_C r_e^2 \tag{4-21}$$

and

$$I_n^2 = 2qI_B \tag{4-22}$$

Two mechanisms limit E_n: the thermal noise of the base resistance and the shot noise of the collector current times the emitter resistance. At low values of source resistance, where E_n dominates, it is desirable that the transistor have a low base resistance, and that operation take place at a high level of collector current. The I_n noise current generator is determined by a single noise mechanism, the shot noise of the base current. For high values of source resistance, where I_n dominates, it is desirable to operate with a low-leakage, high-beta transistor at a low level of direct collector current.

Plots of E_n and I_n versus collector current for certain values of transistor parameters are shown in Fig. 4-4. The E_n curve has a minimum value at high-collector currents determined by the thermal noise of the base resistance. The noise voltage rises steadily with decreasing collector current. The magnitude of the noise current generator I_n decreases with decreasing collector current and with increasing β_o. This illustrates the minor effect of beta and the strong dependence on collector current.

4-6 MINIMIZING THE NOISE FIGURE

Optimum noise factor was defined in Chapter 2. The discussion resulted in the following expression:

$$F_{\text{opt}} = 1 + \frac{E_n I_n}{2kT\,\Delta f} \tag{2-14}$$

Values for E_n and I_n from Eqs. 4-21 and 4-22 can be substituted into Eq. 2-14, and the result is an expression for *the optimum noise factor of a bipolar transistor*:

$$F_{\text{opt}} = 1 + \sqrt{\frac{2r_{bb'}}{\beta_o r_e} + \frac{1}{\beta_o}} \tag{4-23}$$

From this expression for optimum noise factor in terms of base resistance, emitter resistance, and transistor beta, we observe that increasing beta reduces the noise factor; similarly, reducing base resistance and/or collector current reduces the noise. The lowest noise is obtained at low-collector currents because the level is limited by the shot noise of the collector current and not by the thermal noise associated with the base current. The data shown in Fig. 4-5 indicates that the input transistor should be operated at a collector current of less than 100 μA.

To design for lowest noise figure we require that the first transistor operate in the shot noise limited region. Minimum noise is highly dependent on the transistor beta and inversely dependent on that parameter. We can define a limit on F_{opt} as follows:

$$F_{\text{opt}} \simeq 1 + \frac{1}{\sqrt{\beta_o}} \tag{4-24}$$

The optimum value of noise factor is only obtained when the source resistance is optimum. That value of R_o is given by

$$R_o = \sqrt{2\beta_o r_{bb'} r_e + \beta_o r_e^2} \tag{4-25}$$

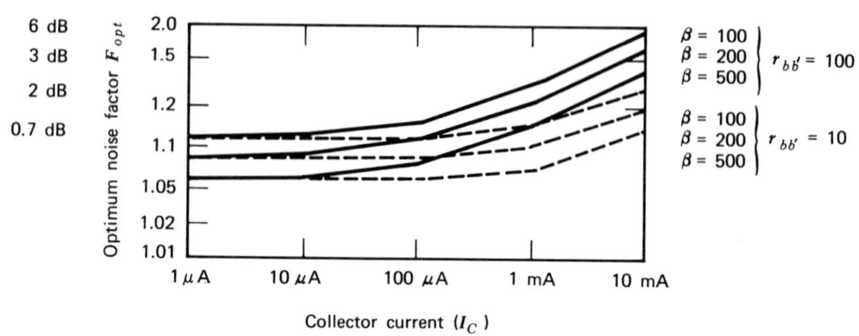

Fig. 4-5. Graph of F_{opt} versus I_C.

A plot of this equation versus I_C for various values of β_o and $r_{bb'}$ is given in Fig. 4-6. It is clear from the figure that decreasing collector current increases the optimum source resistance. Transistor beta and base resistance have only a slight effect on R_o. In the limiting case when base resistance is negligible, the optimum source resistance is linearly related to collector current. Decreasing collector current linearly increases the optimum source resistance according to

$$R_o \simeq r_e \sqrt{\beta_o} \tag{4-26}$$

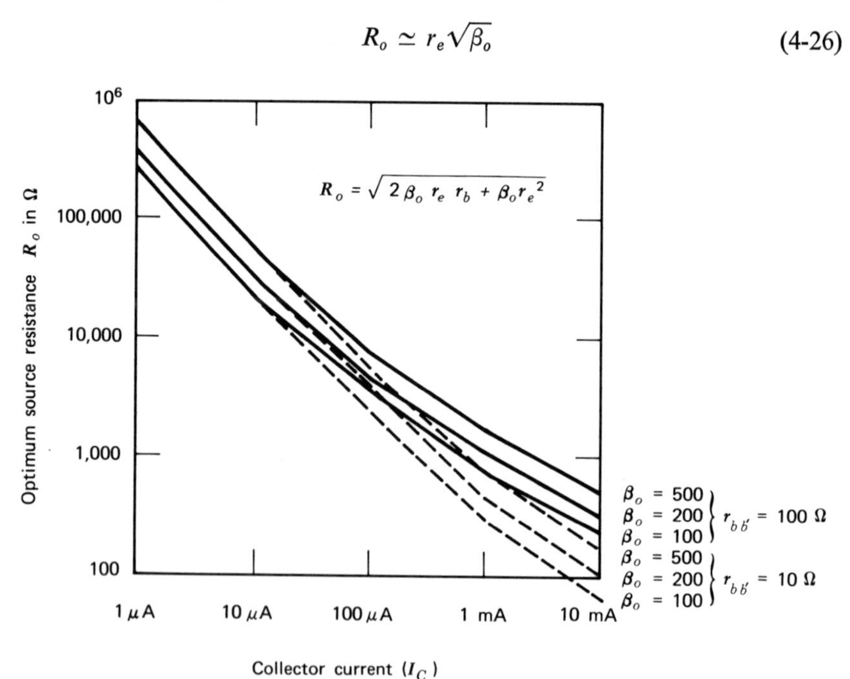

Fig. 4-6. Graph of optimum source resistance versus I_C.

The preceding discussions indicate that lowest noise figure is obtained at low collector currents for source resistances of a few thousand ohms to a few hundred thousand ohms. The lower impedance limit is set by the base resistance. *A low-noise transistor should have a low value of base-spreading resistance, and a high value of* β_o.

4-7 THE 1/f NOISE REGION

Special consideration must be given to operation in the 1/f noise region. The same general design criteria apply, but different conditions exist for optimization. From the theoretical equations for E_n and I_n, the 1/f components can be extracted. They are

$$E_n^2 = \frac{2qf_L I_B^{\gamma} r_b^2}{f^{\alpha}} \tag{4-27}$$

and

$$I_n^2 = \frac{2qf_L I_B^{\gamma}}{f^{\alpha}} \tag{4-28}$$

These two expressions differ only in the 1/f base resistance r_b. The optimum source resistance is E_n/I_n. In this case, R_o *equals the base resistance and is not dependent on operating point.*

The minimum noise factor F_{opt} obtained at source resistance R_o is

$$F_{opt} = 1 + \frac{4kTqf_L I_B^{\gamma} r_b}{f^{\alpha}} \tag{4-29}$$

The minimum noise factor is directly related to base current and base resistance. To minimize the noise factor in the 1/f region we select a good low-noise transistor with a small base resistance, operated at the lowest possible collector current. As we have seen, this criterion also gives low-noise performance in the midband region; however, it does not assure good high-frequency performance.

4-8 NOISE VARIATION WITH OPERATING CONDITIONS

It has been noted that equivalent input noise is dependent on three contributors

$$E_{ni}^2 = E_n^2 + I_n^2 R_s^2 + E_t^2$$

We wish to investigate the effect of static collector current on this noise equivalent. In Fig. 4-4 data are given on E_n versus I_C and I_n versus I_C. The

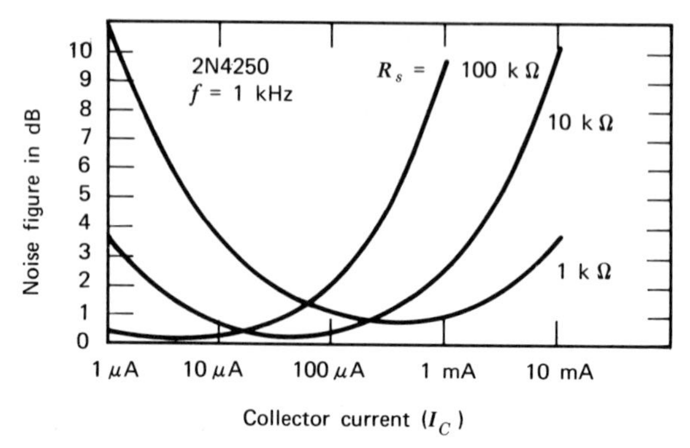

Fig. 4-7. Effect of collector current and source resistance on noise figure.

relative effect of I_n is dependent on R_s. Thermal noise E_t is constant with I_C. When these three terms are squared and added to form a plot of E_{ni}^2 versus I_C, it is clear that the total noise will be high at both extremes of I_C, and that for intermediate values of I_C the curve will perhaps experience a minimum.

The effect noted in the preceding paragraph is apparent in the plot of NF versus I_C shown in Fig. 4-7. It is seen that minima exist for all values of R_s. There is only one value of collector current and source resistance for the lowest noise figure. Since published noise data are often given for one source resistance only, it is well to consider this curve when designing an amplifier for a different source resistance.

Test data on type 2N4250 transistor form the basis for the plot of noise figure versus source resistance shown in Fig. 4-8. Observe that the optimum source resistance decreases with increasing collector current. At a fixed

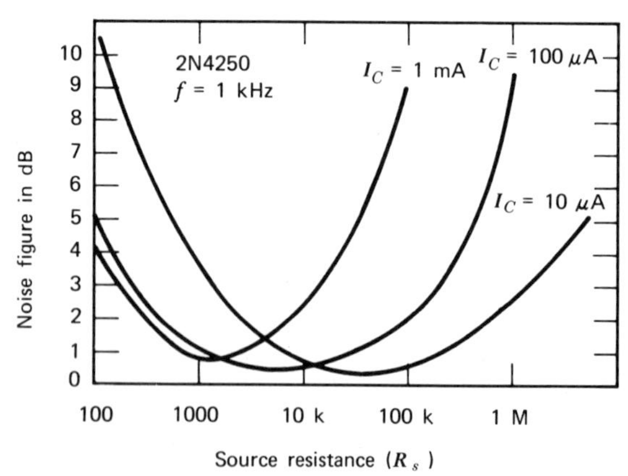

Fig. 4-8. Noise figure variation with source resistance for several values of collector current.

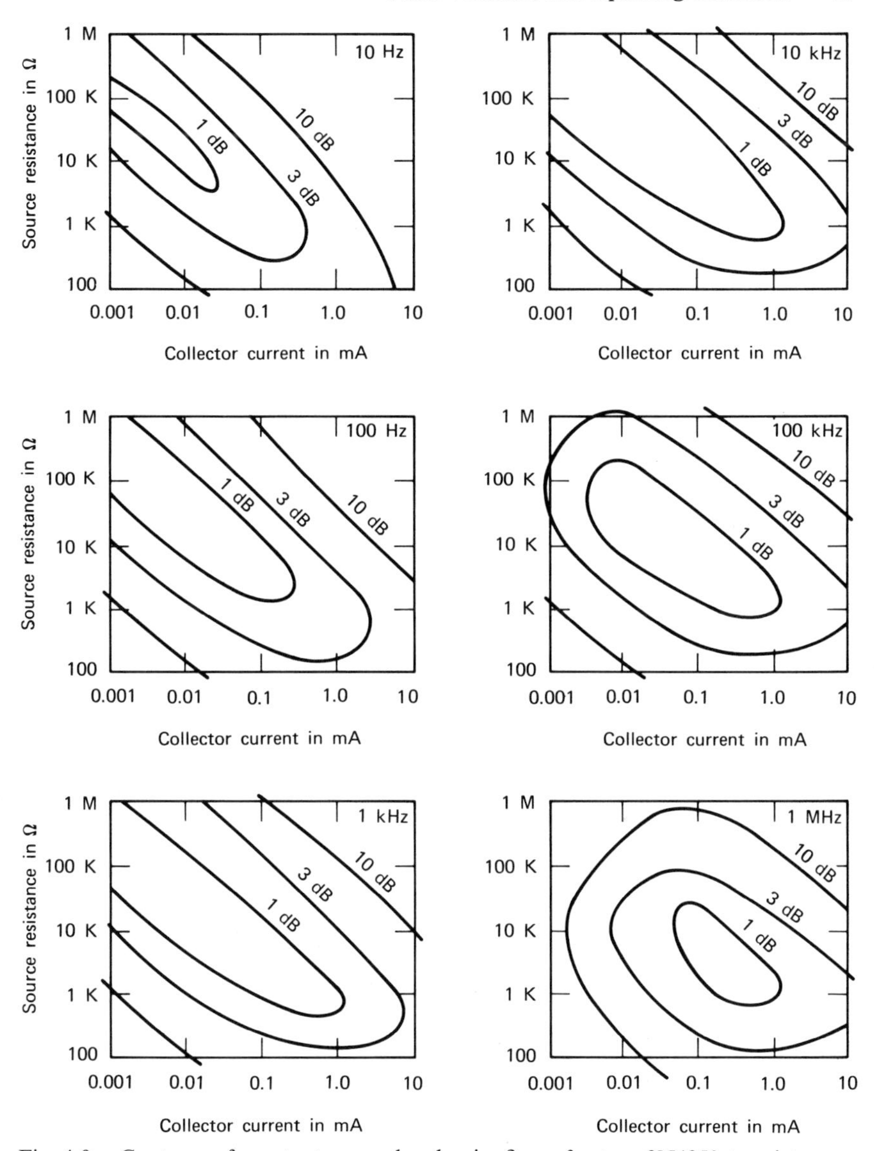

Fig. 4-9. Contours of constant narrowband noise figure for type 2N4250 transistor.

value of collector current the noise is higher for any other source resistance. A minimum value of noise figure results when the E_n noise is equal to the I_nR_s noise. Below the optimum source resistance the amplifier noise E_n is constant even though the source noise is decreasing. Above the optimum value the I_nR_s noise is increasing faster than thermal noise. Both of these conditions cause the ratio of total noise to thermal noise to increase, and result in the minimum in the noise-figure curve.

An alternate method of presenting noise-figure data is shown in Fig. 4-9. These plots are referred to as *contours of constant noise figure*. Each of the graphs in the figure shows data taken at a different frequency. These data pertain to a type 2N4250 silicon transistor at 10, 10^2, 10^3, 10^4, 10^5, and 10^6 Hz.

The contours of constant noise figure are convenient for analyzing noise performance. The major limitation in their use arises in broadband applications. For examination of frequency effects, plots such as those shown in Fig. 4-10 are useful. We see E_n and I_n behavior for the frequency range from 10 Hz to 1 MHz. In both curves there is a flat frequency-independent region with noise increasing at both low and high frequencies. At low frequencies the slope is proportional to $f^{-1/2}$; at high frequencies the slope is proportional to frequency.

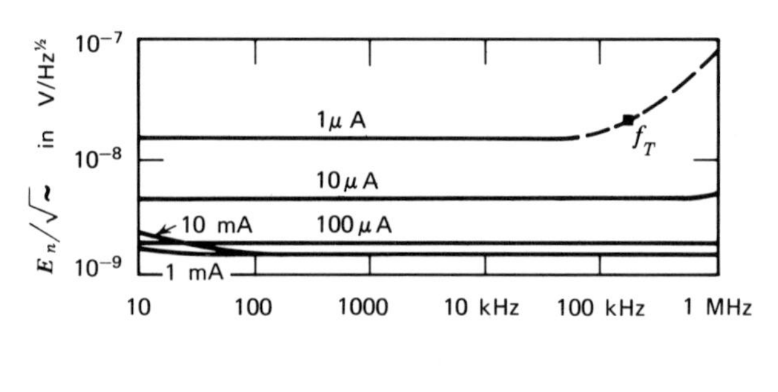

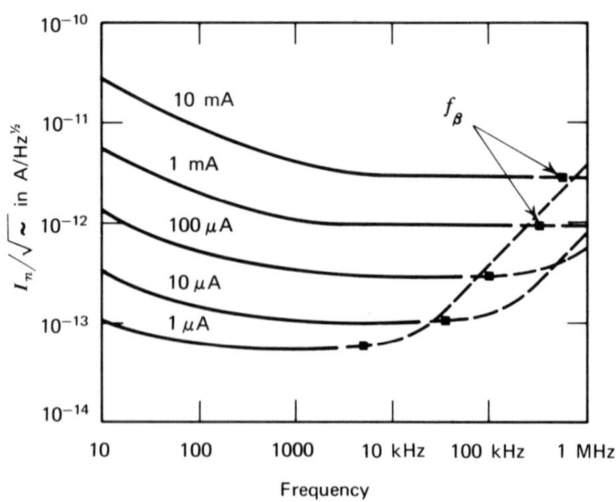

Fig. 4-10. Noise performance of 2N4250.

There is a small amount of 1/f noise at the highest collector currents in the E_n curve. The high-frequency break is located at f_T; therefore, best performance can be obtained at high collector currents where f_T is greatest.

The I_n curves show midband noise increasing with increased collector current. The 1/f noise component is more prominent. For low-frequency operation it is clear that a low-collector current is desirable. The high-frequency corner is approximately $f_T/\sqrt{\beta_o}$. The high-frequency portion of the plot is based on calculations using the simplified hybrid-π version of Fig. 4-3. If the feedback capacity $C_{b'c}$ were included in the model, the noise may be larger than indicated.

4-9 TEST RESULTS

A table of noise parameter values for seven transistor types is given in Table 4-2. These data are based on sample sizes of ten or less. While not statistically significant, these results are of value to show trends in noise parameter variations.

Since E_n is strongly dependent on base resistance, its variations for a given transistor type are not wide. On the other hand, because I_n is proportional to β_o, its variations are wider. Variations in 1/f noise at 20 Hz are wild. It is possible to observe units with values 10 or 100 times larger than nominal. For critical applications in the 1/f region, it is recommended that a larger number of units be tested to determine a dependable manufacturer and type (process). In addition, it is usually necessary to test every unit and discard those that are obviously defective.

Appendix I is devoted to noise characteristics measured on nine bipolar transistor types. Noise data are presented in two forms:

1. Noise parameters E_n and I_n are plotted against frequency and collector current.
2. Contours of constant noise figure are plotted on source resistance, collector current axes.

Conclusions that can be reached from the noise data regarding recommended types include the following:

1. Low broadband noise: 2N4104, 2N4250, and 2N3964
2. Low I_n and 1/f noise: 2N4104, 2N4250, and 2N3964
3. Low R_s applications: 2N4125 and 2N4403
4. High R_s—high-frequency applications: 2N4124
5. Low cost—low noise: 2N4250

Table 4-2. Table of Noise Parameter Variations

Transistor	Type	Number of Units Tested	Nominal E_n nV/Hz$^{1/2}$ 10 kHz at 1 mA	Variation ± %	Nominal I_n pA/Hz$^{1/2}$ 10 kHz at 1 mA	Variation ± %	1/f Nominal I_n at 20 Hz	Variation ± %
2N4250	PNP-Si	5	1.58	13	1.23	10	6.61	33%
2N4403	PNP-Si	10	0.875	9	1.61	18	12.6	37%
2N4058	PNP-Si	10	2.1	24	2.03	50	29.6	87%
2N4124	NPN-Si	10	1.38	12	1.48	18	17.9	47%
2N4935	NPN-Si	10	1.2	17	3.3	33	100.0	67%
2N4044	NPN-Si	8	3.65	10	2.3	32	25.0	24%
2N404	PNP-Ge	6	2.05	7	1.75	14	17.6	64%

4-10 NOISE FIGURE OF A COMMON-BASE STAGE

The information previously presented concerning noise in bipolar transistors and the methods of analysis already discussed are now applied to a new problem. In this example, we wish to *determine the noise figure of a common-base transistor amplifier stage*. To add some spice to this presentation the small-signal T equivalent circuit is used to represent the transistor. That model is shown in Fig. 4-11*a*. The dependent current generator αi_e is linked to the instantaneous direction of emitter current i_e. Elements of this model are of course related to elements of the hybrid-π [1]. For example, $\alpha = \beta/(\beta + 1)$.

The short-circuit current gain α and collector impedance Z_c are frequency dependent.

$$\alpha = \frac{\alpha_o}{1 + jf/f_a}$$

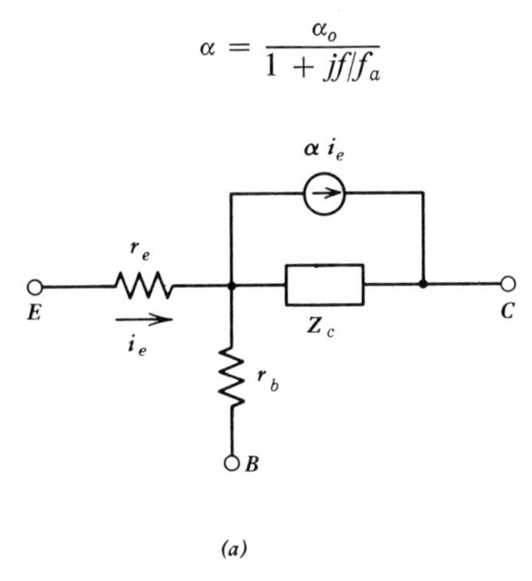

(*a*)

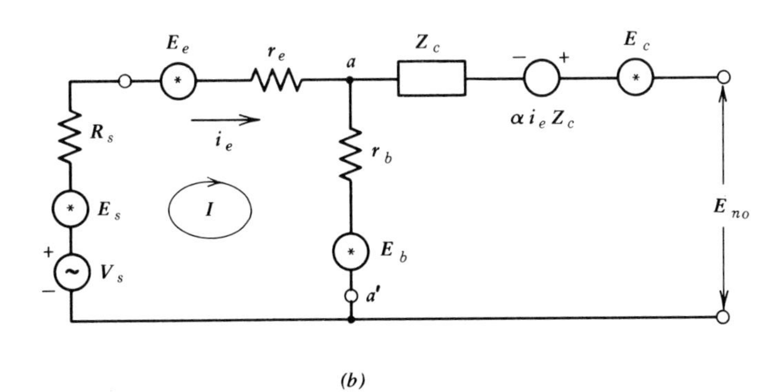

(*b*)

Fig. 4-11. (*a*) Small-signal T model; (*b*) noise equivalent circuit with open-circuit load.

where f_α is the α-cutoff frequency and α_o is the reference value of α. The collector impedance represents parallel r_C and C_c:

$$Z_c = \frac{1}{1/r_c + j\omega C_c}$$

Figure 4-11*b* shows the model with thermal and shot noise sources E_s, E_e, E_b, and E_c added. The signal source V_s and source resistance R_s are also shown. Source interchanges have been made so that all current sources have been replaced by equivalent voltage sources. The noise generators are [4]

$$E_s^2 = 4kTR_s\,\Delta f \quad \text{(thermal noise of source resistance)}$$

$$E_e^2 = 2kTr_e\,\Delta f \quad \text{(shot noise of emitter current)}$$

$$E_b^2 = 4kTr_b\,\Delta f \quad \text{(thermal noise of base resistance)}$$

$$E_c^2 = 2qI_C\left(1 - \frac{|\alpha|^2}{\alpha_o}\right)|Z_c|^2\,\Delta f \quad \text{(recombination noise in the base)}$$

This last effect is carried over into the collector circuit. As α declines with increasing frequency, it becomes more important.

After manipulation using $I_C = \alpha I_E$,

$$E_c^2 = \frac{2kT\alpha_o}{r_e}(1 - \alpha_o)|Z_c|^2 F(f)\,\Delta f \qquad \text{where} \qquad F(f) = \frac{1 + \left(\dfrac{f}{f_\alpha(1 - \alpha_o)^{1/2}}\right)^2}{1 + \left(\dfrac{f}{f_\alpha}\right)^2}$$

We wish to find the output noise voltage designated in the figure as E_{no}. That voltage depends on all the noise sources and the dependent generator. Since the load terminals are open-circuited, the direct contributions of E_s and E_e are simply the result of voltage division around the closed left-hand loop; the fractions of E_s and of E_e across r_b are

$$E_s\frac{r_b}{(r_b + r_e + R_s)} \qquad \text{and} \qquad E_e\frac{r_b}{(r_b + r_e + R_s)} \tag{4-30}$$

Generator E_b contributes its total voltage to E_{no}, minus the correlated voltage drop across r_b caused by the current resulting from that voltage. If we consider E_b to be instantaneously positive at the a' terminal, the voltage from a to a' due to that generator is

$$E_b\frac{r_b}{(r_b + r_e + R_s)} - E_b$$

or

$$\frac{E_b(-r_e - R_s)}{(r_b + r_e + R_s)} \tag{4-31}$$

The contribution of E_c is simply its own voltage:

$$E_c^2 = \frac{2kT\alpha_o}{r_e}(1 - \alpha_o)|Z_c|^2F(f)\,\Delta f \tag{4-32}$$

Now we consider the dependent generator contribution. For calculation purposes, we must assign a direction to noise current I. Let us say that it flows in the direction of i_e in the figure. We can write

$$I^2 = \frac{E_s^2 + E_e^2 + E_b^2}{(r_b + r_e + R_s)^2} \tag{4-33}$$

The dependent generator is, therefore,

$$\alpha^2 I^2 Z_c^2 = \frac{\alpha^2 Z_c^2(E_s^2 + E_e^2 + E_b^2)}{(r_b + r_e + R_s)^2} \tag{4-34}$$

Now the total output voltage E_{no} is the sum of mean square voltages given in Eqs. 4-30, 4-31, 4-32, and 4-34:

$$E_{no}^2 = \frac{E_s^2(r_b + \alpha Z_c)^2 + E_e^2(r_b + \alpha Z_c)^2 + E_b^2(\alpha Z_c - r_e - R_s)^2}{(r_b + r_e + R_s)^2} + E_c^2 \tag{4-35}$$

The absolute magnitude of the impedance terms is utilized in this equation, for the noise voltages cannot be added according to phasor techniques.

Since, in general, $Z_c \gg r_b$ and $Z_c \gg R_s + r_e$, a simplified version of Eq. 4-35 can be written:

$$E_{no}^2 = \frac{E_s^2(\alpha Z_c)^2 + E_e^2(\alpha Z_c)^2 + E_b^2(\alpha Z_c)^2}{(r_b + r_e + R_s)^2} + E_c^2 \tag{4-36}$$

The output S/N ratio is

$$\frac{S_o}{N_o} = \frac{A_{oc}^2 V_s^2}{E_{no}^2} \tag{4-37}$$

where A_{oc} is the open-circuit voltage gain. At the input terminals

$$\frac{S_i}{N_i} = \frac{V_s^2}{E_s^2} \tag{4-38}$$

Noise factor was defined in Eq. 2-10. Here we have

$$F = \frac{S_i/N_i}{S_o/N_o} = \frac{V_s^2/E_s^2}{A_{oc}^2 V_s^2/E_{no}^2} = \frac{E_{no}^2}{A_{oc}^2 E_s^2} \tag{4-39}$$

The denominator of this last equation is the first term in Eq. 4-36: $A_{oc}^2 E_s^2 = E_s^2(\alpha Z_c)^2/(r_b + r_e + R_s)^2$. Therefore

$$F = 1 + \frac{E_e^2}{E_s^2} + \frac{E_b^2}{E_s^2} + \frac{E_c^2(r_b + r_e + R_s)^2}{E_s^2(\alpha Z_c)^2} \tag{4-40}$$

In terms of transistor parameters, we obtain

$$F = 1 + \frac{r_e}{2R_s} + \frac{r_b}{R_s} + \frac{(r_b + r_e + R_s)^2}{2\beta_o r_e R_s}\left[1 + \left(\frac{f}{f_\alpha\sqrt{1-\alpha_o}}\right)^2\right] \tag{4-41}$$

It can be shown that Eq. 4-41 is virtually identical to F obtained from Eq. 4-16 using the hybrid-π representative for a common-emitter stage. *An important conclusion is that noise figure is independent of transistor configuration.* Further discussion appears in Chapter 12.

4-11 NOISE IN UHF-MICROWAVE TRANSISTORS

The amplification of low-level signals in the frequency band above several hundred MHz is often accomplished with bipolar transistors. Lumped representations of device behavior at high frequencies are not necessarily valid because of the distributed nature of the processes involved. A successful representation for determination of gain, stability, and terminal impedances uses the set of four scattering (s) parameters. The s-parameters are not the elements in an equivalent electrical network. They are relatively easy to measure and have gained wide acceptance.

It has been noted that low noise in bipolar transistors is associated with operation at low levels of static collector current. Unfortunately the gain-bandwidth parameter f_T is low-valued at low-current operating points.

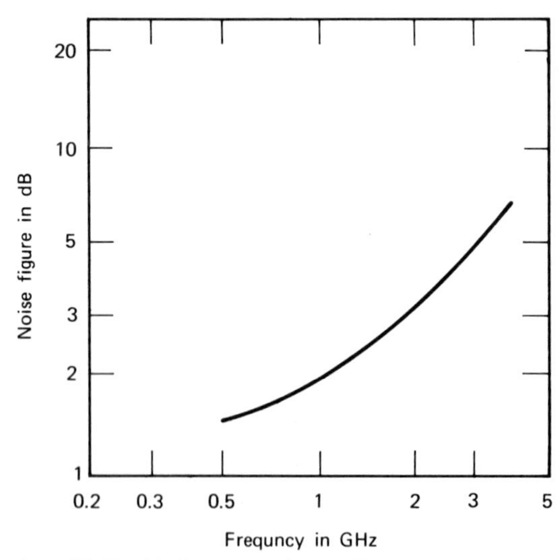

Fig. 4-12. Behavior of NF with frequency for a microwave transistor.

Therefore, it is not possible to achieve lowest noise and highest gain. A compromise is suggested. For the type 2N5651 microwave transistor the maximum f_T is 2.2 GHz and is obtained at $I_C = 8$ mA. For $f_T > 1$ GHz it is necessary for I_C to be greater than 1 mA.

It is customary for engineers involved with UHF and microwave devices and circuits to describe noise behavior with the noise figure. The graph shown in Fig. 4-12 shows the variation in NF with frequency for the type 2N5761 S-band microwave transistor, operating at 3 mA and 10 V from a 50 Ω source. At 2 GHz, NF is a minimum for this transistor at $I_C = 3$ mA. The power gain parameter $|s_{21}|$ is largest for $2.5 < I_C < 8$ mA.

SUMMARY

a. The bipolar transistor contains noise sources whose spectral densities are given by

$$I_f^2 = KI_B^{\gamma}/f \qquad \text{1/f noise}$$

$$I_b^2 = 2qI_B \qquad \text{shot noise}$$

$$E_b^2 = 4kTr_{bb'} \qquad \text{thermal noise}$$

$$I_c^2 = 2qI_C \qquad \text{shot noise}$$

b. Relative to midband reference, parameters E_n and I_n increase at low frequencies because of the 1/f noise source and increase at high frequencies because of reduced gain.

c. Both E_n and I_n are highly dependent on operating-point coordinates.

d. At midfrequencies the noise parameters of a bipolar transistor are

$$E_n^2 = 4kTr_{bb'} + 2qI_C r_e^2$$

$$I_n^2 = 2qI_B$$

e. Optimum noise factor:

$$F_{\text{opt}} = 1 + \left(\frac{2r_{bb'}}{\beta_o r_e} + \frac{1}{\beta_o}\right)^{1/2}$$

f. Optimum source resistance:

$$R_o = (2\beta_o r_{bb'} r_e + \beta_o r_e^2)^{1/2}$$

g. Best noise performance is obtained for $10\ \text{k}\Omega < R_s < 100\ \text{k}\Omega$ and $1\ \mu\text{A} < I_C < 100\ \mu\text{A}$.

h. Noise figure is independent of transistor configuration.

PROBLEMS

1. Derive Eq. 4-4.

2. Explain the Miller effect as it applies to bipolar transistors [1].

3. In deriving Eq. 4-19 from Eq. 4-16, what assumption is made in addition to the omission of element R_s?

4. Use E_n and I_n values for a 2N4044 transistor from Table 4-2 to determine F_{opt} at normal ambient temperature.

5. For $r_{bb'} = 50\ \Omega$, $\beta_o = 100$, and $q/kT = 40$, calculate F_{opt} for $I_C = 10\ \mu\text{A}$, $100\ \mu\text{A}$, 1 mA, and 10 mA.

6. Show that the expression for noise figure of a common-emitter stage is the same as Eq. 4-41 for a common-base stage by interchanging emitter and base branches in the model of Fig. 4-11*a*.

REFERENCES

1. Fitchen, F. C., *Transistor Circuit Analysis and Design*, 2nd ed., Van Nostrand Reinhold, New York, 1966, p. 107.
2. Chenette, E. R., "Low-Noise Transistor Amplifiers," *Solid-State Design*, **5**, 2 (February 1964), 27–30.
3. Plumb, J. L., and E. R. Chenette, "Flicker Noise in Transistors," *IEEE Trans. Elect. Dev.*, **ED-10**, 5 (September 1963), 304–308.
4. Nielsen, E. G., "Behavior of Noise Figure in Junction Transistors," *Proc. IRE*, **45**, 7 (July 1957), 957–963.

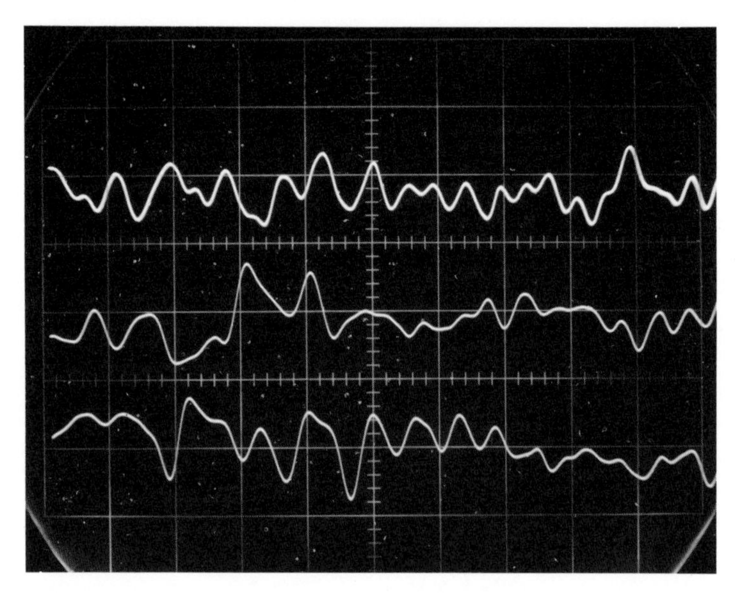

The power content of l/f noise in each decade of frequency is equal, as shown in these traces. Top waveform, $\Delta f = 20$ Hz, horizontal 100 msec/cm; center waveform, $\Delta f = 200$ Hz, horizontal 10 msec/cm; bottom waveform, $\Delta f = 2$ kHz, horizontal 1 msec/cm.

Chapter 5

EXCESS NOISE IN TRANSISTORS

One of the principle noise mechanisms discussed in Chapter 1 was excess or 1/f noise. In Section 4-2 it was noted that this type of excess noise is present in bipolar transistors, and a model was presented for noise analysis of transistor circuits.

We now focus our attention on several additional topics associated with excess noise in transistors. The initial concern in this chapter is the tie between excess noise and the reliability of the device. A means of damaging transistors that results in a high level of excess noise is considered in Section 5-2. The chapter concludes with a discussion of another form of excess noise that is referred to as *burst* or *popcorn* noise.

5-1 RELIABILITY AND NOISE

The noise mechanisms within a transistor have been attributed to thermal and shot effects, and in addition we have noted the presence of l/f noise. It has been shown that *correlation exists between high values of l/f noise and poor reliability*. This correlation follows from the fact that l/f noise is very sensitive to changes in a transistor's surface conditions and the ambient atmosphere within the transistor package, and these changes, if significant, result in transistor failure.

Two types of reliability testing programs can be performed using noise data. Routine noise tests on each transistor constitute one approach. Another means of study involves tests on a device over a period of time to determine *changes* in behavior. Noise tests require that some definition of normal noise levels be made. A normal level of 1/f noise varies among transistor types. A number of units must be tested to establish the norm.

Surface contamination can be detected by testing for 1/f noise current (I_n) at a few hundred μA of collector current and gradually increasing the collector voltage. If the noise current increases in proportion to applied voltage, surface contamination is most probably present and is acting as a semiconductor resistor across the collector-base junction. If the noise increases abruptly or exponentially with the voltage, the collector-base junction is breaking into an avalanche. When the avalanche takes place *within rated operating voltages* trouble is indicated, probably in the form of an irregularity of the collector-base junction.

Faults such as defective contacts, fractures, or irregularities at the emitter-base junction can be detected by measuring noise voltage (E_n) with the transistor operated under high-current conditions. These conditions are indicated by an abnormally high level of 1/f noise. A test of 1/f noise current I_n made at an intermediate value of collector current can indicate base region defects.

Not all potentially defective units show up in the tests previously noted. Certain units show increased noise after several hours or days of normal operation. These "sleepers" cannot be detected by simple E_n or I_n measurements. Fortunately they are not often encountered.

Artificial aging to correlate noise with the probability of early failure can be accomplished by immersing the devices in a 350°C environment for 10-hr periods, followed by 10 hr at room temperature [1]. It has been noted that under these circumstances a large increase in 1/f noise occurs just prior to failure. Further, it has been shown that units with low initial values of noise current have a longer life under artificial aging.

5-2 AVALANCHE BREAKDOWN AND NOISE

Caution! A circuit designer, evaluation engineer, or inspector making routine tests or adjustments can inadvertently ruin a perfectly good low-noise, high-gain transistor. This can happen when the base-emitter junction of a bipolar transistor (discrete or IC) is reverse-biased beyond its avalanche knee. The noise behavior of the device can be increased by ten times because of a circuit turn-on transient, a signal transient, or the testing of V_{EBO}.

Character of the Problem

Figure 5-1 shows the progressive increase in noise current I_n caused by pulses of reverse bias current applied to the emitter-base diode of a sample

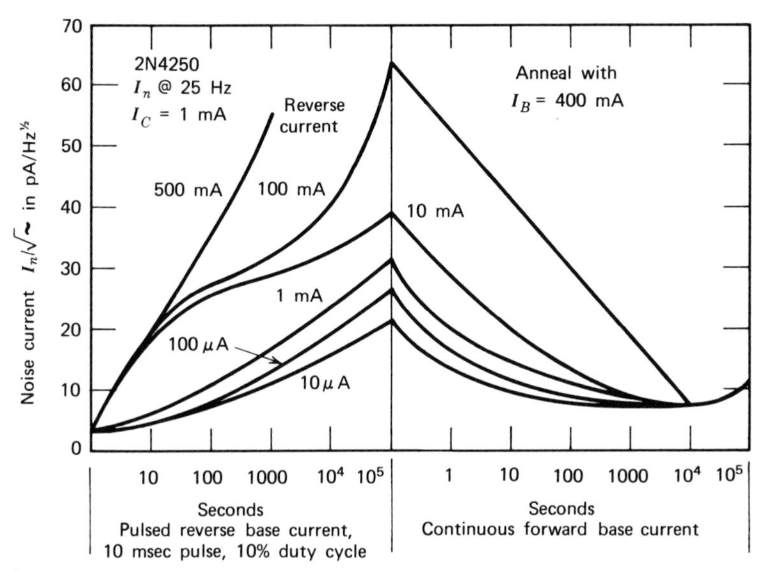

Fig. 5-1. Increase in noise current with avalanching, and the decrease resulting from annealing.

type 2N4250 silicon transistor. In normal operation that diode is forward-biased; the pulses represent avalanching of the junction under a reverse bias. In order to minimize heating of the transistor, the current pulses were 10 msec in duration, and the duty cycle was 10%.

It is found that reverse biasing of the junction below avalanche causes no damage [2]. The damaging effect is observed to be proportional to total charge flow and the logarithm of time [3]; consequently, a steady reverse current causes the same damage in 1/10 the time shown in Fig. 5-1.

Also shown in Fig. 5-1 is the result of a current anneal on noise behavior. It is found that the avalanche-induced damage can be partially repaired by a temperature or a current anneal. McDonald reports that heating the chip to approximately 300°C returns the noise to near original values [2]. The annealing was accomplished on the unit shown in Fig. 5-1 by passing a continuous 400-mA forward current through the emitter-base junction. Whereas this means of repair is of interest, it is not a practical way to overcome the avalanche-caused damage.

The noise observed is mainly excess 1/f noise. Frequently it includes "popcorn" noise (discussed in the next section). The equivalent noise voltage E_n can double in value, whereas noise current I_n increases perhaps ten times.

Along with large increases in noise, the avalanched transistor suffers a severe deterioration in its dc current gain parameter h_{FE} (dc β). Test data are shown in Fig. 5-2. Again, an annealing is possible to repair the damage, at least in part. The degradation in h_{FE} is attributed to fast surface states caused by the hot carriers created during an emitter-base avalanche stress [2]. This increase in base current causes a resulting increase in excess noise proportional to $(\Delta I_B)^{1/2}$, where ΔI_B is the increase in I_B necessary to maintain I_C constant when h_{FE} has decreased in value.

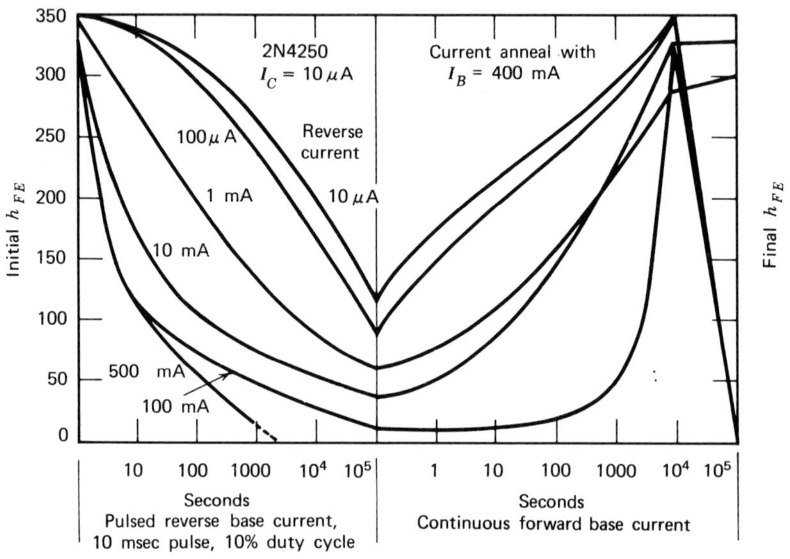

Fig. 5-2. Deterioration in h_{FE} with avalanching, and the increase resulting from annealing.

Noise is probably the most sensitive measure of the quality of a transistor, and h_{FE} is an important measure. A small increase in recombination centers in the base region significantly decreases gain and increases noise. On the other hand, the avalanche voltage is virtually independent of minor defects. For this reason the junction can be operated as a reference or regulator diode without destruction.

In the data shown in Fig. 5-3 we note the effect of 1-sec current pulses on six type 2N4124 transistors. The transistors differed greatly in their noise behavior prior to stressing. It is obvious from the graph that units with high noise levels were less affected by the pulsing than the quieter units.

Circuit Causes of Avalanching

Three causes of base-emitter junction avalanching are circuit design, signal transients, and V_{EBO} testing. When an ac amplifier is turned on it may experience wild voltage variations as the coupling and bypass capacitors charge,

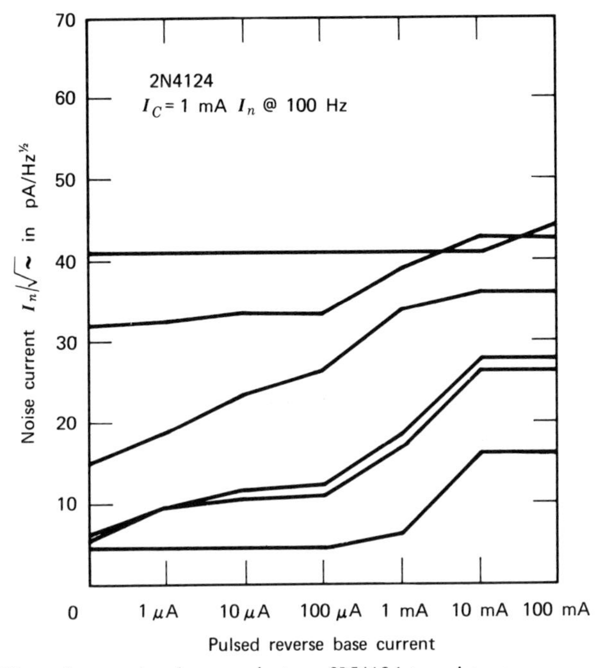

Fig. 5-3. Effect of current pulses on six type 2N4124 transistors.

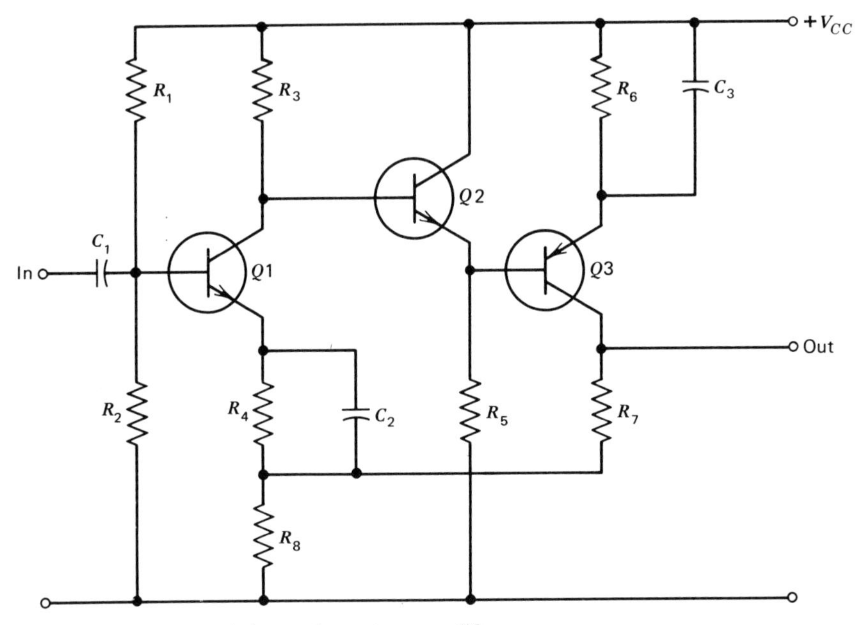

Fig. 5-4. Direct-coupled complementary amplifier.

and there may be avalanching. To examine the conditions that spawn a turn-on transient we consider the common direct-coupled complementary amplifier shown in Fig. 5-4. When the circuit is first turned on by connection to V_{CC}, transistors $Q1$ and $Q3$ are fully on while they charge C_2 and C_3. (The initial voltages across all capacitances are assumed to be zero and they cannot change instantaneously.) Thus the base of $Q2$ is near ground and the emitter near $+V_{CC}$. This can result in avalanching the emitter-base junction of $Q2$ because of the charging currents of C_2 and C_3. Although insufficient to burn out $Q2$, this transient can result in a significant increase in the excess noise current I_n. Since $Q2$ usually operates from a large source resistance, R_3, the avalanching can significantly increase the total noise of the amplifier. A similar effect can result when a signal transient overloads an amplifier.

Three solutions to the turn-on problem are shown in Fig. 5-5. In Fig. 5-5*a*, C_3 is connected to bypass R_6 to ground instead of V_{CC}. A second solution is to place a reverse diode $D1$ across the base-emitter junction of $Q2$ and add a series limiting resistor R_9. The charging transient passes through $D1$. In normal operation, $D1$ has no effect on the gain or noise. In Fig. 5-5*b*, the *npn* $Q2$ is replaced by its *pnp* complement. The transient simply tries to turn on $Q2$ harder, and the avalanching problem is therefore circumvented.

Avalanching can occur when operating from a biased source as shown in Fig. 5-6*a*. When the 100-V supply is turned on, capacitor C_1 is charged to a high voltage through R_B. Now if point A is grounded, or the circuit is turned off, the voltage on C_1 can avalanche the base of $Q1$. Diode $D1$ is added to protect the base from this problem.

A differential amplifier can be protected with two diodes as shown in Fig. 5-6*b*. The voltage between the two input lines is limited to about 1.4 V peak-to-peak. Resistors R_1 and R_2 must be large enough to absorb the energy of the transient yet smaller than the signal source resistance so that they add little thermal noise.

Nonlinear circuits such as the free-running multivibrator in Fig. 5-6*c* suffer from avalanche damage. In the absence of protection diodes $D1$ and $D2$, the transistor base is avalanched every cycle by the stored charge on capacitors C_1 and C_2, if the supply voltage V_{CC} is greater than the base breakdown. Continued operation can decrease the β of the transistors until oscillation stops; that is, circuit failure. Diodes $D1$ and $D2$ must have a breakdown voltage greater than V_{CC} and leakage less than that of the emitter-base diode.

Measurement of V_{EBO}, the emitter-base breakdown voltage of a transistor, is generally considered nondestructive as long as the current is limited. In fact, damage may go unnoticed since a 1-sec, 1-mA pulse only reduces h_{FE} by about 20%. The increased noise, however, is a significant problem in low-level circuits. There are two ways to measure V_{EBO}. One is to avalanche the

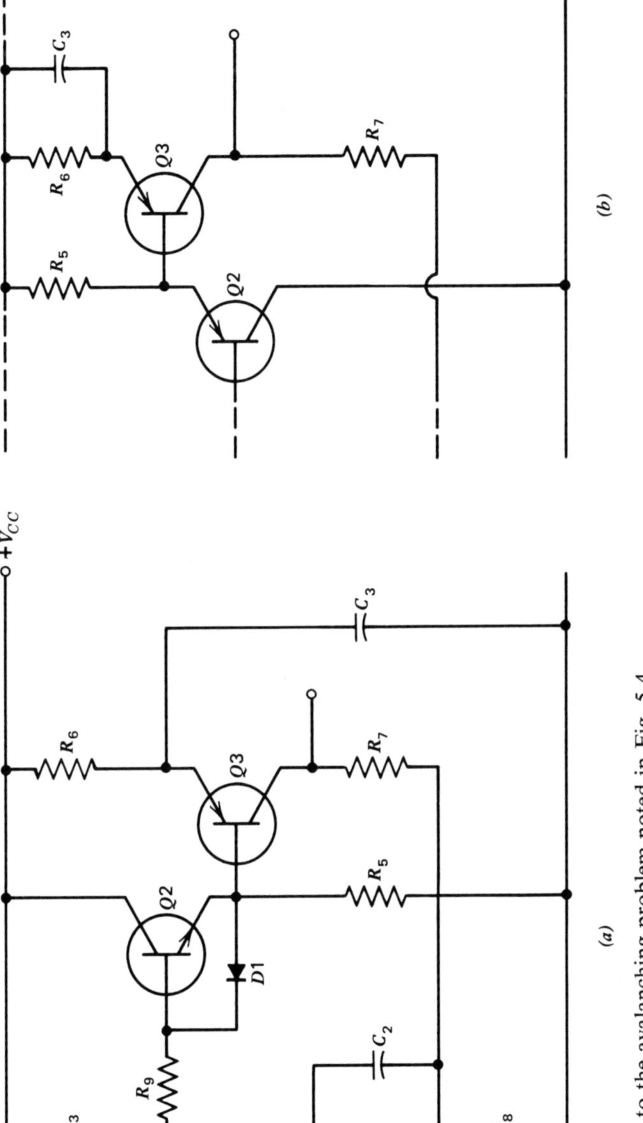

Fig. 5-5. Solutions to the avalanching problem noted in Fig. 5-4.

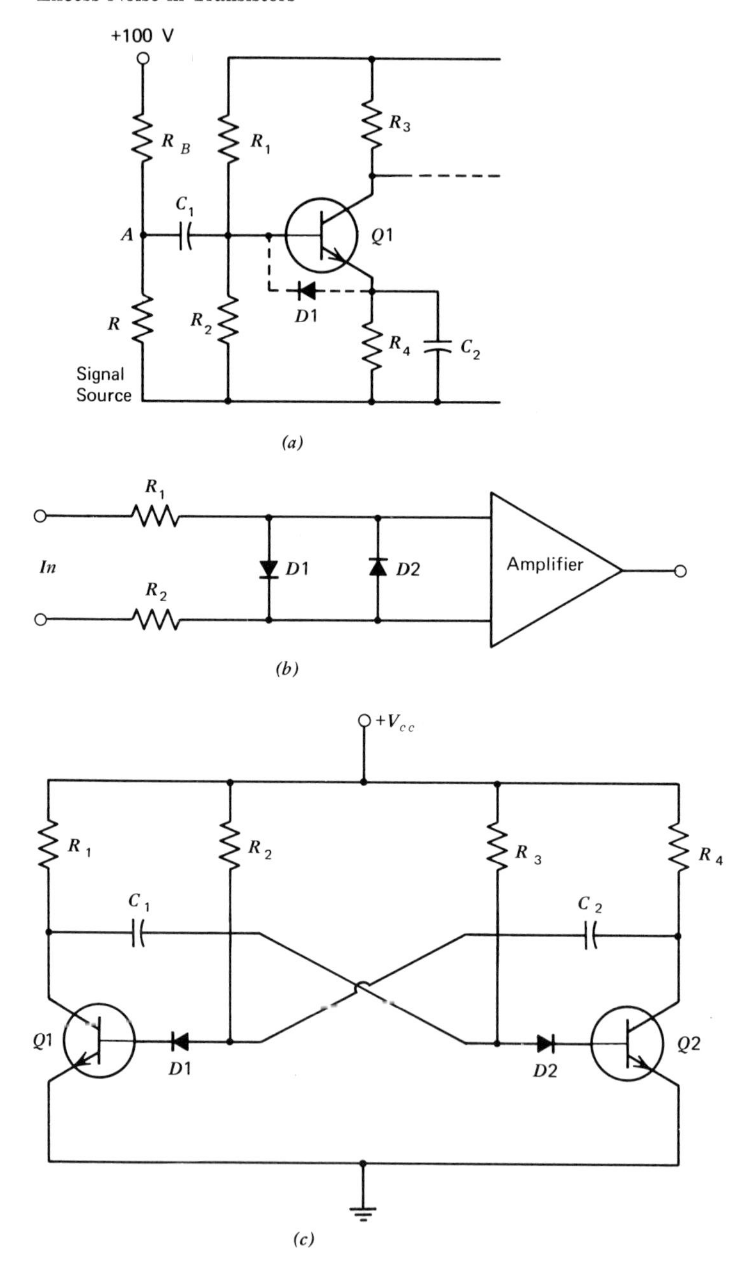

Fig. 5-6. Examples of solutions to the avalanching problem.

base-emitter junction and measure the voltage drop. A better method is to specify a minimum acceptable V_{EBO}, then test for current flow (or breakdown) at that voltage. Frequently it is not necessary to measure V_{EBO} especially for low-level linear circuits.

5-3 POPCORN NOISE

Popcorn noise, also referred to as *burst noise*, was first observed in point-contact diodes. It has since been found in tunnel diodes, juction diodes, certain resistors, junction transistors, and integrated circuits. The name "popcorn" stems from the fact that when it is fed to a loudspeaker, the result sounds like corn popping.

Popcorn-noise waveforms are shown in the frontispiece. Obviously this is not white noise. The power spectral density of this noise is a $1/f^{\alpha}$ function with $1 < \alpha < 2$. It is often found to vary as $1/f^2$. The noise-frequency plot has a roller-coaster effect with one or more plateaus. This noise is masked by other mechanisms such as shot noise, unless frequency selective networks are used to filter out higher-frequency noise.

For a *pn*-junction diode, the amplitude of popcorn pulses was found to be no higher than a few tenths of a μA. The pulse width has been found to vary from a few μsec upward [4,5].

A typical noise burst of 10^{-8} A with a duration of 1 msec represents some 10^8 charge carriers. It is doubtful that such a large number of carriers would be directly involved in the process; it seems more likely that a modulation mechanism is present, and therefore small numbers of carriers are controlling the large carrier flow. A physical model for burst noise in semiconductor junctions has been proposed that is based on the presence of two types of defects. Suppose that a metallic precipitate were to exist within a metallurgical *pn* junction, and in the space charge region adjacent to that defect a recombination-generation center or a trap were located. It can be argued that if the occupancy of the recombination-generation center changes, the current flowing through the defect is modulated. Thus the center in effect controls the current across the potential barrier, and therefore the generation of burst noise.

If we consider that noise power spectral density of popcorn noise is inversely proportional to f^2, the equivalent noise current generator representing this effect has the form

$$I_{bb}^2 = \frac{K' \, \Delta f}{f^2} \tag{5-1}$$

Studies have refined Eq. 5-1 to

$$I_{bb}{}^2 = \frac{KI_B\,\Delta f}{1 + \pi^2 f^2/4a^2} \tag{5-2}$$

where the constant a represents the number of burst/sec. Equation 5-2 predicts a leveling off of popcorn noise at very low frequencies.

The total base resistance ($r_{bb'}$) of a bipolar transistor can be considered to be composed of two sections: the relatively small resistance from the base contact to the edge of the emitter junction, and a larger resistance lying beneath the emitter. These have been called *inactive* (r_i) and *active* (r_a) base resistances, respectively. The total resistance $r_{bb'} = r_i + r_a$. It has been reported that 1/f noise attributable to surface recombination effects should be affected by r_i, whereas 1/f noise caused by recombination in the active base region should be affected by the entire base resistance [6]. It is also found that popcorn noise is associated with the value of inactive base resistance.

A 1/f noise equivalent circuit for a common-emitter connected transistor that has six noise sources, including popcorn noise, two 1/f noise sources, two shot noise sources, and thermal noise, is shown in Fig. 5-7. The values of the noise generators are

Shot: $I_b{}^2 = 2qI_B\,\Delta f$

Shot: $I_c{}^2 = 2qI_C\,\Delta f$

Thermal: $E_b{}^2 = 4kTr_{bb'}\,\Delta f$

Burst: $I_{bb}{}^2 = \dfrac{KI_B\,\Delta f}{1 + \pi^2 f^2/4a^2}$

1/f: $I_{f1}{}^2 = \dfrac{K_1 I_B{}^{\gamma_1}\,\Delta f}{f}$

1/f: $I_{f2}{}^2 = \dfrac{K_2 I_B{}^{\gamma_2}\,\Delta f}{f}$

This noise equivalent circuit can be used for noise analysis. The constants present in the expressions for 1/f and burst noise sources must be determined experimentally for each transistor.

Microplasma noise has been reported to be present in IC operational amplifiers [7]. The amplitude of microplasma noise can be several orders of magnitude larger than popcorn noise, whereas the pulse width can be extremely small. Microplasma noise is a local surface high-field breakdown phenomenon at the collector-base junction. It often can be traced to an input stage, but if originating in a later amplifying stage, it will still seriously degrade performance.

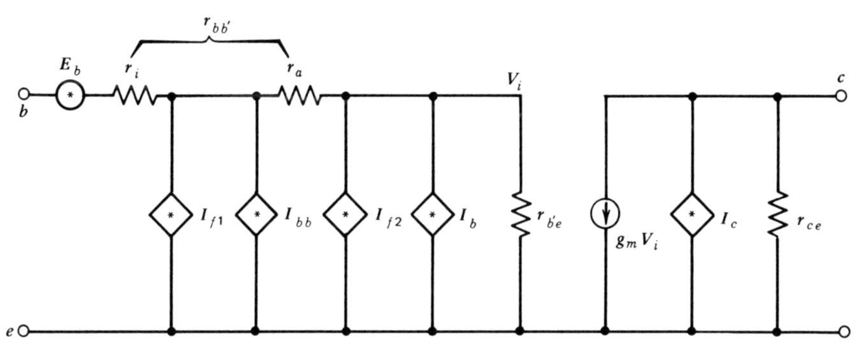

Fig. 5-7. Modified hybrid-π transistor low-frequency equivalent circuit with six noise generators.

SUMMARY

a. Transistors with high values of l/f noise tend to be less reliable than low-noise units.

b. Noise testing can indicate surface contamination and other defects in transistors.

c. The emitter-base junction, after avalanche breakdown, exhibits a great deal of l/f noise, particularly in I_n.

d. Avalanche damage reduces the current gain parameter, h_{FE}.

e. Avalanche damage can occur in development or inspection testing, or it may result from turn-on or turn-off transients in an otherwise well-designed circuit.

f. Popcorn or burst noise is found in some bipolar transistors and in junction diodes. It has been mathematically represented by

$$I_{bb}{}^2 = \frac{KI_B\,\Delta f}{1 + \pi^2 f^2/4a^2}$$

This current generator parallels the l/f noise generator in the transistor noise model.

PROBLEMS

1. When a forward-biased silicon diode is used for protection, we assume for simplicity that its resistance is nearly infinite until the applied voltage becomes equal to about 0.6 V. Then its resistance becomes low. Using $I = I_R\epsilon^{qV/kT}$ to represent a diode's characteristic, derive an expression for the dynamic resistance from

$\partial V/\partial I$, and determine values for that resistance at $V = 0.2$ V, 0.4 V, 0.6 V, 0.8 V. Consider $I_R = 10^{-9}$ A.

2. It has been stated that Eq. 5-2 predicts a leveling off of popcorn noise at low frequencies. Consider $KI_B = 10^{-14}$ and $\pi^2/4a^2 = 10^{-2}$. Plot spot $I_{bb}{}^2$ versus f for these values.

3. Can you explain why the constant a in Eq. 5-2 can be related to the number of bursts/sec present in popcorn noise?

4. Simplify Fig. 5-7 by eliminating all noise sources except I_{f1} and I_{f2}. Show that the model associates I_{f1} with r_i and I_{f2} with $r_i + r_a$ by deriving the expression for generator $g_m V_i$.

REFERENCES

1. Van der Ziel, A., and H. Tong, "Low-Frequency Noise Predicts When A Transistor Will Fail," *Electronics*, **39** (November 28, 1966), 95–97.
2. McDonald, B. A., "Avalanche-Induced 1/f Noise in Bipolar Transistors," *IEEE Trans. Electron Devices*, **ED-17**, 2 (February 1970), 134–136.
3. Collins, D. R., "h_{FE} Degradation Due to Reverse Bias Emitter-Base Junction Stress," *IEEE Trans. Electron Devices*, **ED-16**, 4 (April 1969), 403–406.
4. Hsu, S. T., and R. J. Whittier, "Characterization of Burst Noise in Silicon Devices," *Solid-State Electronics*, **12** (November 1969).
5. Hsu, S. T., R. J. Whittier, and C. A. Mead, "Physical Model for Burst Noise in Semiconductor Devices," *Solid-State Electronics*, **13** (July 1970).
6. Jaeger, R. C., and A. J. Brodersen, "Low-Frequency Noise Sources in Bipolar Junction Transistors," *IEEE Trans. Electron Devices*, **ED-17**, 2 (February 1970), 128–134.
7. Hsu, S. T., "Bistable Noise in Operational Amplifiers," *IEEE J. Solid-State Circuits*, **SC-6**, 6 (December 1971), 399–403.

Chapter 6

NOISE IN FIELD-EFFECT TRANSISTORS, SUPER-BETA TRANSISTORS, AND INTEGRATED CIRCUITS

In addition to bipolar transistors and passive components, the modern electronic system can contain other components. Both the field-effect transistor (FET) and the super-beta transistor (SBT) are being used as input stages in high-quality amplifiers. The noise behavior of these devices is examined in the present chapter.

In the search for acceptable low-noise devices and systems, the electronic designer must be aware of the capabilities of commercial integrated circuits (IC). Although a linear IC can contain dozens of circuit elements, it may be treated as an off-the-shelf device by the design engineer, and utilized in a system as one employs any electronic component. When it it possible to use noncustom ICs, significant savings in time, costs, space, and improvement in reliability can result.

This chapter concludes Part I, Noise Mechanisms and Models. Additional information on the noise behavior of passive components is given in Chapter 9.

6-1 NOISE IN FETs

The FET can be fabricated in two distinctly different ways. In the junction FET (JFET), the voltage applied to its reverse-biased silicon *pn* junction modulates the conductivity of a channel below the junction by varying the size of the depletion region. The insulated-gate FET (MOSFET) accomplishes control of current flow by a capacitor type of action. Charge in the conductive channel in the vicinity of the gate (input) electrode is attracted or repelled by the potential applied to the gate. This charge accumulation can discourage current flow through the channel.

Both the JFET and the MOSFET can be assembled in either conductivity type depending on whether the channel is *p*- or *n*-type silicon. The output characteristics curve for all FET types is similar to the corresponding curve for the bipolar transistor. However, the important input quantity is gate-to-source voltage instead of base current. The input resistance of the JFET is that of a reverse-biased diode; for the MOSFET, input resistance is even higher. Additional information is available in the literature [1].

The noise equivalent circuit for common-source operation shown in Fig. 6-1 applies to both the JFET and the MOSFET. Small-signal operation is simulated by C_{gs}, C_{gd}, C_{ds}, and the dependent current generator $g_m V_1$ with parallel resistance r_{ds}. Noise current generator I_g is the result of three physical processes: shot noise of the current flowing through the gate, 1/f noise, and thermal fluctuations in the drain circuit that are coupled into the gate circuit. Noise generator I_d is the result of thermal excitation of carriers in the channel of the device. Generators I_g and I_d are found to be correlated at high frequencies.

The E_n-I_n representation is applicable to both FET types. The curves shown in Fig. 6-2 are typical of a JFET. Figure 6-2*a* shows the typical characteristic of low E_n noise at high frequencies (region 2). Behavior of

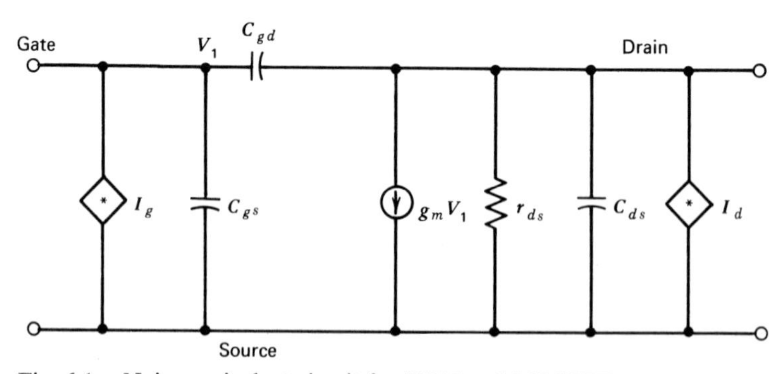

Fig. 6-1. Noise equivalent circuit for JFET and MOSFET.

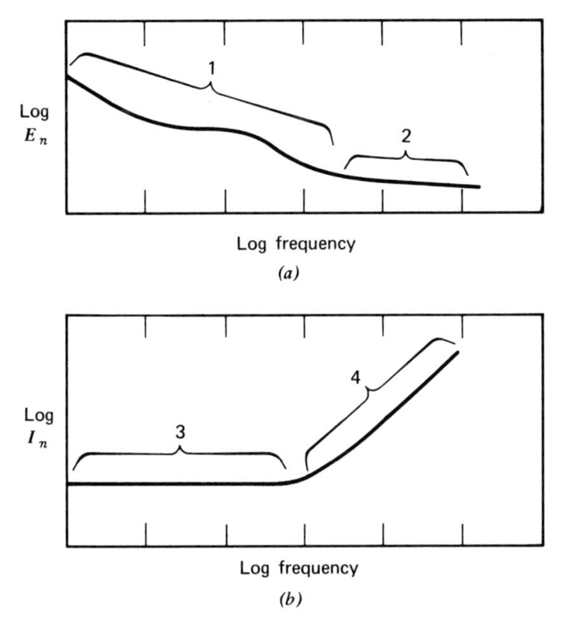

Fig. 6-2. Typical noise behavior of a FET.

open-circuit noise current I_n is given in Fig. 6-2*b*. Noise current is low in region 3 and increases with frequency.

The observed behavior in region 2 is due to the thermal noise of the channel. This has been shown by van der Ziel [2,3] to be dependent on the resistance of the channel and follows the expression

$$R_n \simeq \left(\frac{2}{3}\right)\left(\frac{1}{g_m}\right) \tag{6-1}$$

Transconductance g_m is conventionally defined. This value of noise resistance R_n is related to E_n according to the thermal noise equation

$$E_n = \sqrt{4kTR_n\,\Delta f} \tag{6-2}$$

In order to minimize the region 2 noise it is necessary to operate the FET where its g_m is large. The largest g_m values can be found at high values of static drain current. Usually g_m is highest in the vicinity of I_{DSS}, the value of I_D where gate-to-source bias voltage V_{GS} is zero.

From measured data on g_m, we can determine R_n from Eq. 6-1 and calculate E_n using Eq. 6-2. These calculated values can be compared with noise measurements of E_n. Such a comparison is given in Table 6-1. Only the minimum values for calculated and measured E_n are shown in the table. It is

Table 6-1. Comparison between Calculated and Measured Values of E_n

FET Type	Calculated E_n in nV	Measured E_n in nV	FET Type	Calculated E_n in nV	Measured E_n in nV
2N2609	1.7	2.7	2N5266	2.8	3.5
2N3460	2.6	3.3	2N5394	1.5	2.0
2N3631	2.6	5.0	C413N	0.6	0.7
2N3684	2.4	3.3	FN88-8	1.8	2.0
2N3821	1.9	2.2	FT0653	1.3	1.4
2N4221A	1.8	2.2	FT0654C	1.5	5.8
2N4416	1.6	2.0	HRN1030	2.2	5.8
2N4869	2.2	2.7	UC410	1.8	5.0
2N5116	1.1	2.2			

evident that in all instances the measured noise is larger than that predicted by calculations.

At the lower frequencies in Fig. 6-2*a*, region 1, there is excess 1/f noise. This noise arises from the trapping of carriers in the so-called Shockley-Read-Hall (SRH) generation-recombination centers in the junction depletion region [4]. These centers are the main sources of reverse-bias leakage current in silicon diodes. A diagramatic cross-section of a junction FET is shown in Fig. 6-3.

Generation centers, represented by *X*s, alternately emit a hole and an electron, and simultaneously fluctuate between a charged and neutral state. This fluctuating charge looks like a change in gate voltage or a true input signal. Thus the channel current varies or "flickers." These fluctuations are the principal source of 1/f noise in the JFET.

The generation centers are due to crystal defects or impurities. Since these centers are one of the sources of reverse-bias current, it is observed that devices with exceptionally low gate leakage current (I_{GSS}) have low 1/f noise.

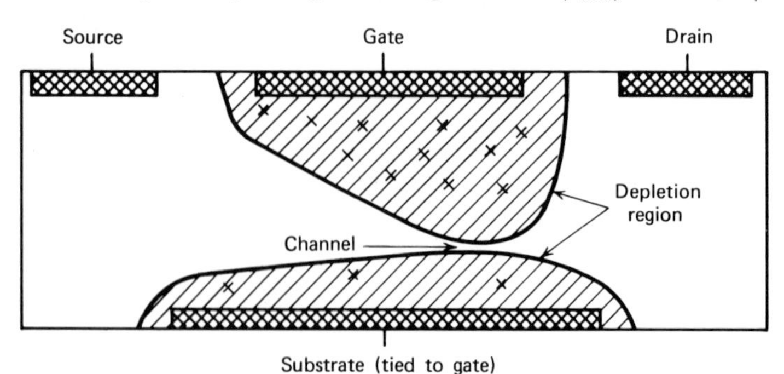

Fig. 6-3. FET cross-section.

High gate leakage does not always imply a high-noise device since the currents can be of surface origin. Charge fluctuations at the surface do not modulate the width of the channel nor do they necessarily give the type of noise being discussed here. Since the MOSFET does not have a *pn* junction, this noise mechanism does not necessarily apply.

Region 1 in Fig. 6-2*a* has bumps or a roller-coaster effect. This is due to the distribution of trapping centers with a variety of time constants.

There does not appear to be any theoretical way of reducing this 1/f noise. Several methods that increase 1/f noise, such as gold doping, are known. The noise is quite process-dependent and varies from run-to-run and supplier-to-supplier. To select a FET for a 1/f application it is necessary to measure the characteristics of a significant sample of devices.

We now refer to the curve of equivalent open-circuit noise current I_n in Fig. 6-2*b*. Region 3 is usually independent of frequency and is determined by the shot noise of the gate leakage. In a few transistors with very high 1/f noise in E_n, the I_n curve has a slight rise at the lowest frequencies. Shot noise is a basic characteristic of current flow across a *pn* junction.

The noise in region 4 of Fig. 6-2*b* is caused by an interesting mechanism. This is the thermal noise of g_{11}, the real part of the input admittance. In FETs, the noise of the channel is conducted to the input at high frequencies by the gate-to-drain capacitance. This appears as a shunt resistance in parallel with the input capacitance, C_{gs}, in the equivalent electrical circuit. Using the model given in Fig. 6-1 with load quantities R_L and C_L to represent r_{ds}, C_{ds}, and external elements connected between drain and source, the equivalent shunt resistance between gate and source can be shown to be

$$R = \frac{1 + \omega^2 C_L R_L}{\omega^2 g_m C_{gd} C_L R_L{}^2}$$

This "reflected" resistance, in general, decreases with increasing frequency. From the discussion given in Section 3-3 pertaining to the effect of a parallel noisy resistance, we can conclude that the coefficient of E_n is reduced with increasing frequency, whereas the effective I_n term is increased for a reduction in the value of R.

The noises in regions 2 and 4 are correlated. This results in an additional noise term as shown in Eq. 1-8.

6-2 MEASURING FET NOISE

FET noise measurements cover such wide ranges of impedance and frequency that four separate measurement methods are used. The noise voltage E_n at low and high frequencies can be measured directly as will be described. To determine low-frequency noise current I_n, two methods are used. The noise can be measured directly, or the gate leakage current can be measured and the noise calculated. High-frequency I_n noise need not be measured directly. Instead, the shunt input resistance can be measured and the I_n thermal noise calculated. Accurate measurement of I_n is difficult at the midfrequencies.

For the low-frequency end of the spectrum (10 Hz–50 kHz), the system shown in Fig. 6-4*a* was used to determine E_n. The noise voltage spectral density is

$$E_n/\sqrt{\sim} = \frac{E_{no}(1/\sqrt{\Delta f})1.13}{K_t} \tag{6-3}$$

where E_{no} = dc output voltage at the recorder

$1/\sqrt{\Delta f}$ = bandwidth factor of wave analyzer

1.13 = meter correction factor (corrects average reading to rms)

K_t = system gain = 10,000 in this system.

For the higher-frequency portion of the spectrum (50 kHz–1.5 MHz), the setup of Fig. 6-4*b* can be used. Gain K_t is measured at each frequency

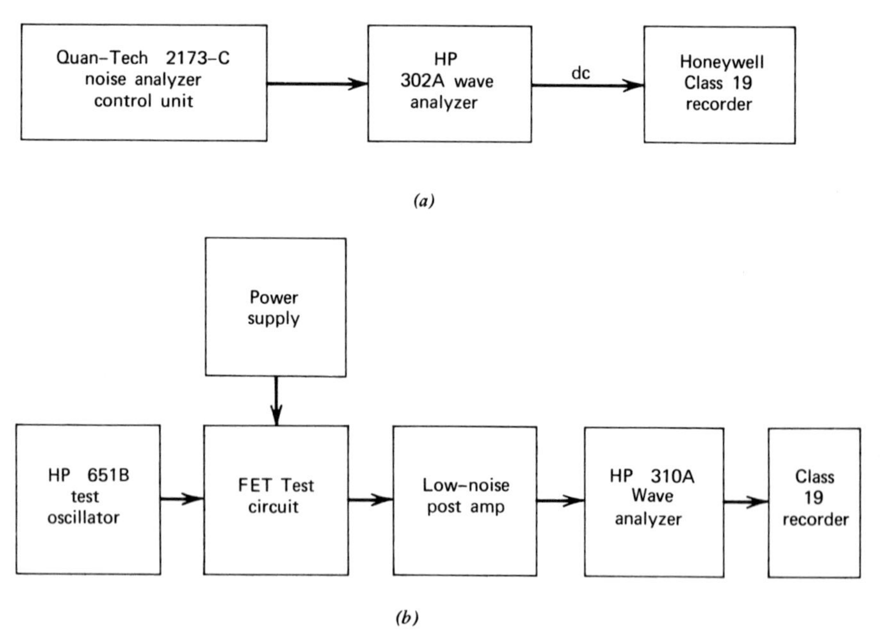

Fig. 6-4. Measuring systems for FET noise.

by supplying a low-level signal from the 651B oscillator and measuring the output on the 310A wave analyzer. The oscillator is then removed, the input to the FET test box is shorted, and the output noise is recorded. Using Eq. 6-3, E_n is then determined.

The normal method of measuring I_n in a bipolar transistor is to measure noise with a large source resistance, and then subtract the thermal noise of R_s from the total measured noise. This procedure is difficult with FETs because of the very low values of I_n. For example, a device with an I_n value of 2×10^{-15} A, and R_s of $10^8\ \Omega$, would yield a noise voltage (I_nR_s) of 2×10^{-7} V. The thermal noise of a 10^8-Ω resistor is 1.26×10^{-6} V, six times greater than the I_nR_s value. To obtain accurate and repeatable I_n measurements the I_nR_s values should be at least three times the thermal noise of R_s The problem of high R_s thermal noise values is solved by using a high-quality (mica) capacitor for R_s. Since the capacitor has no thermal noise, the measured output noise consists of I_nX_c plus E_n. Then $I_nX_c = \sqrt{E_{ni}^2 - E_n^2}$ and $I_n = I_nX_c/X_c$. This method is valid only at frequencies below 100 to 200 Hz, as X_c decreases rapidly with increasing frequency. It was found that on units with a high value of E_n, such as the MOSFET, the E_n component was still the dominant noise source and this made I_n difficult to measure. To verify the I_n low-frequency measurement, an electrometer was placed in series with the gate lead to measure the gate leakage current (I_{GSS}). The measured I_{GSS} is then converted to shot noise current from the formula

$$\frac{I_s}{\sqrt{\Delta f}} = (2qI_{GSS})^{1/2} \tag{6-4}$$

The two methods of determining low-frequency I_n were in close agreement on all units measured.

To determine the high-frequency behavior of I_n, the circuit shown in Fig. 6-5 can be used. The Wayne-Kerr bridge is a high-frequency bridge that measures equivalent parallel R_p and C_p values from 15 kHz to 5 MHz. However, it does not indicate R_p values above 10 MΩ. Since most FETs have an R_p value much greater than 10 MΩ below 500 kHz, we are limited to the 500-kHz to 5-MHz range. From data measured over this range we are able to determine the high-frequency I_n corner and slope. The 556 oscilloscope was used above 1.5 MHz to observe the null on the bridge because the 310A wave analyzer is limited to 1.5 MHz. From the R_p values measured, I_n is determined from the formula

$$\frac{I_n}{\sqrt{\Delta f}} = \left(\frac{4kT}{R_p}\right)^{1/2} \tag{6-5}$$

Noise data, taken using the methods discussed, are presented in Appendix

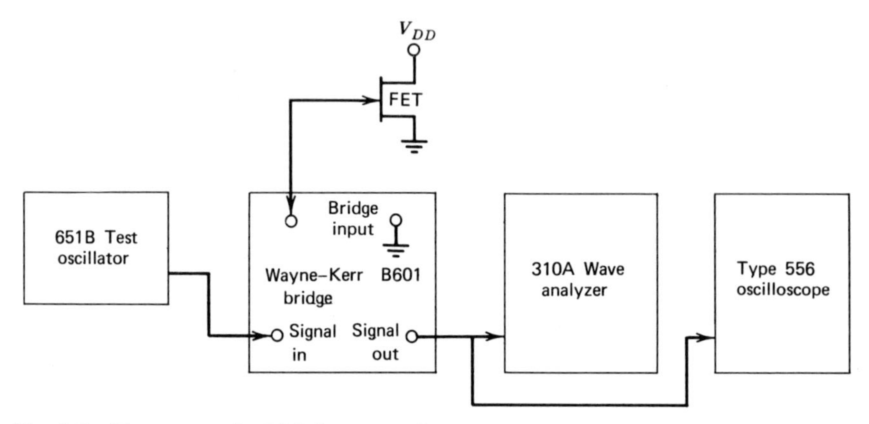

Fig. 6-5. Test system for high-frequency I_n measurements.

II and are summarized in Table 6-2. The values given in the figure are minima, or the best performance obtained at the best operating point. As we may predict using Eq. 6-1, operation of the JFET at drain current equal to I_{DSS} provides superior noise behavior. The measured noise data confirm that prediction. However, operation at values of $0.1I_{DSS}$ or $0.2I_{DSS}$ does not cause

Table 6-2. Measured Noise Data on FETs

FET Type	E_n @ Midband $\text{nV}/\sqrt{\sim}$	E_n @ 10 Hz $\text{nV}/\sqrt{\sim}$	I_n @ Midband $\text{fA}/\sqrt{\sim}$	I_n @ 1 MHz $\text{fA}/\sqrt{\sim}$
2N2609	2.8	31	13	130
2N3460	4	40	11	70
2N3631 (MOS)	5	800	0.2	20
2N3684	3.5	55	8.2	34
2N3821	2.2	7	3.5	19
2N4221A	2.2	50	8.5	18
2N4416	2	20	13	23
2N4869	3	10	6	65
2N5116	3	160	35	210
2N5266	4	80	3.5	60
2N5394	2	30	14	85
C413N	0.8	6	31	230
FN88-8	2	3	8	80
FT0653	1.4	7	6.5	100
FT0654C	6	40	14	140
HRN1030 (MOS)	6	800	0.7	40
UC410	5	230	14	120

Note: fA ≡ femto-ampere ≡ 10^{-15} A.

a very serious increase in E_n. It is also observed that I_n is not affected by the value of reverse-gate bias.

For values of source resistance below 500 kΩ the E_n mechanism dominates over I_n. Thermal noise of a source of 500 kΩ is 90 nV/Hz$^{1/2}$. As can be seen from Table 6-2, all of the units are well below that value at midfrequencies. For systems with large values of source resistance (>1 MΩ), the I_nZ_s noise term is dominant. Among the JFET types, the 2N3821 and 2N5266 performed well. The value of I_n is, however, proportional to the square root of gate leakage. MOSFETs, with no junction, have low I_n values.

The results of noise tests on a type 2N5266 JFET are shown in Fig. 6-6.

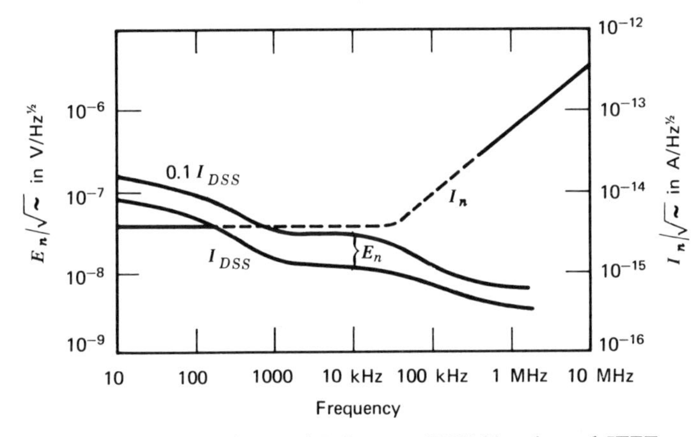

Fig. 6-6. Behavior of E_n and I_n for type 2N5266 *p*-channel JFET.

Both parameters E_n and I_n are plotted against frequency. The roller-coaster effect and the dependence on operating point are obvious in the E_n curves. Clearly I_n is very low valued at low frequencies.

6-3 NOISE IN SBTs

The *super-beta transistor* (SBT), also referred to as a *super-gain transistor*, differs from conventional bipolar transistors primarily in the value of its short-circuit current-gain parameter. The β of a typical SBT is in the 1000 to 10,000 range, at collector-current levels as low as 1 μA. The SBT has the disadvantage of its collector-to-emitter voltage breakdown being less than 5 V.

The SBT can be fabricated on the same chip as conventional monolithic bipolar transistors, but requires a second emitter diffusion to achieve an extremely narrow base width. The narrow base width results in the high-β and low-voltage breakdown characteristics. In a monolithic IC, the SBT can be most useful as the input stage of an amplifier. The SBT is also available as a discrete circuit element.

Whereas the SBT does have a large current gain, its input resistance is also large. Previously, it was noted that voltage gain can be expressed as

$$A_v = \frac{A_i R_L}{R_i} \tag{6-7}$$

An SBT fabricated to obtain $\beta = 2000$ has an input resistance of perhaps 1 MΩ. Using the value of β as an approximation to A_i, and when feeding a 20,000-Ω load, the voltage gain of the stage is only 40. Thus the SBT is not superior to the conventional bipolar transistor when considering voltage gain alone.

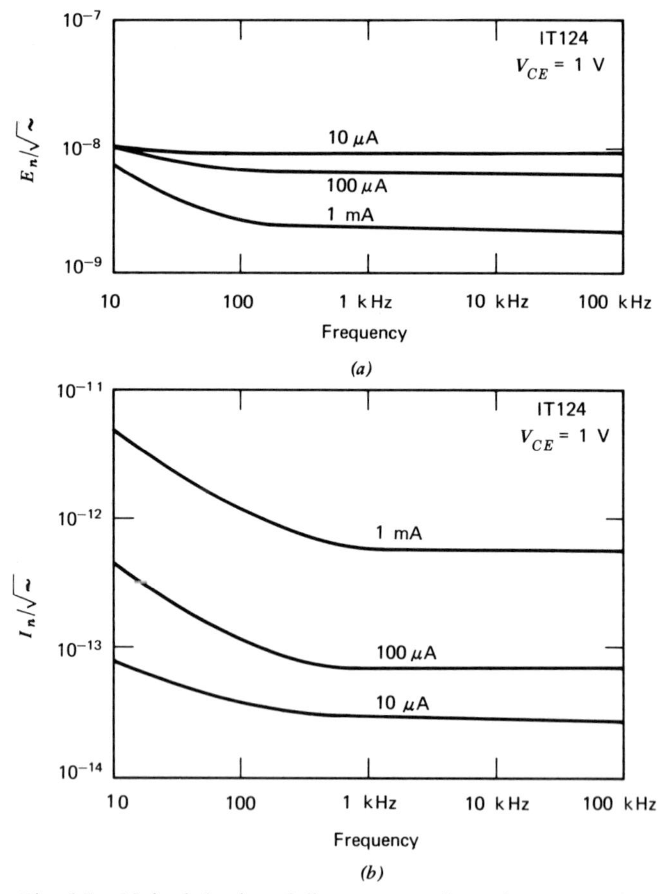

Fig. 6-7. Noise behavior of discrete super-β transistor.

Noise tests were performed on a number of discrete SBTs. The curves shown in Fig. 6-7 are representative. In (*a*) of the figure we see a reasonable noise level for E_n and little 1/f noise. An upturn in the spot current noise was apparent in test data on all devices with the break frequency occurring around 200 Hz. The behavior of E_n and I_n with I_C is similar to that reported earlier for conventional bipolar devices.

Popcorn noise is observed in some SBT units.

6-4 NOISE IN ICs

There are two basic types of linear IC: the hybrid IC and the monolithic IC. Our major concern is with the noise behavior of the input stage. If the input stage in a hybrid IC is composed of discrete elements, the discussions of transistor noise in Chapters 4 and 5, and of component noise in Chapter 9, are applicable. On the other hand, if we are considering the behavior of monolithic transistors in monolithic or hybrid ICs, this section will supplement the previously presented information.

Many linear ICs make use of the differential pair as shown in Fig. 6-8

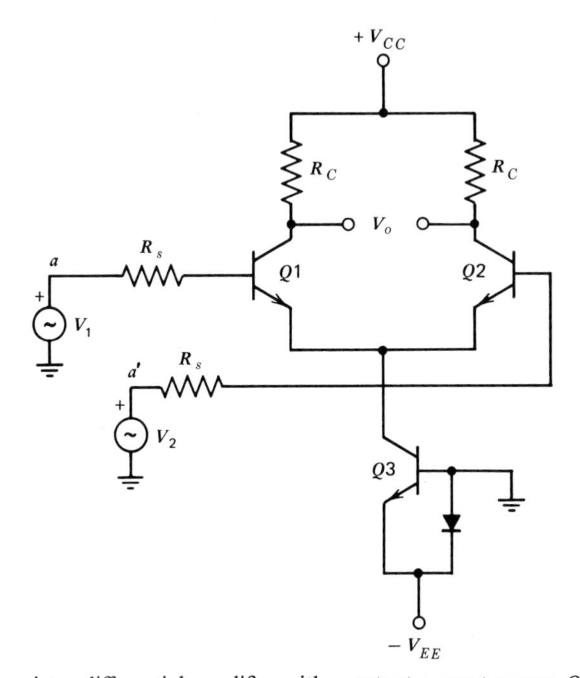

Fig. 6-8. Transistor differential amplifier with constant-current source $Q3$.

and discussed in Ch. 12. In that circuit, transistors $Q1$ and $Q2$ provide the desired amplification. Transistor $Q3$ forms part of a dc constant-current source to provide $Q1$ and $Q2$ with a constant level of emitter current to form a stable operating point for those transistors.

In normal operation the differential amplifier has a voltage gain approximately given by

$$A_v = \frac{V_o}{V_1 - V_2} \simeq \frac{\beta R_L}{R_s + r_{bb'} + r_{b'e}} \tag{6-6}$$

Parameters of $Q1$ and $Q2$ are assumed equal, and V_2 is normally equal in magnitude and opposite in phase to V_1. Therefore, in a practical case, the effective input voltage is equal to $2V_1$. If V_1 and V_2 are a single voltage source V_s connected between points a and a', the V_o/V_s gain is as given in Eq. 6-6.

It has been shown that the noise sources of the two transistors in the monolithic differential amplifier are statistically independent [5]. Consequently, the pair exhibits about twice the level of noise of a single transistor.

Noise mechanisms in monolithic transistors do not differ from the presentation of Chapter 4. Shot noise sources are dependent on the dc base current and the dc collector current. The ohmic portion of the base resistance is responsible for thermal noise. Low-frequency noise is also present. Expressions used to represent these mechanisms are the same for monolithic transistors and discrete transistors.

Popcorn or burst noise can also be present in some units. The popcorn noise behavior of the IC transistor is as discussed in the preceding chapter. To represent the popcorn noise source a noise-current generator as given in Eq. 5-2 can be connected to the junction between active and inactive base resistances. The model is shown in Fig. 5-7.

In a differential input amplifier, popcorn noise would appear as a differential mode signal, and would be located at the input of either transistor in the pair. The operation of any system would be severely affected by any popcorn noise generated in the initial stage. Popcorn noise in the second stage of an amplifier can seriously degrade performance.

6-5 IC NOISE PERFORMANCE

The E_n-I_n noise model and the concept of total equivalent input noise are useful in describing the noise behavior of IC amplifiers. These representations are used in this discussion of noise in selected linear ICs. Measured noise data are available in Appendix III.

Generally it is found that ICs are inherently higher-noise devices than discrete transistors. IC transistors have lower βs because of tradeoffs in device design that result in not being able to optimize fully for low noise. The noise performance of an IC depends strongly on the manufacturer, and not exclusively on the type number.

The μA741C is an internally compensated operational amplifier with differential input stage. When supplied with ± 15 V power it exhibits the following typical characteristics:

Voltage gain: 200,000 ($R_L \geq 2$ kΩ)
Input resistance: 2 MΩ
Input capacitance: 1.4 pF
Output resistance: 75 Ω
Common-mode rejection ratio: 90 dB ($R_s \leq 10$ kΩ)
Power consumption: 50 mW

Since op amps are seldom used open loop, no frequency response data are

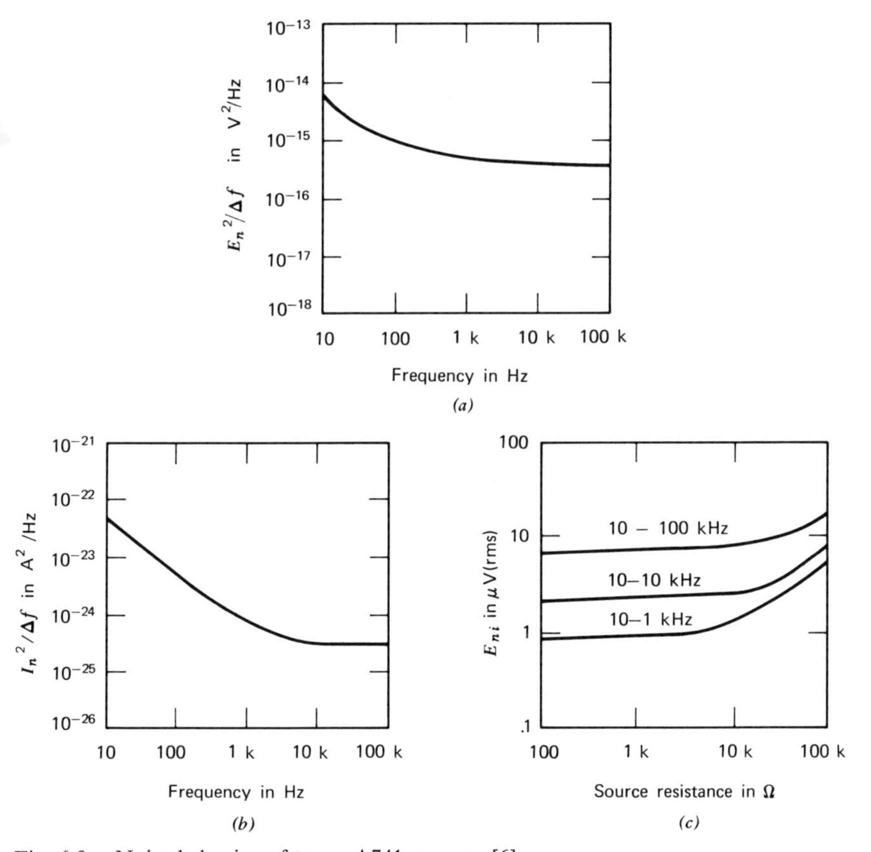

Fig. 6-9. Noise behavior of type μA741 op amp [6].

presented. However, when connected as a unity-gain amplifier by using appropriate feedback, the transient rise time is 0.3 μsec for $R_L = 2\ \text{k}\Omega$ and $C_L \leq 100$ pF.

Noise behavior of the μA741 is given in Fig. 6-9. In *a* and *b* of the figure the narrowband E_n^2 and the narrowband I_n^2 are shown as functions of frequency. Figure 6-9*c* depicts the dependence of E_{ni} on R_s and Δf.

The type μA725 is an operational amplifier specifically for instrumentation service. Typical specifications for operation from ± 15-V power supplies are

Voltage gain: 3×10^6 ($R_L \geq 2\ \text{k}\Omega$)
Input resistance: 1.5 MΩ
Output resistance: 150 Ω
Common-mode rejection ratio: 120 dB ($R_s \leq 10\ \text{k}\Omega$)
Power consumption: 80 mW

Spot noise parameters given by the manufacturer are

$f_o = 10$ Hz: $E_n = 15\ \text{nV/Hz}^{1/2}$, $I_n = 1.0\ \text{pA/Hz}^{1/2}$
$f_o = 100$ Hz: $E_n = 9.0\ \text{nV/Hz}^{1/2}$, $I_n = 0.3\ \text{pA/Hz}^{1/2}$
$f_o = 1$ kHz: $E_n = 9.0\ \text{nV/Hz}^{1/2}$, $I_n = 0.15\ \text{pA/Hz}^{1/2}$

Noise behavior of the μA725 is shown in Fig. 6-10. In *a* and *b* we see E_n^2 and I_n^2 versus frequency. The broadband E_{ni} is given in *c* of the figure. Contours of constant spot-noise figure are plotted in Fig. 6-10*d* against frequency. From the curve in *e* it is clear that to minimize the 1-kHz value of *F*, a source resistance of greater than 50 kΩ is necessary. Power supplies are ± 15 V.

In Fig. 6-10*f*, the effect of a common-mode noise voltage is shown on a system with unity closed-loop voltage gain. The common-mode rejection ratio is not infinite, and consequently common-mode voltages can be expressed in terms of an equivalent input noise voltage. Further discussion is given in Section 12-17.

Micropower linear IC amplifiers are available. A micropower amplifier can be defined as a system in which the standby power dissipation (product of dc supply voltage and current) is less than a mW. The type μA735 is an example of such an amplifier. It has a voltage gain of 20,000, 10-MΩ input resistance, and the power consumption is but 100 μW with ± 3-V supplies. Because of low-current densities, this type of device has low excess noise.

Tests on several μA735 samples gave the following results when operated at ± 3 V: $E_n = 20\ \text{nV/Hz}^{1/2}$ with a 1/f corner of 50 Hz; $I_n = 1 \times 10^{-13}$ A/Hz$^{1/2}$ with a 1/f corner of 100 Hz. These figures represent best performance of a small number of samples and are presented simply as a guide to typical values. Although the 1/f noise is low, it is greater than for certain types of

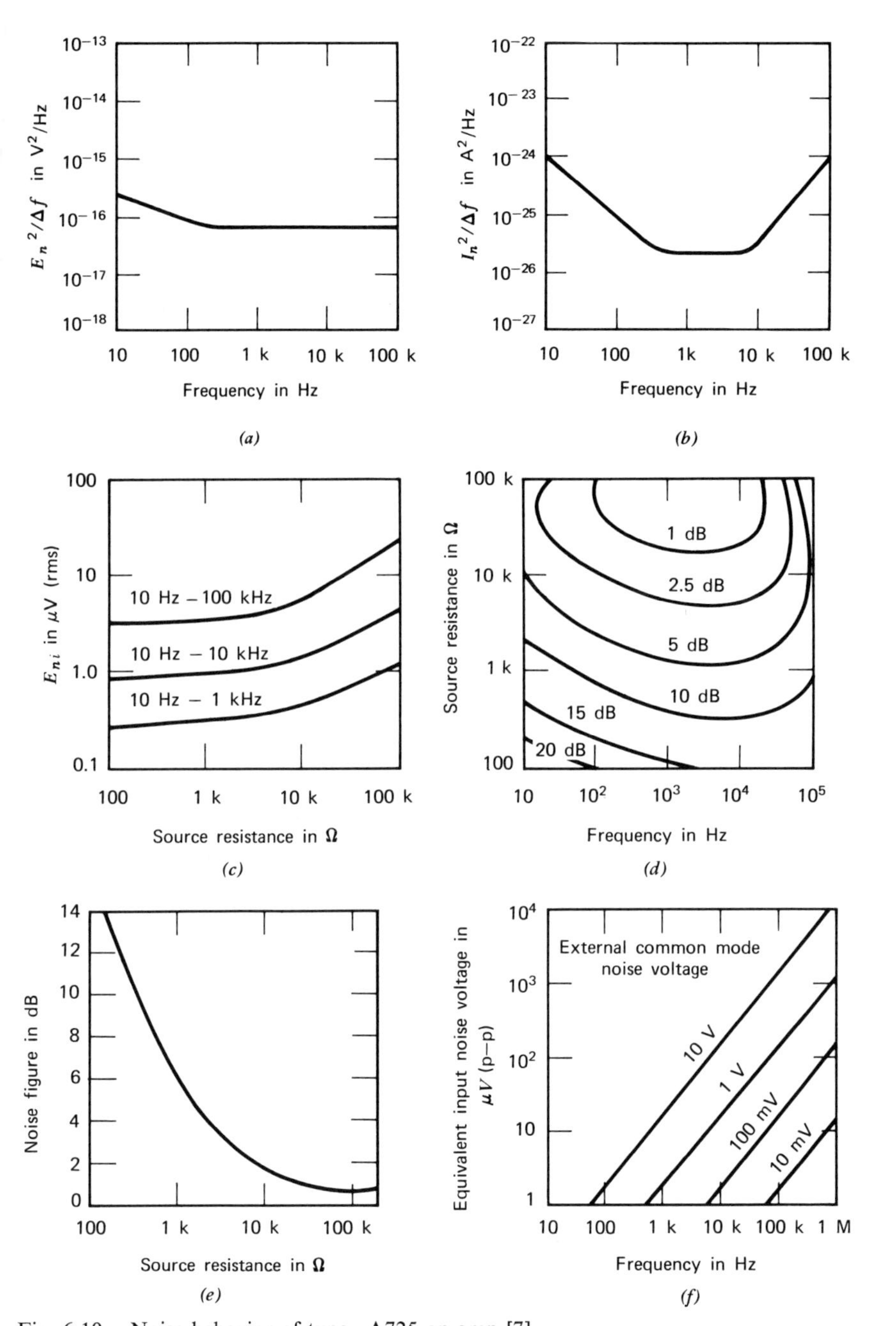

Fig. 6-10. Noise behavior of type μA725 op amp [7].

discrete transistors. An improvement in noise performance is possible if the user adds an ac-coupled *npn* stage operating at 2 μA ahead of the μA735.

The diagram of Fig. 6-11 shows connections made to the μA735 for noise measurements. Because the signal source is fed to the + input terminal, the

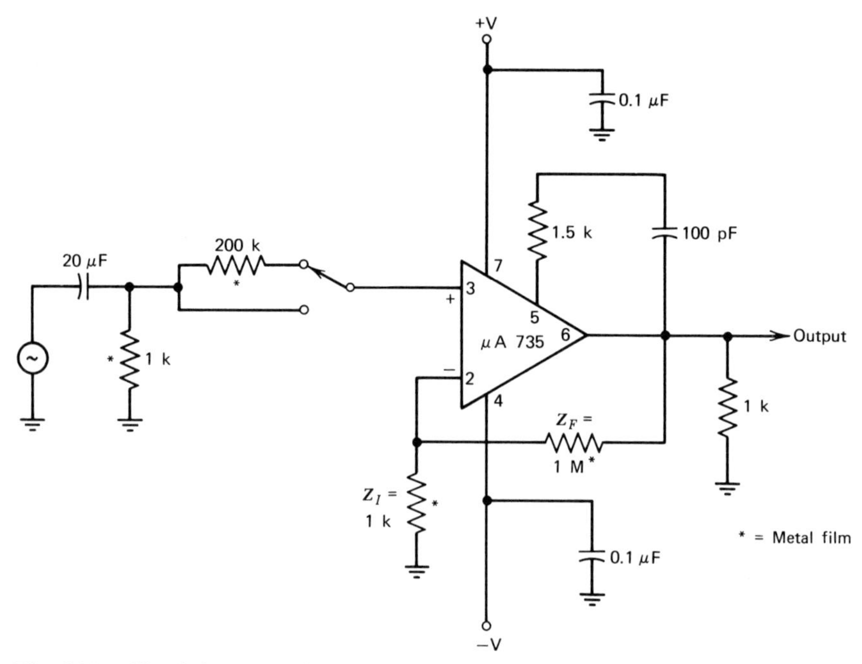

Fig. 6-11. Circuit for measuring the noise of a micropower op amp.

connection is noninverting. Feedback elements Z_F and Z_I set the gain at about 1000. The 1.5-kΩ resistor and 100-pF capacitor perform the function of frequency compensation. Other capacitors are for blocking and filtering. The 1-kΩ resistor is a load. For I_n measurement, a 200-kΩ metal film resistor is employed as shown.

SUMMARY

a. Noise currents in the JFET are caused by shot noise of gate leakage current, 1/f mechanisms, and thermal noise. The MOSFET has no shot-noise component.

b. Because field-effect devices have low I_n, they are useful in systems with sensors of high internal resistance.

c. The SBT can be considered for an input stage. Because of the extremely low-base bias current required, they are attractive for input stages in IC operational amplifiers. Noise behavior is similar to that of conventional bipolar transistors.

d. Noise performance of monolithic IC amplifiers is generally not as good as can be accomplished by custom design using discrete devices.

e. A bipolar stage preceding an IC amplifier can provide low noise along with the advantages of the IC.

PROBLEMS

1. Consider a type 2N3821 JFET biased at I_{DSS}. The resistance of the signal source is 10 MΩ, and gate biasing elements can be neglected.

(a) Derive an expression for noise figure for this simple circuit in terms of R_s, E_n, I_n, and correlation factor C.

(b) Using midband data from Table 6-2, calculate the noise figure. Consider that the correlation is unity.

2. Convert the 1 kHz data given in Fig. 6-9*c* at 100 kΩ and 100 Ω to spot-noise values of E_n^2 and I_n^2. How do your values compare with those shown in *a* and *b* of the figure?

3. Calculate CMRR from Fig. 6-10*f* at $f = 1$ kHz and a common-mode voltage of 1 V. Does the CMRR calculated at 1 kHz and 10 V differ from your answer at 1 V?

4. Predict the wideband noise voltage to be expected at the load terminals of an IC amplifier with voltage gain of 10,000. Consider $R_s = 0$, $E_n = 10$ nV/Hz$^{1/2}$, and $I_n = 0.2$ pA/Hz$^{1/2}$ over the band from 1 to 50 kHz.

5. Use the information given in the preceding problem and the fact that $R_s = 10^4$ Ω to predict the wideband output noise voltage over the same frequency band.

REFERENCES

1. Fitchen, F. C., *Transistor Circuit Analysis and Design*, 2nd ed., Van Nostrand Reinhold, New York, 1966, pp. 18–22.
2. van der Ziel, A., "Thermal Noise in Field Effect Transistors," *Proc. IEEE*, **50**, 8 (August 1962), 1808–1812.
3. van der Ziel, A., "Gate Noise in Field Effect Transistors at Moderately High Frequencies," *Proc. IEEE*, **51**, 3 (March 1963), 461–467.
4. Lauritzen, P. O., "Low-Frequency Generation Noise in Junction Field Effect Transistors," *Solid-State Electronics*, **8**, 1 (January 1965), 41–58.

5. Brodersen, A. J., E. R. Chenette, and R. C. Jaeger, "Noise in Integrated Circuit Transistors," *IEEE J. Solid-State Circuits*, **SC-5**, 2 (April 1970), 63–66.
6. "Internally Compensated Operational Amplifier, μA741C," Bulletin of Fairchild Semiconductor, Mountain View, California, April 1969.
7. "Instrumentation Operational Amplifier, μA725B," Bulletin of Fairchild Semiconductor, Mountain View, California, June 1970.

PART II

Design Techniques and Examples

Chapter 7

LOW-NOISE DESIGN

We have discussed the noise mechanisms present in electronic devices, and have shown models to represent the noise behavior of those devices. Furthermore, we have treated the methods of circuit analysis peculiar to noise problems. Now we turn our attention to system design. The information contained in the preceding chapters is used to create new low-noise systems that perform according to preassigned requirements.

Low-noise design from the system designer's viewpoint usually is concerned with the following problem: given a sensor with known signal, noise, impedance, and response characteristics, how do we optimize the amplifier design to achieve the lowest value of equivalent input noise?

The amplifying portion of the system must be *matched* to the sensor. This matching is the essense of low-noise design.

7-1 DESIGN PHILOSOPHY

When designing an amplifier for a specific application, there are many constraints to be met and decisions to be made. These include gain, bandwidth, impedance levels, feedback, stability, cost, and, as expected, noise. The amplifier designer can elect one of two paths. Typically he worries about the gain and bandwidth first and later in the design process he checks for noise. We strongly urge the reverse approach, with initial emphasis on noise performance. First, the input stage device is considered, discrete or IC, bipolar or FET. The operating point is then selected. If preliminary analysis shows that the noise specification can be met, a circuit configuration (CE, CC, or CB) can be selected and the amplifier designed to meet the remainder of the circuit requirements. A critical noise specification can be one of the most serious limiting factors in a design; it is best to meet the most stringent requirement head-on.

Noise is essentially unaffected by circuit configuration and overall negative feedback; therefore the transistor and its operating point can be selected to meet the circuit noise requirements, and then the configuration or feedback can be determined to meet the gain, bandwidth, and impedance requirements. This approach allows the circuit designer to optimize for the noise and for the other circuit requirements independently.

After selecting a circuit configuration, analyzing it for nonnoise requirements may indicate that it will not meet all the specifications. If the bandwidth is too narrow, more stages and additional feedback can be added, the collector current of the input transistor can be increased, or a transistor with a larger f_T can be selected. Then noise can be recalculated to see if it is still within specifications. This iterative procedure insures obtaining satisfactory noise performance and prevents locking in on a high-noise condition at the very start of the design. Once the design is frozen and you find that you have too much noise, it is difficult to make minor changes that lower the noise.

The question can be asked, "In a low-level application do you always design the amplifier for the lowest possible noise?" The answer is "no." Usually, there is not much value in designing for a noise figure of less than 3 dB. A NF of 3 dB is equivalent to a signal-to-noise ratio of 1:1. In other words, the amplifier and sensor are each contributing equal noise, and the total system noise is $\sqrt{2}$ times either component. If, from supreme design effort, the amplifier noise is reduced to $\frac{1}{10}$ of the source noise, the total system noise is now $1/\sqrt{2}$ or about 0.707 of the 3-dB condition. Hence there is little to be gained by such a reduction. It is impressive to say that your amplifier has a NF of 0.3 dB, but really, in terms of signal-to-noise ratio, little is gained with a NF of less than 3 dB.

When a system operates from a reactive source that has a small resistive component, the amplifier must be carefully designed. In this application, the NF is not very meaningful. With an inductance and small series resistance, thermal noise can be negligible, but in reality the system noise is dominated by $I_n X_L$ at the higher frequencies. *For best performance do not design for minimum NF, but design for lowest equivalent input noise* E_{ni} *over the bandwidth.*

7-2 DESIGN PROCEDURE

The ultimate limit on equivalent input noise is determined by the sensor impedance and the first stage $Q1$ of the amplifier. The source impedance $Z_s(f)$ and noise generators $E_n(f)$ and $I_n(f)$ representing $Q1$ are each a different function of frequency. Initial steps in the design procedure are the selection of the type of input device, such as bipolar transistor, FET, or IC, and the associated operating point to obtain the desired noise characteristic. In the simplest case of a resistive source, match the amplifier's optimum source resistance R_o to the resistance of the source or sensor. This minimizes the equivalent input noise at a single frequency. If the amplifier is operated over a band of frequencies, the noise must be integrated over this interval. Since the noise mechanisms and sensor impedance are functions of frequency, the computer program NOISE was developed to perform this integration. The NOISE program is the subject of Chapter 8. The resulting analysis may indicate that there is too much noise to meet the requirements or it may suggest a different operating point. By changing devices and/or operating points, theoretical performance can approach an optimum. If the initial analysis indicates that the specifications cannot be met, then no amount of circuit design is going to reduce the equivalent input noise to an acceptable level.

After selecting the input stage the circuit is designed. Set up the biasing, determine the succeeding stages, the coupling networks, and the power supply; then analyze the total noise of the entire system, including the bias-network contributions to insure that you can still meet the noise specifications. Finally, add the overall negative feedback to provide the desired impedance, gain, and frequency response. Of course, these requirements should have been kept in mind while doing the initial noise design. Input impedance, frequency response, and gain are not basic to one type of transistor or to one particular operating point. Overall negative feedback can be used to increase or decrease input impedance or to broaden the frequency response and set the gain.

If it appears at this time that one or more of these circuit requirements cannot be met, then readjust the circuit to suit these needs and make another noise analysis to insure that the design continues to meet the noise specifications. Continuing around the loop of noise analysis, circuit synthesis, and back to noise analysis finally establishes the best design for each specific application.

The diagram shown in Fig. 7-1 symbolizes quantities of interest in this design discussion. Equivalent noise generators $E_n(f)$ and $I_n(f)$ represent the sum of all of the noise contributions of the amplifier stages and components. However, at this point in the design procedure the stages following $Q1$ may not be determined, and thus their contributions are not known.

It was shown in Chapter 2 that the influence on NF from stages following the first one is often negligible for a high-gain first stage, and if not negligible the attenuation is caused by available power gain of the first stage. Chapter 12 examines cascading and provides for evaluation of the gain multipliers.

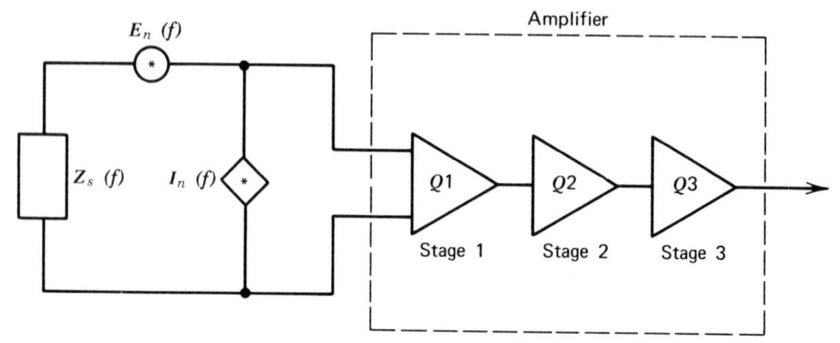

Fig. 7-1. Sensor-amplifier system.

When using bipolar transistors connected common-emitter, approximate relations are available to suggest biasing. In midband, design for $I_{C2} = \beta_2 I_{C1}/10$, and in the 1/f region, select $I_{C2} = I_{C1}$. Further discussion is available in Chapter 12.

7-3 SELECTION OF ACTIVE DEVICES

The active device can be a bipolar or a FET, or it can be an IC with bipolar or FET input stage. Selection depends primarily on source resistance and frequency range. It is difficut to say exactly where each type of device should be used. To assist in decisions of this type, a general guide is shown in the chart of Fig. 7-2. At the lowest values of source resistance it is usually necessary to use transformer coupling at the input to match the source resistance to the amplifier R_o. Bipolar transistors are most useful in a mid-range. Some valuable adjustment can be made by changing the transistor collector current. There is a slight difference between *pnp* and *npn* transistors; the *pnp* can have a lower base resistance due to higher mobility in its base region. The *pnp* thus has a lower thermal noise voltage and can be used with smaller source resistances. On the other hand, *npn* transistors often have a

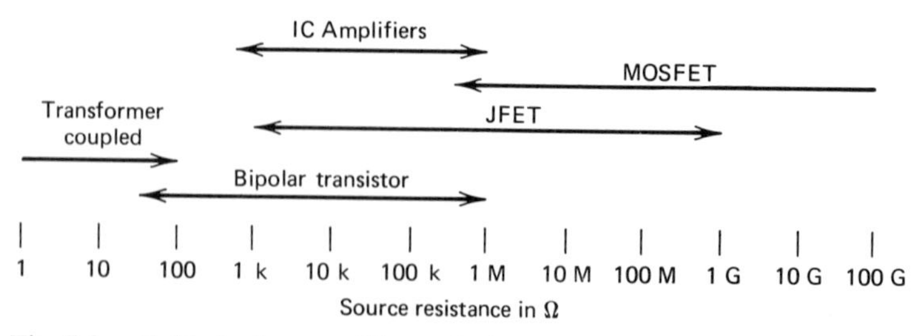

Fig. 7-2. Guide to the use of input devices.

slightly larger β_o and f_T, and therefore can be more useful at the higher end of the resistance range.

At higher values of source resistance, FETs are more desirable because of their very low noise current I_n. In some instances they are even preferred when a low E_n is desired. For instance, at frequencies of less than 10 Hz, a JFET can be selected that has lower noise than available bipolar transistors because some units do not show as much 1/f noise. In fact, when operating with a very wide range of source resistances, such as in an instrumentation amplifier application, a FET is generally preferred for the input stage. A good FET has E_n slightly larger than that of a bipolar transistor, and its I_n is significantly lower. This is of particular value when operating from a reactive source over a wide frequency range because the source impedance is linearly related to frequency. Another advantage of the FET is its high-input resistance and low-input capacitance; thus it is particularly useful as a voltage amplifier. FET input impedance is highly frequency sensitive because of the high-input resistance. The low-noise FET can be several times more expensive than a comparable bipolar transistor.

In general, the JFET has low excess noise; with high source resistances the MOSFET with its extremely low values of I_n has an advantage, although the MOSFET may have 10 to 1000 times the 1/f noise voltage. As processing techniques improve, the MOSFET becomes increasingly more attractive.

Often, because of their small size and low cost, we would like to use IC amplifiers. In general, if the lowest obtainable noise is not required, ICs can be used to good advantage. Frequently an IC can be selected that exhibits a noise level of only two to five times that of a discrete transistor circuit. If maximum state-of-the-art performance is desired, it is possible to use one or two discrete (bipolar or FET) stages ahead of the IC. This option is often selected in practical design. One caution must be observed when applying ICs: if a low gain, less than 10, is desired, frequently the feedback resistors are a serious additional source of noise.

7-4 DESIGN WITH FEEDBACK

After determination of the input device, its operating point and configuration, we can add overall multistage negative feedback to achieve the required input impedance, amplifier gain, and frequency response. In effect, single-stage negative feedback is utilized if the design employs either the common-base or common-collector configurations because these stages have 100% current and voltage feedback, respectively. An additional dimension is added by using negative feedback around several cascaded stages. As discussed in Chapter 12, negative feedback does not increase or decrease the equivalent input noise except for the additional noise contribution of the feedback resistors themselves.

If an amplifier with a low input resistance is desired, the common-base configuration can be used as an input stage. The input resistance can be further reduced by the addition of overall negative feedback to the emitter input of the common-base stage. See Fig. 7-3*a*. The input resistance is reduced

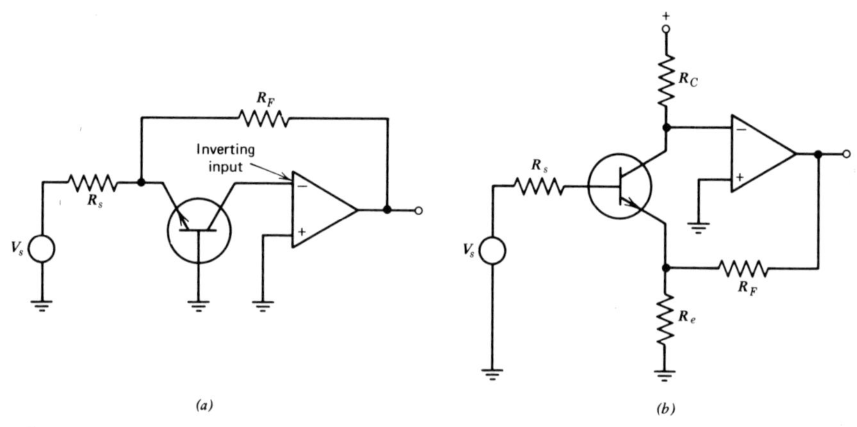

Fig. 7-3. Feedback circuits.

in proportion to the feedback factor, the ratio of the open-loop to closed-loop amplifier gains. A low input resistance is required when an amplifier must respond to signal current rather than voltage. The CB stage is also useful to minimize the effects of shunting capacitance on frequency response.

A high impedance is desirable when the amplifier must respond to the voltage signal from the source. A high input impedance can be obtained with either the common-emitter or the common-collector configurations. The FET, however, potentially offers the highest input resistance. Overall negative feedback to the emitter of a common-emitter input stage raises the amplifier input impedance. An example of this connection is shown in Fig. 7-3*b*.

Determining the input impedance of an amplifier with overall negative feedback can be confusing. If the input impedance is measured with the typical low-resistance impedance bridge, the measured values will agree well with calculations. On the other hand, the input impedance can be measured by inserting a variable resistor in series with the input and increasing its value until the output signal halves; the input impedance then should equal the value of the series resistor. Using this method on a negative feedback amplifier often produces a lower measured value of input impedance than that measured with the impedance bridge. Which value is correct? That depends on the source resistance. For low-source resistances the bridge value is correct. For very large resistances, greater than open loop R_i, the second method is correct. If we do not simulate actual system conditions when making measurements of this type, the test results obtained can be erroneous.

The question arises, "How can we design an amplifier with a 600-Ω input resistance to operate from a 600-Ω source without doubling the system noise?" If a 600-Ω resistor were connected in parallel with the amplifier input terminals to provide the proper input resistance, the noise of that resistor would equal the noise of the source and give a minimum NF of 3 dB. The answer is to use negative feedback. As was pointed out, for a common-emitter input stage overall negative feedback to the emitter increases the input impedance, and overall negative feedback to the base decreases the input impedance. By the use of negative feedback to both emitter and base simultaneously it is possible to obtain the desired input resistance without substantially increasing the noise. One such arrangement is shown in Fig. 7-4.

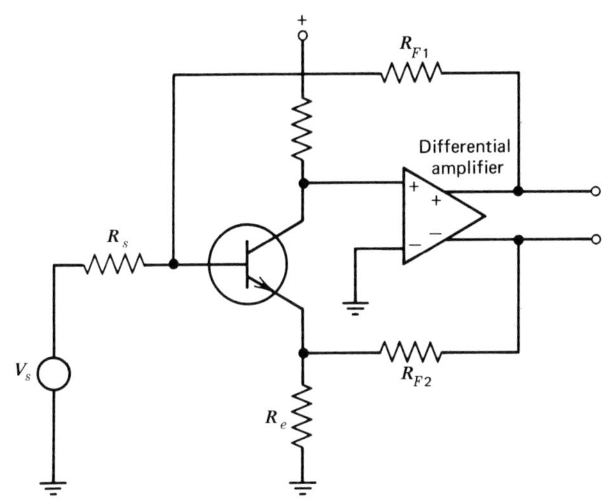

Fig. 7-4. Common-emitter stage with feedback to base and emitter.

7-5 BANDWIDTH AND SOURCE REQUIREMENTS

An important point to remember when designing low-noise amplifiers is not to overdesign for wide bandwidth. Be sure that the amplifier has definite low- and high-frequency roll-offs and that these are set as narrow as possible to pass only the signal spectrum required. The amplifier has a certain amount of noise in each hertz of bandwidth, and the greater the amplifier bandwidth the greater the output noise. There is no value in a response wider than the spectrum of the signal.

The noise curves of most devices, such as bipolar transistors, indicate that the noise spectral density is constant at the midfrequencies, but increases at both low and high frequencies. In general, increasing the transistor quiescent collector current increases the 1/f noise and decreases the high-frequency noise, so there is no one current that always provides the best performance for all applications. To determine the best performance for a specific application it is necessary to integrate the total noise over the frequency range of interest. Although it is possible to approximate the total noise, usually it is better to use a computer program with the exact circuit model as described in Chapter 8, especially if the source is reactive.

The impedance of an inductive source rises with frequency. This means that the amplifier E_n is important at low frequencies, and I_n is important at high frequencies. For this type of application, either a bipolar transistor operating at a low value of collector current, or a FET having a low E_n at low frequencies, may be recommended. It was shown in Chapter 3 how the series inductance of the source influences the amplifier noise current I_n contribution to E_{ni}.

For a sensor that has shunt capacitance, the noise voltage E_n of the amplifier is the most critical. This is discussed in Chapter 3. Thus the best performance can be obtained with a bipolar transistor operated at a high collector current to minimize E_n. If the source resistance is large it may be desirable to use a FET having a low-noise voltage at high frequencies.

A sensor that has both inductance and capacitance and is resonant in the passband can be a problem because of the opposite operating conditions required. In this case it is difficult to generalize because of the wide range of impedance encountered over the frequency range of interest. Since both low E_n and I_n are required a good FET would seem to be suggested. As shown in Chapter 2 the sensor resonance decreases the effect of the amplifier noise voltage in proportion to the sensor Q factor. This can be used to reduce the amplifier noise contribution at the resonant frequency, but can make signal conditioning more difficult because of the resonant peak. The best improvement is generally obtained if the resonance is placed at the high-frequency corner near cutoff. A current amplifier such as a common-base stage can be used for signal linearization since the output signal now is unaffected by the parallel resonance of the sensor.

To design a low-noise amplifier to operate at high frequencies from a high-source resistance requires a low value of I_n at the frequencies of interest, and, in addition, a low input capacitance to assure adequate frequency response. These requirements can be met with a FET operated in the source-follower configuration or with an emitter-follower operated at low collector current. The input capacitance can be further reduced by the use of overall negative feedback.

7-6 TRANSFORMER COUPLING

In order to couple the sensor to the amplifier in an electronic system, it is sometimes advantageous to utilize a coupling or input transformer. When it is necessary but impractical to achieve the optimum NF using operating-point adjustment of the input transistor, transformer coupling can be considered. Such a situation can arise with a low-resistance source. According to Fig. 4-6, this source characteristic suggests that the transistor be operated at a collector current near 10 mA to assure low-noise operation. However, this high value of I_C is shown in Fig. 4-5 to result in relatively poor F_{opt}. If transformer coupling is employed, a lower value of I_C could be selected, and the noise performance improved.

A transformer has the ability to transform impedance levels. To prove this statement consider the ideal transformer shown in Fig. 7-5*a*. It is lossless;

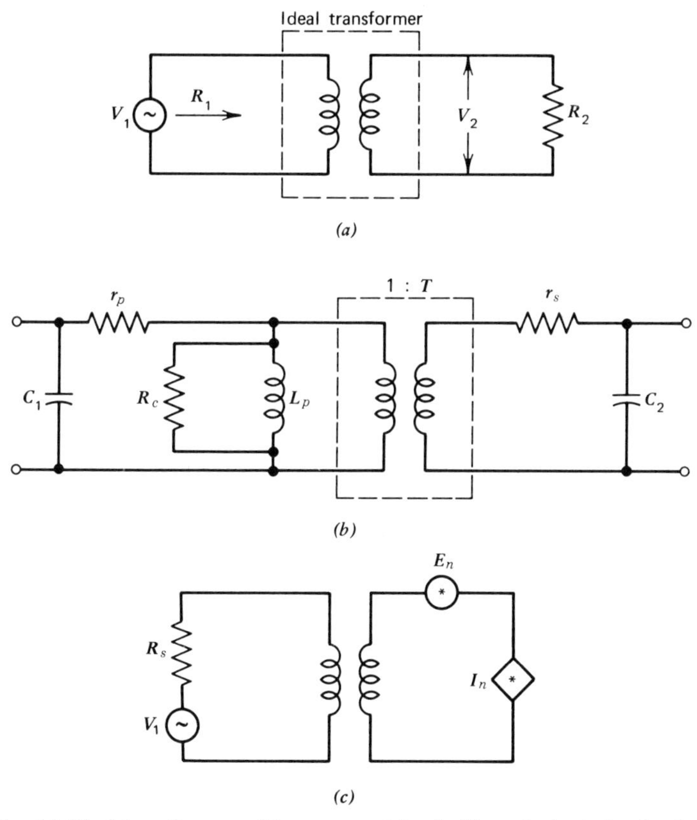

Fig. 7-5. (*a*) Ideal transformer with source and load; (*b*) equivalent circuit of practical coupling transformer; (*c*) amplifier noise parameters located at secondary side.

therefore the power levels at the primary and secondary are equal:

$$\frac{V_1^2}{R_1} = \frac{V_2^2}{R_2} \tag{7-1}$$

where R_2 is the resistive load connected to the secondary terminals, and R_1 is the effective (reflected) load at the primary terminals (not the dc resistance of the transformer winding, for that resistance is assumed negligible when we state that the device is ideal). Let the ratio of primary to secondary turns be 1:T. Then the voltages are related according to

$$V_2 = TV_1 \tag{7-2}$$

Substitution of Eq. 7-2 into Eq. 7-1 yields

$$R_1 = \frac{R_2}{T^2} \tag{7-3a}$$

If R_1 is a fixed resistance, and we wish to find the resistance reflected to the secondary port, we obtain

$$R_2 = T^2 R_1 \tag{7-3b}$$

By an appropriate choice of T, a desired impedance level can be realized.

The assumption was made that the transformer is ideal. Practical transformers have ohmic resistance in both primary and secondary windings that results in signal power loss. These are shown as r_p and r_s in the simplified equivalent circuit of Fig. 7-5*b*. The primary winding inductance L_p shunts the ideal portion of the practical transformer in the figure, and is the cause of poor performance at low frequencies. High-frequency performance is deteriorated by wiring capacitances C_1 and C_2. Element R_c is a representation of core losses.

The use of an input transformer between sensor and amplifier improves system noise performance by matching sensor resistance with amplifier optimum source resistance R_o. Consider the secondary circuit as shown in Fig. 7-5*c* containing noise parameters E_n and I_n. We have defined $R_o = E_n/I_n$. When these quantities are reflected to the primary, we obtain

$$E_n' = \frac{E_n}{T} \qquad \text{and} \qquad I_n' = TI_n$$

from Eq. 7-2. The ratio of these reflected parameters we call R_o':

$$R_0' = \frac{E_n'}{I_n'} = \frac{E_n}{T^2 I_n} = \frac{R_o}{T^2}$$

Let us proceed to match R_o' to source resistance R_s. Since we are free to select T, it follows that

$$T^2 = \frac{R_o}{R_s} \tag{7-4}$$

This assures us that the amplifier sees the optimum source resistance.

In addition to being useful to achieve an optimum source resistance level. transformer coupling can have other uses such as to maximize signal power transfer by matching resistance levels, or to isolate portions of a system. Coupling transformers are further discussed in Section 8-10, and the noise behavior of transformers is considered in Section 9-5.

7-7 QUALITY OF PERFORMANCE

To a system engineer, low-noise operation implies an acceptable signal-to-noise ratio at the system output terminals. To the designer this means designing for the minimum equivalent input noise E_{ni}. In effect E_{ni} is a normalized reciprocal of signal-to-noise ratio referred to the point where the signal originates. The general expression for equivalent input noise was given as Eq. 2-1:

$$E_{ni}^2 = E_t^2 + E_n^2 + I_n^2|Z_s|^2 \tag{7-5}$$

The equivalent input noise for any amplifier, whether designed from discrete transistors or purchased as a complete unit such as an IC op amp, can be described in terms of noise voltage E_n and noise current I_n, as shown in Fig. 7-6. Values used for E_n and I_n can be either spot noise at a single frequency

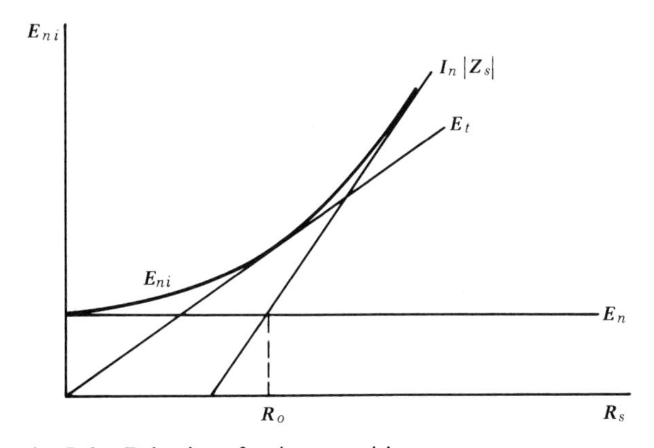

Fig. 7-6. Behavior of noise quantities.

or total integrated noise over a known bandwidth. From this curve the total equivalent input noise can be determined for any source impedance Z_s. Remember that the thermal noise curve E_t is determined by the real part of the impedance only. On the other hand, E_{ni} is dependent on the absolute value of the total source impedance.

When designing for maximum signal-to-noise ratio and the source impedance is not specified, select a sensor having the highest internal signal-to-noise ratio with internal impedance equal to the amplifier optimum noise source resistance R_o, if such a selection is possible. If the sensor resistance is much less than R_o, adding a resistor in series with the sensor does not reduce the equivalent input noise. In fact, it probably increases it. Only when an increase in sensor resistance increases the signal proportionally should the total resistance value be adjusted.

In the amplifier design for maximum signal-to-noise ratio, for a sensor whose impedance is already determined, the values of E_n and $I_n Z_s$ are adjusted by active device selection so that the lowest E_{ni} can be obtained. As an example, consider the E_{ni} curves on two amplifiers given in Fig. 2-2. For a 1000-Ω source resistance, E_{ni} for amplifier (*a*) is the lowest. If the source resistance is 100 kΩ, the amplifier represented in (*b*) clearly gives the lowest equivalent input noise. Methods for optimizing E_n and I_n are described in Chapters 4 and 6.

It should be obvious that the design engineer is responsible for all aspects of his design. When using transistors and ICs it is convenient to accept information from manufacturer's data sheets as typical or nominal. Often this is not the case. Also, noise performance from such data may not be the lowest obtainable; perhaps data have been given for a convenient or standard value of I_C, rather than for the best performance point. For intelligent design, additional information must be obtained by laboratory testing. Information such as that presented in Chapter 4 can be extremely useful in zeroing-in on an optimal design.

7-8 DESIGN EXAMPLES

The design examples presented in this section apply the information presented here and in preceding chapters. The examples are not complete in all respects; however, they should prove to be of assistance to the reader who is unfamiliar with low-noise design procedures.

Design Example 1—Amplifier Comparison

OBJECTIVE. To compare first-stage active devices for an amplifier operating at 10 kHz from a resistive source of 100 kΩ.

SOLUTION. Figure 7-2 indicates that an IC, FET, or a bipolar could be used for this application. We consider each of these devices in this example.

The μA741 op amp is a likely candidate for this job. According to the curves in Appendix III, $E_n = 20$ nV and $I_n = 0.4$ pA at a frequency of 10 kHz. The optimum source resistance, therefore, is 50 kΩ.

Noise figure can be obtained from

$$NF = 10 \log \left[\frac{E_t^2 + E_n^2 + I_n^2 R_s^2}{E_t^2} \right] \tag{7-6}$$

For this device, we find that $E_{ni} = 60$ nV and NF $= 3.5$ dB.

Now let us consider a bipolar transistor. The curve of optimum source resistance R_o versus collector current I_C shown in Fig. 4-7 indicates minimum noise at a collector current of 2 μA. From the contours of constant NF given in Appendix I for the *npn*-type 2N4124 transistor, the 10-kHz curve indicates that for $R_s = 100$ kΩ, I_C for minimum noise should be 2 μA. The resulting NF is less than 1 dB. Extrapolating between the 1 and 10 μA curves for the 2N4124 suggests that $E_n = 12$ nV and $I_n = 0.12$ pA. This gives E_{ni} of 44.5 nV and a NF of 0.9 dB.

The type 2N4869 JFET has $E_n = 3$ nV and $I_n = 6$ fA. A calculation shows the equivalent input noise voltage to be 40.1 nV and the NF is 0.05 dB.

Thus all three types of devices can be used depending on the amount of noise tolerable and the cost. Other specifications would be required to determine lowest system cost.

Design Example 2—Transformer-Coupled Amplifier

OBJECTIVE. To design an amplifier for a sensor test instrument that measures signal and noise performance at 100 Hz for sensors with resistances of from 5 to 500 Ω.

SOLUTION. From Fig. 7-2 we conclude that it is usually desirable to use an input transformer with low values of source resistance. Then a transistor can be selected for operation at its minimum NF, and the transformer matches the source resistance to R_o of the transistor.

From the noise curves for bipolar transistors given in Appendix I we select the type 2N4250. The NF contour at 100 Hz suggests that $I_C = 30$ μA for the case of $R_o = 10$ kΩ. The resulting NF is less than 1 dB.

At the selected operating point, $E_n = 3$ nV and $I_n = 0.3$ pA.

The turns ratio of the transformer makes the sensor resistance look like R_o. Since resistances from 5 to 500 Ω are to be used, we choose 50 Ω as a compromise value. The turns ratio determined from Eq. 7-4 is $(R_1/R_2)^{1/2}$, or in this instance $(R_s/R_o)^{1/2}$. When evaluated, $T = 1{:}14$.

To completely specify the coupling transformer, additional information is necessary. The primary dc resistance plus reflected secondary resistance should be much less than the smallest source resistance, 5 Ω. The primary resistance can be about 0.1 Ω and secondary resistance 20 Ω. To minimize the noise contribution of the shunting core loss equivalent, R_c should be much greater than the largest source resistance, 500 Ω. A value of 10 kΩ would certainly be satisfactory. Both frequency response and noise dictate that the primary inductance L_p must be much larger than the 500-Ω source resistance at the lowest frequency of interest, 100 Hz. This requirement is met if the primary inductance is greater than 16 H. This inductance can be achieved in shielded transformers having a volume of a few cubic inches.

The reflected noise voltage of the transistor can be divided by the turns ratio in order to refer these quantities to the primary or sensor side of the transformer. This results in $E_n' = 0.21$ nV and $I_n' = 4.2$ pA. Transistor input resistance, reflected to the primary, is 50 Ω.

If we assume that the transformer does not contribute noise, the following summary is valid for three values of sensor resistance:

R_s	5 Ω	50 Ω	500 Ω
E_t	0.28 nV	0.89 nV	2.8 nV
E_{ni}	0.35 nV	0.93 nV	3.5 nV
NF	1.19 dB	0.40 dB	1.19 dB

Design Example 3—Magnetic Tape Recorder System

OBJECTIVE. To design a preamplifier and equalizer for use with a magnetic tape recorder read-head. The system output voltage should be flat from 400 Hz to 1.6 MHz.

SOLUTION. Electrically, the read-head is a complex sensor. It is necessary initially to determine the characteristics of that device in order to design the system intelligently. The result of the analysis of the read-head is the equivalent circuit given in Fig. 7-7*a*. The elements of this model are

- L_H the inductance of the head
- R_H the total resistive loss of the head
- C_H the head, cable, and stray shunt capacitance
- E_{ns} the thermal noise voltage of R_H
- V_s the voltage signal from the read-head
- E_p stray pickup and tape noise

L_H, R_H, E_{ns}, and V_s are functions of frequency. There are two noise sources:

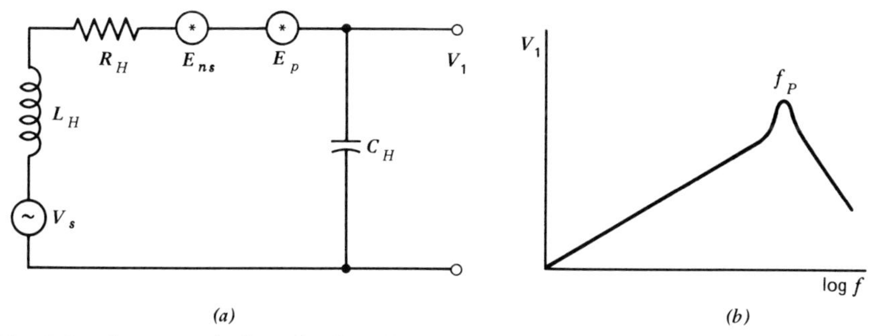

Fig. 7-7. Sensor equivalent circuit and response.

E_{ns} represents thermal noise in the real portion of the sensor impedance, and E_p represents collected pickup and tape noise.

Signal voltage V_s is proportional to the rate of change of magnetic flux on the tape; this signal increases linearly with frequency. Depending on the value of C_H, output voltage falls off at high frequencies. Signal behavior is shown in Fig. 7-7*b*.

To achieve a flat overall response, the system requires an equalizer to shape the response. The equalizer, located in the main amplifier, has a response that is the inverse of the head output.

We wish to maximize S/N, or achieve the equivalent of minimizing total input noise E_{ni}. To determine the equivalent input noise E_{ni}, the total output noise E_{no} is determined from the network of Fig. 7-8. Then, the total output signal V_o is calculated. The output signal-to-noise ratio is

$$\frac{S_o}{N_o} = \frac{V_o^2}{E_{no}^2} \tag{7-7}$$

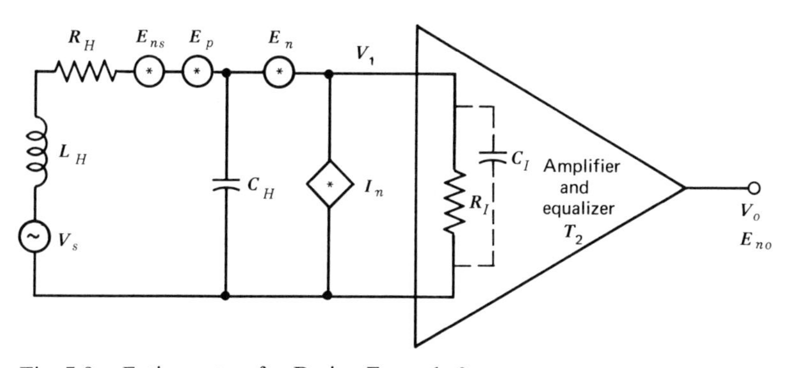

Fig. 7-8. Entire system for Design Example 3.

and the equivalent input noise is

$$E_{ni} = \frac{E_{no}}{K_t} \tag{7-8}$$

Since $V_o = K_t V_s$, Eq. 7-7 can be written

$$\frac{S_o}{N_o} = \frac{V_s^2}{E_{ni}^2} \tag{7-9}$$

The system can be partitioned into two sections in order to write

$$V_o = K_t V_s = T_1 T_2 V_s \tag{7-10}$$

where T_1 is the voltage transfer function of the input network from V_s to V_1, including the amplifier input impedance, and T_2 is the voltage transfer function of the amplifier and equalizer.

The signal voltage from the read-head is found to be

$$V_s = A\eta g\omega\phi N \tag{7-11}$$

where A is a gain constant, η is the transducer efficiency with a maximum value of unity, g is the gap loss, ω is the angular frequency $2\pi f$, ϕ is the tape flux, and N is the number of turns on the head. The signal at the amplifier input is a nonlinear function of frequency and must be equalized so that the output is flat from 400 Hz to 1.6 MHz. This means that the total system response should behave according to

$$K_t(A\eta g\omega\phi N) = \text{constant}$$

The system gain K_t can be written in terms of the impedances in the circuit of Fig. 7-8 and the amplifier-equalizer response T_2:

$$K_t = T_2 \frac{R_I}{(R_H + j\omega L_H)[1 + j\omega R_I(C_H + C_I)] + R_I} \tag{7-12}$$

To obtain the S/N ratio or the equivalent input noise we require an expression for total noise in a similar form. Writing a term for each of the three noise mechanisms at E_{no} gives

$$E_{no}^2 = T_2^2 \left\{ E_{ns}^2 \left(\frac{V_1}{V_s}\right)^2 + E_n^2 \left[\frac{R_I}{R_I + \dfrac{(R_H + j\omega L_H)(1 + j\omega R_I C_I)}{1 - \omega^2 L_H C_H + j\omega R_H C_H}}\right]^2 + I_n^2 \left[\frac{R_I(R_H + j\omega L_H)}{R_I + (R_H + j\omega L_H)[1 + j\omega R_I(C_I + C_H)]}\right]^2 \right\} \tag{7-13}$$

Note that $(V_1/V_s)^2 = T_1^2$

By comparing Eq. 7-13 with Eq. 7-12 it is apparent that the coefficient of the I_n^2 term is

$$\left[\frac{(R_H + j\omega L_H)K_t}{T_2}\right]^2$$

After much manipulation it can be shown also that the coefficient of the E_n^2 term is

$$\left[\frac{[1 + j\omega C_H(R_H + j\omega L_H)]K_t}{T_2}\right]^2$$

Therefore the absolute magnitude of the total noise at the output of the amplifier can then be simply written as

$$E_{no}^2 = K_t^2\{E_{ns}^2 + [(1 - \omega^2 C_H L_H)^2 + \omega^2 C_H^2 R_H^2]E_n^2 + (R_H^2 + \omega^2 L_H^2)I_n^2\} \tag{7-14}$$

and from Eq. 7-9 the S/N ratio (independent of amplifier gain) is

$$\frac{S_o}{N_o} = \frac{(A\eta g\omega\phi N)^2}{E_{ns}^2 + [(1 - \omega^2 C_H L_H)^2 + \omega^2 C_H^2 R_H^2]E_n^2 + (R_H^2 + \omega^2 L_H^2)I_n^2} \tag{7-15}$$

It can be seen that the signal-to-noise ratio is independent of the amplifier input resistance and capacitance. The noise current I_n contribution is dependent on the head resistance and inductance, but not capacitance. The E_n contribution is increased by $R_H L_H$ and C_H, but may have a minimum value when $\omega^2 C_H L_H$ is unity.

We recall that both E_n and I_n are dependent on f. Mathematical expressions for those quantities are given in Section 8-3. Evaluation of an equation such as (7-14) over a bandwidth requires integration of noise voltage that can be accomplished easily with the aid of a digital computer.

To equalize the system for a flat response, function T_2 must behave in the manner shown in Fig. 7-9. The curve is in four convenient sections. Representations are

Section I $\left(\dfrac{f/f_1}{[1 + (f/f_1)^2]^{1/2}}\right)^{N1}$

Section II $\left(\dfrac{(f/f_3)^2 + 1}{(f/f_2)^2 + 1}\right)^{1/2}$

Section III $\left(\dfrac{[(f/f_4)^2 - 1]^2 + (f/f')^2}{[(f/f_4)^2 - 1]^2 + (f/f'')^2}\right)^{1/2}$

Section IV $\left(\dfrac{1}{[(f/f_5)^2 + 1]^{1/2}}\right)^{N2}$

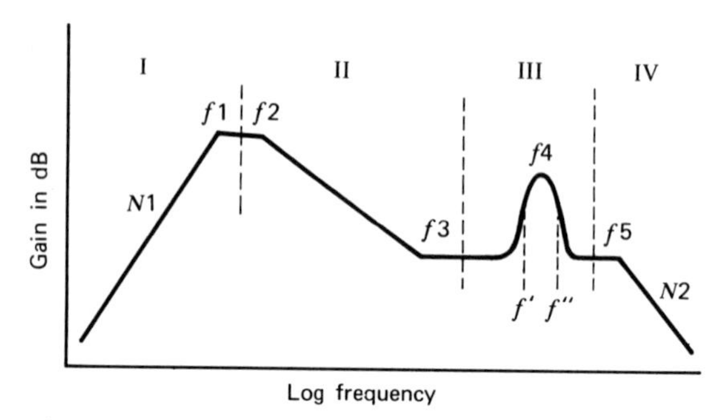

Fig. 7-9. Equalization curve for design example.

Quantity N represents the slope in Sections I and IV. $N = 1$ corresponds to 6 dB/octave; $N = 2$ is equivalent to 12 dB/octave. Further discussion is included in Section 8-5.

There is a slight resonance in the head output as noted by the peak at f_P in Fig. 7-7*b*. Section III of Fig. 7-9 would have to be opposite to that shown in order to compensate that peaking. The peak at $f4$ is actually to extend the system response to higher frequencies; $f4$ occurs at a higher frequency than f_P.

SUMMARY

a. Low-noise design implies attaining a low value of equivalent input noise E_{ni}. Since

$$\frac{S_o}{N_o} = \frac{V_s^2}{E_{ni}^2}$$

minimizing E_{ni} maximizes S_o/N_o.

b. The noise specification is often the most important and may be the most difficult to meet. An NF below 3 dB may be unnecessary.

c. A recommended low-noise design procedure with sensor characteristics fixed:

1. Select input device and configuration (Chapter 7).
2. Decide on operating point (Chapters 4–5).
3. Design bias circuit (Chapter 10).
4. Determine succeeding stages (Chapter 12).
5. Consider coupling (Chapter 7).
6. Power supply requirements (Chapter 11).

7. Location of equalizers.
8. Overall feedback (Chapter 12).

An iterative analysis-synthesis procedure determines the final system. Digital computer assistance is valuable.

d. Input devices considered are
   the bipolar transistor (conventional or SBT),
   the FET (JFET or MOSFET), and
   the IC (bipolar or FET input stage).

e. Because it can transform impedance levels, a coupling transformer can be used with low-resistance sensors.

PROBLEMS

1. Specify the turns ratio required to couple a 200-Ω source to a transistor with $R_o = 5$ kΩ. What is the signal level at the transistor input terminal if the sensor signal is 2 mV?

2. Show that the primary inductance in a transformer determines the lower-cutoff frequency f_L. By definition, f_L is the frequency where signal transmission has been reduced to 0.707 of the midfrequency reference value.

3. A given sensor has series resistance of 1 kΩ. Select an operating point for the 2N4403 transistor from Appendix I. Determine the wideband noise figure of your design for the 10 to 100 kHz band, and find E_{ni} for this band.

4. Sketch S/N versus f from Eq. 7-15. Consider that the numerator is Kf, $E_n = E_{ns} = 10^{-6}$ V/Hz$^{1/2}$, $C_H = 10^{-10}$, $L_H = 2.5 \times 10^{-4}$, $R_H = 10^3$, and $I_n = 10^{-11}$ A/Hz$^{1/2}$.

Chapter 8

COMPUTER-AIDED DESIGN OF LOW-NOISE SYSTEMS

The preceding chapters discussed noise models for the amplifiers, sensors, and components that make up an electronic system. In the design of low-noise systems, that information is used to predict performance. However, it should be noted that the associated calculations can be very time consuming. Network and sensor impedances can be complex. Transistor noise contains all frequencies. To predict the total noise in a usable bandwidth it is necessary to calculate the noise at many frequencies and integrate the mean-square noise voltage over the bandwidth of interest.

To simplify the design process a digital computer can be utilized for many of the calculations. This chapter considers a computer program that has been found useful in design. It is called "NOISE."

8-1 "NOISE" PROGRAM

The NOISE program provides a noise analysis of an electronic system. The system is considered to be composed of three subsystems:

1. The sensor and associated circuitry.
2. The amplifier, or the first transistor stage.
3. The gain and frequency response of the electronics.

NOISE must be provided with data concerning sensor characteristics, amplifier or transistor characteristics, response information, and frequency ranges. The program performs the calculations requested by the user. The calculations are designated 1 to 6. The choices are

1. Total equivalent input noise over a band.
2. Input network frequency response.
3. Input noise at one frequency.
4. Input noise versus frequency.
5. Total noise at the output.
6. Total system gain.

Calculation 1 integrates and prints E_{ni} for each frequency interval selected. This is the total noise at the input, independent of amplifier gain and input impedance.

Calculation 2 determines the transfer function from the sensor to the amplifier input terminals, including the amplifier input impedance. The frequency range must be selected. The program calculates at 10 frequencies/decade stepping by $\sqrt[10]{10}$.

Calculation 3 determines the equivalent input noise E_{ni} in a 1-Hz bandwidth at any single frequency. The frequency must be specified.

Calculation 4 determines the noise voltage spectral density at the input as a function of frequency, $E_{ni}(f)$. The frequency range must be specified; then the computer steps through the band printing the noise at 10 frequencies/decade as in calculation 2.

Calculation 5 integrates the total noise at the output of the amplifier over the intervals selected. The equivalent input noise is calculated and multiplied by the input network transfer function and the amplifier transfer function. This gives the total noise at the output after it has been amplified and equalized. After analyzing the input noise with calculations 1, 3, and 4, we can determine the output noise with calculation 5 to find the signal-to-noise ratio after equalization.

Calculation 6 prints the total system gain versus frequency. For the selected frequency range, it prints results at 10 frequencies/decade. The total gain is the product of the input transfer function as determined in calculation 2 and the amplifier response as supplied in the data.

By coupling-in a plot routine, calculations 2, 4, and 6 plot gain and noise versus frequency.

NOISE has been found useful by practicing engineers. The version of the program discussed here is written in FORTRAN and listed in Appendix IV. A BASIC language version used with a time-shared system is available to interested readers [1].

A sample computer run of all six calculations is given in Section 8-13.

8-2 SUBROUTINES

The system to be analyzed is depicted in Fig. 8-1. The three major subdivisions were given in the preceding section as sensor, amplifier noise, and gain and response.

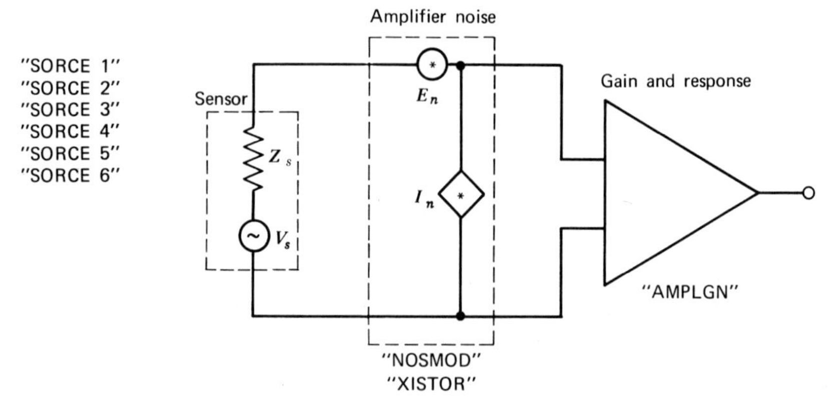

Fig. 8-1. Subprogram names in NOISE.

In order to accommodate different types of signal sources, *the user selects a subroutine from the six available to describe his specific source.* He must enter data describing the source. The subroutines are representative of sensors according to the following table:

Subroutine	*Sensor Model*
SORCE1	Resistive source
SORCE2	Biased resistive source
SORCE3	Coil-RLC source
SORCE4	Biased diode source
SORCE5	Transformer model
SORCE6	Piezoelectric sensor

The models used are discussed in Sections 8-6 through 8-11.

To represent the amplifier or first transistor, the user selects either the NOSMOD subroutine or the XISTOR subroutine. NOSMOD is the E_n-I_n representation for the active device. Provision is made to accommodate the variation in these parameters with frequency. Detailed discussion is given in Section 8-3.

If the user prefers to perform noise calculations using a transistor noise model as given in Chapter 4 rather than the E_n-I_n model, he uses the XISTOR subroutine instead of NOSMOD. XISTOR is described in Section 8-4.

The third major section of the system is referred to as AMPLGN in Fig. 8-1. *The AMPLGN subroutine is concerned with the voltage transfer function of the post amplifier and equalizer.* Provision is made for simulating many types of networks. This is considered in Section 8-5.

In summary, then, the major sections of the system to be analyzed are represented by three major subroutines in the computer program:

SORCE1 to SORCE6
NOSMOD or XISTOR
AMPLGN

Other subroutines are noted in Section 8-13.

8-3 "NOSMOD" UNIVERSAL MODEL

The amplifier noise model using parameters E_n and I_n is shown in Fig. 8-2. Total equivalent input noise is given by Eq. 2-7 for a simple sensor-amplifier:

$$E_{ni}^2 \simeq E_{ns}^2 + E_n^2 + I_n^2 R_s^2 \tag{2-7}$$

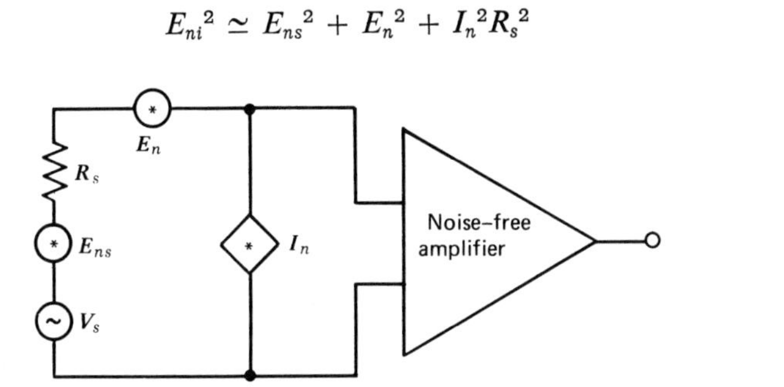

Fig. 8-2. Noise model.

When the sensor network is more complex, terms on the right-hand side of Eq. 2-7 pick up coefficients as described in Chapter 3. These coefficients are determined in the SORCE subroutines.

Noise parameter behavior commonly varies with frequency as shown in Fig. 8-3. At midband the noise is independent of the frequency. At low frequencies the noise rises at 3 dB/octave because of the 1/f mechanism. At high frequencies the noise rises at 6 dB/octave because of the transistor gain roll-off.

To use NOSMOD, values for midband noise parameters E_n and I_n and the noise corners are necessary. Symbols F1 through F4 represent the low- and high-frequency corners; the mean-squared noise voltage as a function of

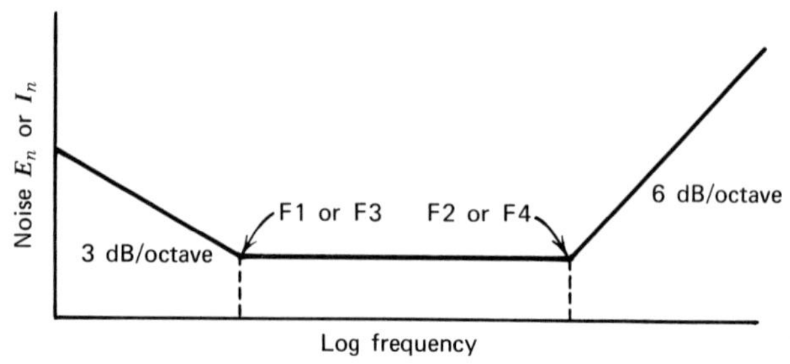

Fig. 8-3. NOSMOD.

frequency is

$$E_n^2(f) = E_n^2\left[\left(\frac{\text{F1}}{f} + 1\right) + \left(\frac{f}{\text{F2}}\right)^2\right] \tag{8-1}$$

Similarly for I_n:

$$I_n^2(f) = I_n^2\left[\left(\frac{\text{F3}}{f} + 1\right) + \left(\frac{f}{\text{F4}}\right)^2\right] \tag{8-2}$$

In the program, the symbols given to these quantities are

E_n = E
I_n = I
F1, F2 = low-and high-E_n noise corners
F3, F4 = low-and high-I_n noise corners

The required noise values can be obtained from the curves of device noise in Chapters 4, 5, and 6, or from measurements on a device or a complete amplifier.

8-4 BIPOLAR TRANSISTOR "XISTOR" MODEL

It was shown in Chapter 4 that the noise of a bipolar transistor can be calculated from its hybrid-π values and certain measured noise constants. The model used is shown in Fig. 8-4. The derived expression for total equivalent input noise of a bipolar transistor is given in Eq. 8-3, and the equivalent noise voltage and current generators E_n and I_n are presented in Eqs. 8-4 and 8-5:

$$E_{ni}^2 = 4kT(r_{bb'} + R_s) + 2qI_B(r_{bb'} + R_s)^2 + \frac{2qI_C}{\beta_o^2}(r_{bb'} + R_s + r_{b'e})^2$$
$$+ \frac{2qf_L I_B^{\gamma}(R_s + r_b)^2}{f^{\alpha}} + 2qI_C(r_{bb'} + R_s)^2\left(\frac{f}{f_T}\right)^2 \tag{8-3}$$

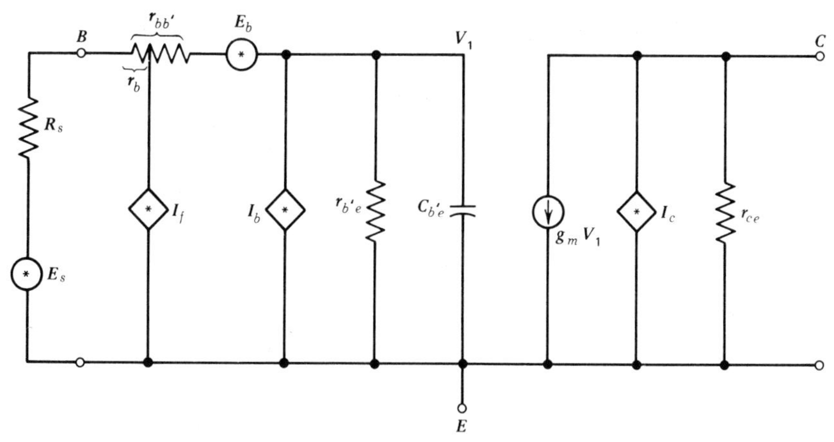

Fig. 8-4. Low-frequency hybrid-π noise model.

Also

$$I_B \simeq I_{CBO} + \frac{I_C}{\beta_o}$$

The noise voltage generator is

$$E_n^2 = 4kTr_{bb'} + 2qI_C r_e^2 + \frac{2qf_L I_B^{\gamma} r_b^2}{f^{\alpha}} + 2qI_C r_{bb'}^2 \left(\frac{f}{f_T}\right)^2 \tag{8-4}$$

and the current generator is

$$I_n^2 = 2qI_B + \frac{2qf_L I_B^{\gamma}}{f^{\alpha}} + 2qI_C \left(\frac{f}{f_T}\right)^2 \tag{8-5}$$

These equations are valid for a 1-Hz frequency band.

Symbol equivalents used in XISTOR are

$r_{bb'}$ = N5 = base resistance
r_b = N6 = the l/f base resistance
I_C = I4 = collector current
I_{CBO} = I7 = collector-to-base leakage
f_L = F7 = l/f corner
f_T = F8 = frequency at which $|\beta| = 1$
γ = Q1
α = Q2
β_o = B = reference value of transistor short-circuit current gain

The subroutine calculates E_n and I_n according to Eqs. 8-4 and 8-5. It is necessary to supply data values for all of the quantities given in the preceding list. Table 4-1 provides such information.

8-5 AMPLIFIER AND EQUALIZER RESPONSE "AMPLGN"

Noise at the load depends on the transfer function of the post amplifier and equalizer. The transfer functions for five frequency-shaping networks are stored in subprogram AMPLGN. An asymptotic plot of the composite of these functions is illustrated in Fig. 8-5. By providing the proper values for

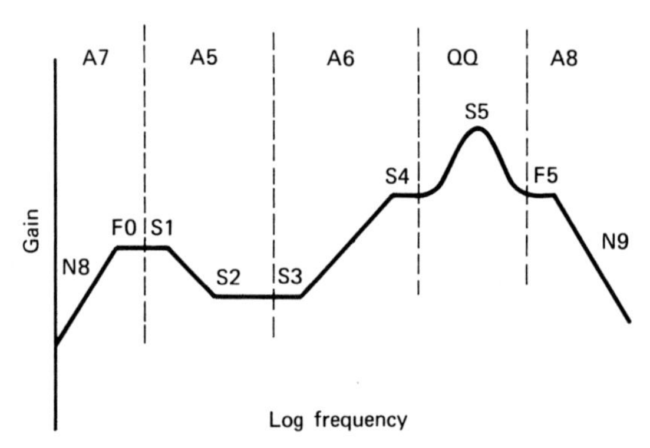

Fig. 8-5. Universal equalization curve.

the frequency corners, these curves can match most amplifiers and equalizers. Midband gain is referred to as K2 and must also form part of the data.

The lowest frequency region of the universal equalization curve starts at a negative value of gain and rises linearly to 0 at frequency F0. This simulates the low-frequency roll-off of one or more stages. The rate-of-rise, N8, is expressed in numbers of time constants. N8 = 1 corresponds to a slope of 6 dB/octave, and N8 = 2 produces a 12-dB/octave roll-off. The transfer function for region A7 is

$$A7 = \left(\frac{F/FO}{\sqrt{1 + (F/FO)^2}}\right)^{N8} \qquad \text{N8} \quad \text{FO} \tag{8-6}$$

Symbol F is being used in this section and in the program to represent the variable f.

Region A5 in Fig. 8-5 starts at 0-dB gain and rolls-off with break frequency S1. The gain decreases linearly at 6 dB/octave (20 dB/decade) to frequency S2. It remains constant for frequencies higher than S2. The transfer function A5 is

$$A5 = \left(\frac{(F/S2)^2 + 1}{(F/S1)^2 + 1}\right)^{1/2} \qquad \text{S1} \quad \text{S2} \tag{8-7}$$

Region A6 starts at 0-dB gain and breaks at frequency S3. It then rises at 6 dB/octave to frequency S4, beyond which the gain remains constant. This is a type of equalization curve used to linearize a sensor high-frequency roll-off. The transfer function for region A6 is virtually the reciprocal of region A5. Thus

$$A6 = \left(\frac{(F/S3)^2 + 1}{(F/S4)^2 + 1}\right)^{1/2} \tag{8-8}$$

S4

S3

Region QQ of the equalization curve is a resonant peak. The hump in the frequency response curve peaks at frequency S5. The circuit Q is entered as Q0. This curve can describe an RLC resonant circuit or peaking as can be found in a feedback amplifier. Mathematical relations for the QQ region are

$$QQ = \frac{A0}{\sqrt{(1 - A0^2)^2 + [2(A0)\rho]^2}} \tag{8-9}$$

$$A0 = \frac{F}{S5}$$

$$\rho \simeq \sqrt{\left(1 + \frac{1}{Q0}\right)} - 1$$

S5

The inverse of this curve may be of interest in some equalizers.

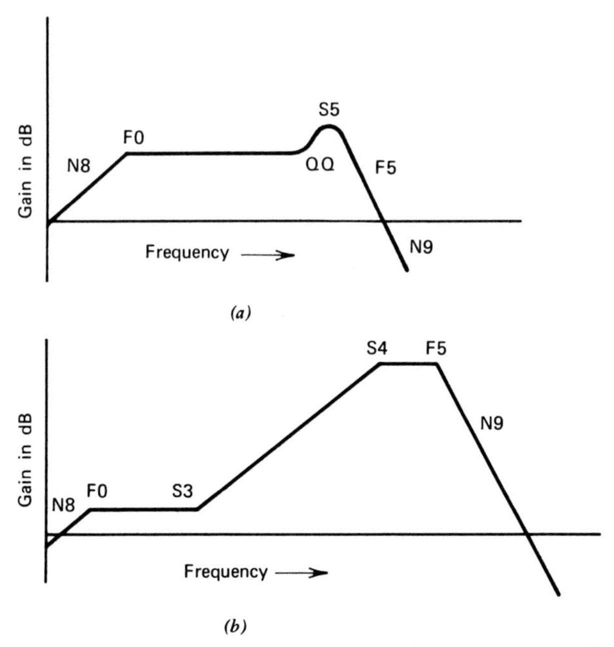

Fig. 8-6. (*a*) Response curve of typical feedback amplifier with peaking; (*b*) equalization curve for sensor with high-frequency roll-off.

Region A8 starts with 0-dB gain and breaks at frequency F5 to roll-off with N9 time constants. If N9 is 1, 2 or 3, the rate of roll-off will be 6, 12, or 18 dB/octave. This describes the high-frequency roll-off region of an amplifier or cascaded amplifiers. The transfer function of region A8 is described by

$$\mathrm{A8} = \left(\frac{1}{\sqrt{1 + (\mathrm{F}/\mathrm{F5})^2}}\right)^{N9} \qquad (8\text{-}10)$$

N9

F5

The total amplifier and equalization gain is the product of these five transfer functions and gain constant K2, the reference gain. In a practical system we would not require all these functions. For some systems only one function would be used.

With the five transfer functions, one can achieve a variety of responses. Figure 8-6*a* illustrates the frequency response curve of a feedback amplifier with peaking. Equations A7 and A8 produce the low- and high-frequency roll-offs and QQ gives the peaking. Figure 8-6*b* is an equalization curve for a sensor with high-frequency roll-off. This uses Eq. A6 to equalize the sensor roll-off and curves A7 and A8 produce the low- and high-frequency roll-offs.

8-6 RESISTIVE SENSOR "SORCE1"

We consider as the first example the case of a resistive sensor. Earlier discussion was given in Section 3-3. Resistive detectors include the thermocouple, thermopile, and PEM infrared cell. In Fig. 8-7 the sensor is symbolized by signal source V_s and series resistance R_s. A coupling capacitor C_C can be used if we are interested exclusively in the time-varying output of the sensor. Element R_L may be useful for impedance matching.

A noise model of the sensor-amplifier system is shown in Fig. 8-8. Shunt capacitance C_P can be the result of the sensor assembly or it may represent the capacitance between wires. The amplifier is represented by noise parameters E_n and I_n, and by its input resistance R_I and input capacitance C_I.

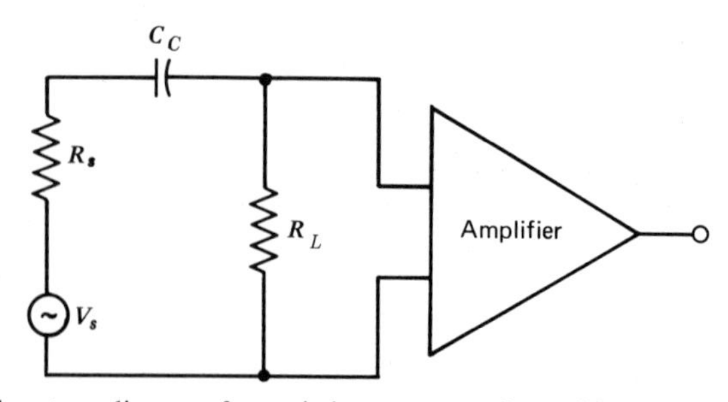

Fig. 8-7. General system diagram for resistive sensor and amplifier.

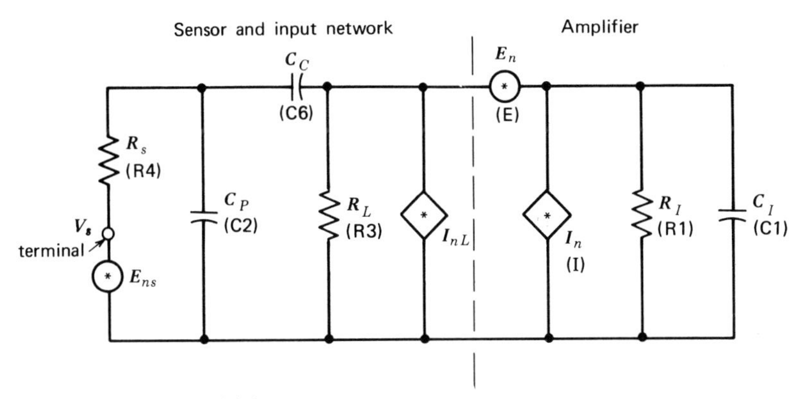

Fig. 8-8. Noise model for computer.

Shown in Fig. 8-8 in parentheses are the symbols used in the computer program. The symbol equivalents are summarized in the following tabulation.

R_s = R4 = sensor resistance
R_L = R3 = bias or load resistance
R_I = R1 = amplifier input resistance
C_I = C1 = amplifier input capacitance
C_C = C6 = coupling capacitance
C_P = C2 = shunt capacitance of sensor or wiring
$E_{ns} = \sqrt{4kTR_s\,\Delta f}$ = thermal noise voltage of R_s
$I_{nL} = (4kT\,\Delta f/R_L)^{1/2}$ = thermal noise current of R_L
E_n and I_n = E and I = amplifier equivalent noise generators

Subroutine SORCE1 requires that values be supplied for C1, C2, C6, R1, R3, and R4, as well as E and I. Generators E_{ns} and I_{nL} are calculated from resistance values. The program calculates equivalent input noise E_{ni}^2, and K_t, the voltage transfer function between amplifier input terminal and sensor signal generator (V_s) terminal. The expressions are

$$E_{ni}^2 = 4kTR_s + E_n^2\left[1 + \frac{R_s}{R_L}\left(1 + \frac{C_P}{C_C}\right) + j\left(\omega R_s C_P - \frac{1}{\omega R_L C_C}\right)\right]^2 + \left(I_n^2 + \frac{4kT}{R_L}\right)\left[\frac{1 + j\omega R_s(C_P + C_C)}{j\omega C_C}\right]^2 \qquad (8\text{-}11)$$

and

$$K_t = \frac{j\omega R_L C_C R_I}{R_I + R_L - \omega^2 R_I R_L R_s(C_I C_P + C_I C_C + C_C C_P) + j\omega[(C_I + C_C)R_I R_L + (C_P + C_C)R_s(R_I + R_L)]} \qquad (8\text{-}12)$$

These equations are somewhat difficult to comprehend unless simplifying assumptions are made. If we omit C_C and R_L, the expression for E_{ni}^2 becomes

$$E_{ni}^2 = E_{ns}^2 + (1 + \omega^2 R_s C_P)^2 E_n^2 + I_n^2 R_s^2 \tag{8-13}$$

We see that Eq. 8-13 is identical to Eq. 3-9, previously presented. For low noise, the noise contribution of R_L is kept low if R_L is large. Similarly, the effect of C_P is minimized if the stray capacitance is as small as possible.

8-7 BIASED RESISTIVE SOURCE "SORCE2"

Biased resistive sources include photoconductive cells, piezoresistive strain gauges, and other elements whose resistance changes with the sensed parameter. As shown in Fig. 8-9, biasing resistance R_B differentiates this sensor type from the type discussed in the preceding section.

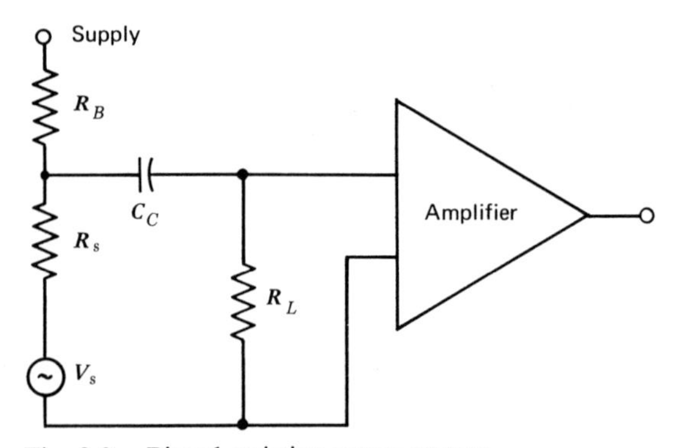

Fig. 8-9. Biased resistive source system.

The noise model shown in Fig. 8-10 includes an additional noise source, I_{nB}, to represent the thermal noise generated in R_B. In parenthesis in the figure are the symbols used in the computer program subroutine. Symbol equivalents are

R_s = R4 = source resistance
R_B = R2 = bias resistance
R_L = R3 = load resistance
R_I = R1 = amplifier input resistance
C_I = C1 = amplifier input capacitance
C_C = C6 = coupling capacitance
C_P = C2 = the sum of sensor and shunt wiring capacitances
F_L = F6 = 1/f noise corner

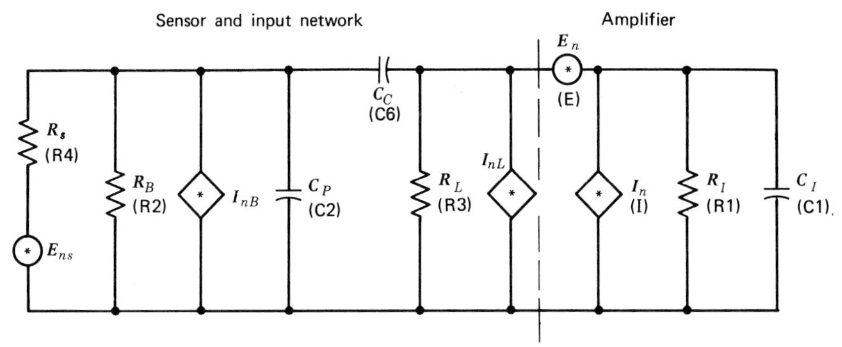

Fig. 8-10. Noise model for computer.

$E_{ns} = (4kTR_s\,\Delta f)^{1/2}$ = thermal noise voltage of R_s
I_{nB} and $I_{nL} = (4kT\,\Delta f/R)^{1/2}$ = thermal noise current of R_B and R_L
E_n and I_n = E and I = equivalent noise generators of the amplifier

SORCE2 requires that values be supplied for C1, C2, C6, F6, R1, R2, R3, and R4, as well as E and I. Parameter F_L is the 1/f noise corner frequency associated with excess noise in the sensor resulting from its bias current. Generators E_{ns}, I_{nB}, and I_{nL} are calculated from resistance values. The program calculates E_{ni}^2 and K_t according to the following equations:

$$E_{ni}^2 = 4kTR_s\left(\frac{F_L}{F} + 1\right) + E_n^2\left\{\frac{C_CR_L(R_B + R_s) + R_BR_s(C_P + C_C) + j[\omega C_PC_SR_sR_BR_L - (R_s + R_B)/\omega]}{R_BR_LC_C}\right\}^2 + \left(I_n^2 + \frac{4kT}{R_L}\right)\left[\frac{R_s(C_P + C_C)}{C_C} - j\frac{R_s + R_B}{\omega R_BC_C}\right]^2 + \left(\frac{4kT}{R_B}\right)\left[R_s + \frac{R_s}{j\omega C_CR_L}\right]^2 \tag{8-14}$$

and

$$K_t = \frac{R_BR_PC_C}{(R_B + R_s)(R_PC_C + R_PC_I) + R_BR_s(C_P + C_C) + j\{\omega C_CC_PR_sR_BR_P - [(R_s + R_B)/\omega] + \omega C_IR_sR_BR_P(C_P + C_C)\}} \tag{8-15}$$

where $R_P = R_LR_I/(R_L + R_I)$.

The computer prints out each of the four terms of Eq. 8-14 separately. The first term in that equation is called SENSOR EN, the thermal noise of the sensor. The second term and its multiplying coefficient is called AMPL EN and is the contribution of the noise voltage of the amplifier. The third term has both the noise current of the amplifier and the thermal noise of the load resistor. The computer separates these and prints the noise current part as AMPL IN*ZS. The thermal noise current of the load R_L and bias resistor R_B are combined and printed as LOAD EN. Thus, by looking at the computer printout, we can see the noise of each section: sensor noise, coupling network noise, and amplifier noise. Total noise is also printed.

Both R_B and R_L contribute thermal noise. Again we wish to make R_L as large as possible. Element R_B affects both the sensor signal and its noise. It is sometimes possible to replace R_B by a noise-free inductance for bias and load. This gives an additional low-frequency roll-off term.

8-8 RLC SOURCE "SORCE3"

The RLC source models coils, inductive pickups, dynamic microphones, linear variable differential transformers, and various other inductive sensors. They may be resonant devices, but their principal characteristic is that the signal is in series with the inductive component. A general system diagram is shown in Fig. 8-11.

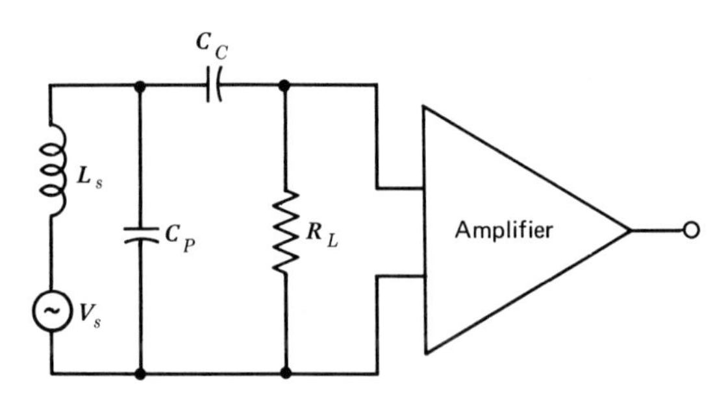

Fig. 8-11. System diagram for RLC sensor.

The noise model shown in Fig. 8-12 contains noise source E_{ns} to represent the thermal noise that may be present in the real part of the sensor impedance. The sensor is assumed to have series inductance represented by L_s; C_P represents any capacitance purposely added to resonant the sensor, as well as wiring and other stray capacitances. Symbol equivalents are

R_s = R4 = sensor series resistance or real part of the impedance
R_L = R3 = load resistance
R_I = R1 = amplifier input resistance
L_s = L2 = sensor series inductance
C_P = C2 = shunt capacitance or resonating capacitance

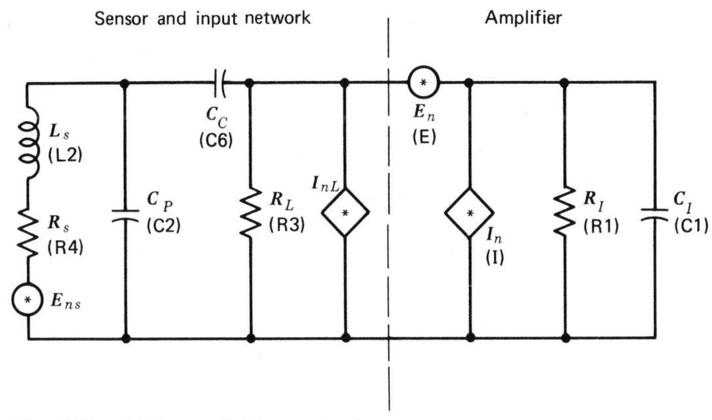

Fig. 8-12. Noise model for computer.

C_C = C6 = coupling capacitance
C_I = C1 = amplifier input capacitance
$E_{ns} = (4kTR_s\,\Delta f)^{1/2}$ = thermal noise voltage of R_s
$I_{nL} = (4kT\,\Delta f/R_L)^{1/2}$ = thermal noise current of R_L
E_n and I_n = E and I = equivalent noise generators of the amplifier

Equations for the equivalent input noise voltage and amplifier input transfer function can be derived.

$$E_{ni}^2 = 4kTR_s + E_n^2\left[\frac{[1 - \omega^2 L_s(C_P + C_C) - \omega^2 C_C C_P R_L R_s] + j\omega[C_C(R_L + R_s) + C_P R_s - \omega^2 C_P C_C L_s R_L]}{j\omega R_L C_C}\right]^2 + \left(I_n^2 + \frac{4kT}{R_L}\right)\left[\frac{1 - \omega^2 L_s(C_P + C_C) + j\omega R_s(C_P + C_C)}{j\omega C_C}\right]^2 \quad (8\text{-}16)$$

$$K_t = \frac{j\omega C_C R_I R_L}{\omega^2\{-R_I R_L R_s(C_I C_C + C_C C_P + C_I C_P) + [(R_I + R_L)/\omega^2] - L_s(C_C + C_P)(R_I + R_L)\} + j\omega\{R_I R_L(C_I + C_P) + R_L R_s(C_P + C_C) - \omega^2 L_s R_I R_L(C_I C_C + C_C C_P + C_I C_P)\}} \quad (8\text{-}17)$$

Necessary data are values for C1, C2, C6, R1, R3, R4, L2, E, and I.

In the computer printouts of noise, the first term in Eq. 8-16 is called SENSOR EN, the second term is AMPL EN, the third term is AMPL IN*ZS, and the noise of the load elements R_L is called LOAD EN in the program.

A similar model, but without R_L and C_C, is studied in Chapter 3 for the circuit of Fig. 3-5. The effect of resonance is more easily seen in Eq. 3-9:

$$E_{ni}^2 = E_{ns}^2 + [(1 - \omega^2 C_P L_s)^2 + \omega^2 C_P^2 R_s^2]E_n^2 + (R_s^2 + \omega^2 L_s^2)I_n^2 \quad (3\text{-}9)$$

At resonance the E_n value has a minimum determined by the Q of the circuit. The I_n term is dependent only on the impedance of the series inductance and resistance.

Equation 8-16 gives the equivalent input noise voltage. The signal is usually a voltage, proportional to the rate-of-change of flux linkage. The coil L_s and resistor R_s also can be expressed as a parallel resistance and inductance. Then the voltage generator E_{ns} could become I_{ns}, a thermal noise current generator. For conversions between parallel and series connections, Eqs. 8-18 and 8-19 are useful:

$$L_s = \frac{L_P R_P^2}{R_P^2 + \omega^2 L_P^2} \qquad R_s = \frac{\omega^2 L_P^2 R_P}{R_P^2 + \omega^2 L_P^2} \tag{8-18}$$

$$L_P = \frac{R_s^2 + \omega^2 L_s^2}{\omega^2 L_s} \qquad R_P = \frac{R_s^2 + \omega^2 L_s^2}{R_s} \tag{8-19}$$

Magnetic-core coils show a decreasing inductance and increasing series resistance at high frequencies due to eddy-current losses. Since the real part of the coil impedance is a thermal noise generator, it may be necessary to calculate the inductance and resistance at each frequency. The real and reactive parts of impedance as a function of frequency can be measured on Boonton or Wayne-Kerr impedance bridges.

The design of an inductive sensor can be optimized for maximum signal-to-noise ratio. The signal is proportional to the number of turns. The coil resistance is proportional to turns, and the noise is proportional to square root of turns up to a point. The signal increases faster than the noise as the number of turns increases. As the coil becomes large in diameter, the resistance increases faster than the square of turns, and the signal-to-noise ratio begins to fall-off.

8-9 BIASED DIODE SENSOR "SORCE4"

An example of the reverse-biased diode sensor is the photodiode. This type of sensor was discussed in detail in Section 3-7. A dc voltage reverse biases the diode and direct current flows through a bias resistor. Incoming radiation causes the current flow to increase. The shot noise of the dc bias current is an extra noise mechanism. A system diagram is shown in Fig. 8-13.

The noise model shown in Fig. 8-14 includes noise generator I_{sh} to represent the shot noise of the photocell direct current. An inductance has been included, although one can not always be used with this type of sensor. Load R_L has been included for completeness; C_C has been omitted for simplicity.

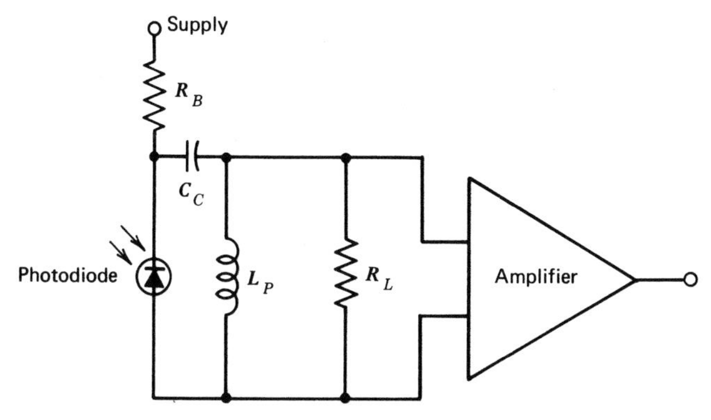

Fig. 8-13. System diagram for diode sensor.

The location of the signal source I_s is clearly indicated in the figure. Symbol equivalents are

R_I = R1 = amplifier input resistance
R_B = R2 = real part of sensor leakage resistance
R_L = R3 = load resistance
C_I = C1 = amplifier input capacitance
C_S = C2 = cell capacitance
C_W = C3 = wiring capacitance
L_P = L1 = shunt inductance
F_L = F6 = 1/f excess-noise corner
I_{dc} = I2 = sensor leakage current
I_{nB} and $I_{nL} = (4kT\,\Delta f/R)^{1/2}$ = thermal noise of R_B and R_L
$I_{sh} = (2qI_{dc}\,\Delta f)^{1/2}$ = shot noise of cell current
E_n and I_n = E and I = amplifier equivalent noise mechanisms

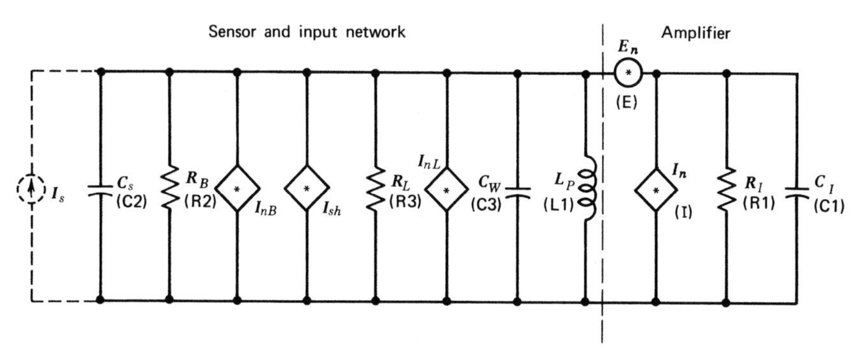

Fig. 8-14. Noise model for computer.

Values for R1, R2, R3, C1, C2, C3, L1, F6, I2, E, and I must be supplied to the program.

For this circuit it is most convenient to derive the expression for equivalent input noise current I_{ni}^2. That equivalent is the sum of noise currents entering the amplifier:

$$I_{ni}^2 = \frac{4kT}{R_L} + 2qI_{dc}\left(\frac{F_L}{F} + 1\right) + \frac{4kT}{R_B} + I_n^2 + \frac{E_n^2}{Z_A^2} \tag{8-20}$$

where

$$\frac{1}{Z_A^2} = \frac{1}{R_P^2} + \frac{(1 - \omega^2 C_P L_P)^2}{(\omega L_P)^2}$$

and

$$R_P = \frac{R_L R_B}{R_L + R_B} \qquad C_P = C_S + C_W$$

The system gain is the ratio of signal voltage at the amplifier terminals to signal source current I_s:

$$K_t = \frac{1}{\dfrac{1}{R_T} - j\left[\dfrac{1 - \omega^2 L_P C_T}{\omega L_P}\right]} \tag{8-21}$$

with

$$R_T = \frac{R_I R_B R_L}{R_I R_B + R_I R_L + R_B R_L} \qquad C_T = C_S + C_W + C_I$$

The second and third terms of Eq. 8-20 are printed out by the subroutine as SENSOR IN. The first term is referred to as LOAD IN, and amplifier noise AMPL EN/ZS is the final term. Amplifier noise current AMPL IN is simply the noise-current generator, the fourth term in Eq. 8-20.

The shunt inductance L_P makes the model more complete. This external inductance resonates with the cell and wiring capacitances thereby reducing the effect of the E_n noise component. If the inductance is not used, make L1 = 10^4 H. If a resonating inductance is used, use an opposite equalization in the amplifier for overall flat frequency response.

A photodiode signal is a current proportional to light intensity; therefore, an equivalent input noise *current* is used in Eq. 8-20. Note the simplicity of the expression. There are five noise current generators. The diode has a shot noise current term, I_{sh}, proportional to the square root of the dc leakage current. All the circuit impedances shunt the signal source. As the shunting impedance decreases, E_n/Z_A rises. Both the cell resistance and the bias resistance are thermal noise current generators. They should be large to minimize the noise contribution. In other words, the cell should have a low leakage and large bias voltage, and a bias resistor should be used.

Although the amplifier input capacity and resistance, C_I and R_I, drop out of the noise expression, they do affect the amplifier gain. This gives us a mechanism for optimizing frequency and noise responses separately.

An unbiased diode or a photovoltaic diode could use a similar model. The biasing resistor R_B would be removed. If there is dc flowing due to the ambient light level, there will still be a shot noise component. The resistance of the diode can be determined from the diode equation. The noise of the diode is either the shot noise, if there is current flowing, or thermal noise of the diode resistance, if there is no current flowing. Diode noise is discussed near the end of Chapter 1.

8-10 TRANSFORMER MODEL "SORCE5"

There are three main reasons for using an input transformer to couple the signal source to the amplifier. The first is to transform the impedance of the source to match the noise resistance R_o of the amplifier and therefore minimize the system noise figure. The second is to provide isolation between the source and amplifier. A third reason is for impedance matching to obtain maximum signal power transfer. Although the transformer can potentially reduce the equivalent input noise of the amplifier, its own noise mechanisms can contribute to the overall system noise.

A system diagram of a transformer-coupled input stage is shown in Fig. 8-15.

The circuit diagram given in Fig. 8-16*a* is a small-signal ac equivalent of the system. The transformer is represented by primary winding resistance r_p and primary inductance L_P, and the reflected secondary series resistance r_s'. This "*T*" equivalent circuit is valid for low-frequency analysis. The load resistance R_L' and input quantities R_I' and C_I' carry the prime designation to indicate they have been "reflected."

A transformer with a turns ratio *T* has a primary to secondary turns ratio

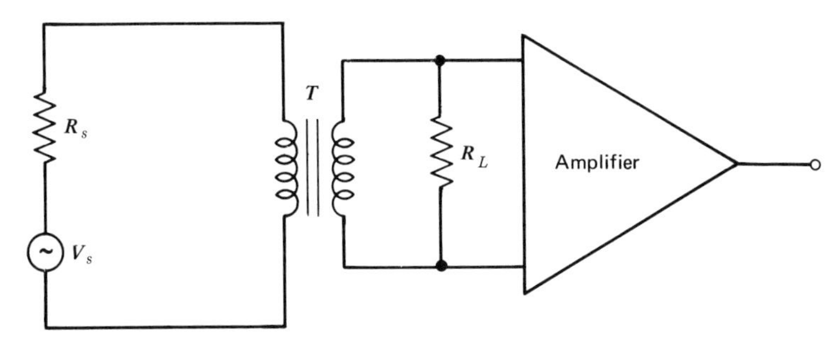

Fig. 8-15. System diagram for transformer-coupled source.

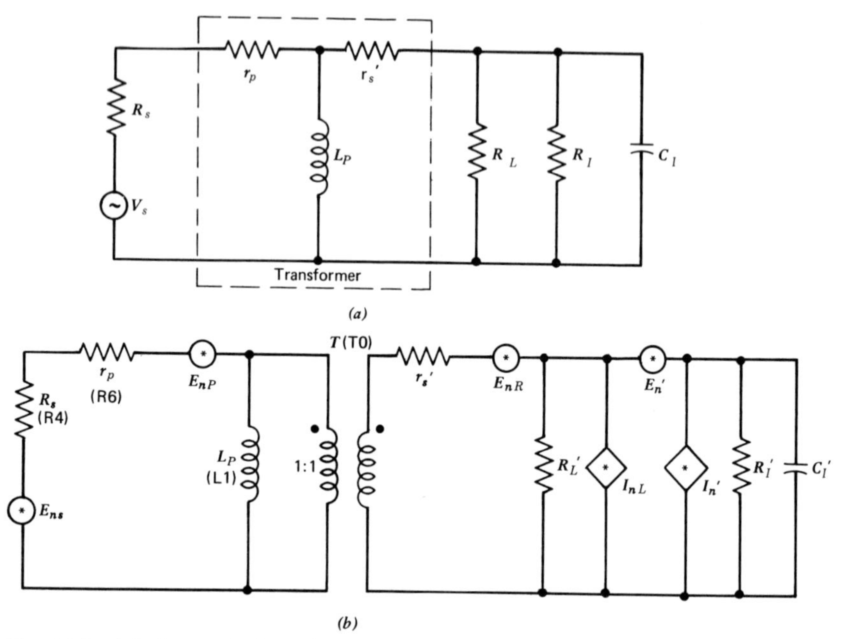

Fig. 8-16. (*a*) Ac equivalent circuit with transformer model; (*b*) noise model for computer.

of $1:T$. If T is a number such as four, there are four times more turns on the secondary winding than on the primary. This results in a stepup of voltage; the secondary voltage would be four times the voltage impressed on the primary winding.

When an impedance Z is connected across the secondary winding, the effect of that load in the primary circuit is Z/T^2. This is referred to as a reflected impedance. If the secondary voltage is high ($T > 1$), the secondary current must be low ($I_2 = I_1/T$) in order to maintain equal power levels in both windings. A transformer cannot amplify power since there is no bias supply. The primed designations in Fig. 8-16*a* represent elements reflected to the primary side by multiplying their values by $1/T^2$. It is of course possible to reflect everything to the secondary. When that is done, primary elements pick up the T^2 multiplier.

The noise model shown in Fig. 8-16*b* includes thermal noise generators associated with R_s, r_p, r_s', and R_L'. Since the turns ratio is used to reflect all secondary quantities to the primary side, the transformer is represented by a 1:1 turns ratio. In circuit analysis this can be omitted entirely.

The symbol equivalents are

r_p = R6 = resistance of the transformer primary
$r'_s = r_s/T^2$ = reflected secondary resistance of the transformer
R_s = R4 = source resistance
R_L = R3 = secondary load resistance
R_I = R1 = amplifier input resistance
C_I = C1 = amplifier input capacitance
L_P = L1 = transformer primary inductance
T = T0 = transformer secondary to primary turns ratio
E_n and I_n = E and I = amplifier equivalent noise mechanisms
$E'_n = E_n/T$
$I'_n = I_nT$
$E_{nP} = \sqrt{4kTr_p\,\Delta f}$ = thermal noise voltage of r_p
E_{nR} = thermal noise voltage of r'_s
I_{nL} = thermal noise current of R'_L

The program assumes that $r_p = r'_s$ as is true with most transformers. Therefore there is only one value for R6. To include the thermal noise of core loss, consider it to parallel the reflected R_L.

The expression for K_t has been derived based on the assumption that C_I is negligible. The results of circuit analysis are

$$E_{ni}^2 = 4kTR_s + 4kTr_p + 4kTr'_s\left[1 + \left(\frac{R_s + r_p}{\omega L_P}\right)^2\right]$$
$$+ (E'_n)^2\left\{\left[\frac{(R'_L + r'_s)(R_s + r_p)}{\omega L_P R'_L}\right]^2 + \left[\frac{R'_L + r'_s + r_p + R_s}{R'_L}\right]^2\right\}$$
$$+ [(I'_n)^2 + I_{nL}^2]\left\{\left[\frac{r'_s(R_s + r_p)}{\omega L_P}\right]^2 + (r'_s + r_p + R_s)^2\right\} \tag{8-22}$$

The voltage gain from V_s terminal to amplifier input terminal is

$$K_t = \frac{j\omega L_p R_P T}{R_A(r_p + R_P) + j\omega L_p(R_A + R_P + r_p)} \tag{8-23}$$

where, for simplicity, we have used

$$R_P = \frac{R'_L R'_I}{R'_L + R'_I} \qquad \text{and} \qquad R_A = R_s + r'_s$$

In the computer program, the value for the SENSOR EN is the first term of Eq. 8-22. The LOAD EN includes the transformer and load resistance noise and is comprised of the second and third terms, as well as I_{nL} and its bracketed coefficient. AMPL EN and AMPL IN*ZS include the fourth term and remainder of the fifth term, respectively.

In addition to the resistive noise mechanisms there are second-order effects when using a transformer that must be reduced. The transformer must be magnetically well shielded with balanced windings to minimize pickup of external ac fields. Also, the transformer should be tightly packaged to minimize its microphonism. The high permeability core in a low-level instrument transformer tends to be highly magnetostrictive and sensitive to flexing and mechanical motion. For high common-mode rejection, interwinding electrostatic shields should be used. When testing a low-level transformer, an ohmmeter should not be used. It could magnetize the core and therefore decrease the inductance and increase the microphonics. It is difficult to degauss the windings once they are magnetized. The inductance and resistance can be measured by using a bridge. Additional discussion is given in Section 9-5.

8-11 PIEZOELECTRIC SENSOR "SORCE6"

Ferroelectric ceramic elements and quartz crystals are examples of piezoelectric transducers. These elements are used in microphones, hydrophones, seismic detectors, vibration sensors, and other devices where there is mechanical to electrical energy conversion. A system diagram is shown in Fig. 8-17.

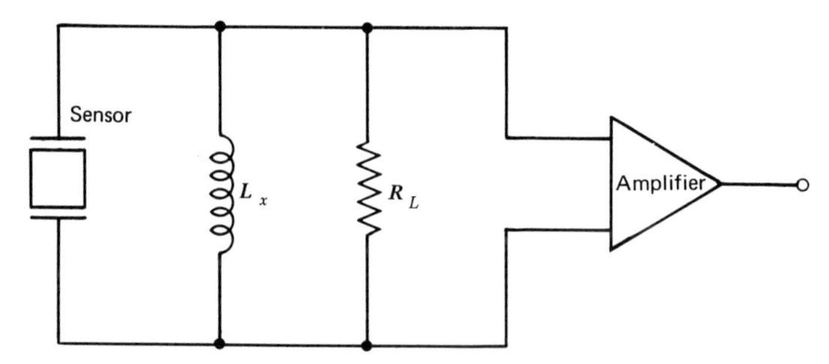

Fig. 8-17. System diagram for piezoelectric sensor.

A noise model for the system is shown in Fig. 8-18. The sensor is modeled by series resistance R_s, and so-called "mechanical" inductance and capacitance, L_M and C_M, respectively. The resulting electrical parameters of the crystal are shown in the figure. In addition, the capacitance of the block or bulk of the crystal parallels the L_M-C_M circuit. At frequencies below L_M-C_M resonance, C_M and C_B form a voltage divider, and since C_B may be 10 C_M, the signal to the amplifier is only $V_s/10$. External inductance L_x can be useful to resonate with sensor and wiring capacitances and therefore improve the signal transmission.

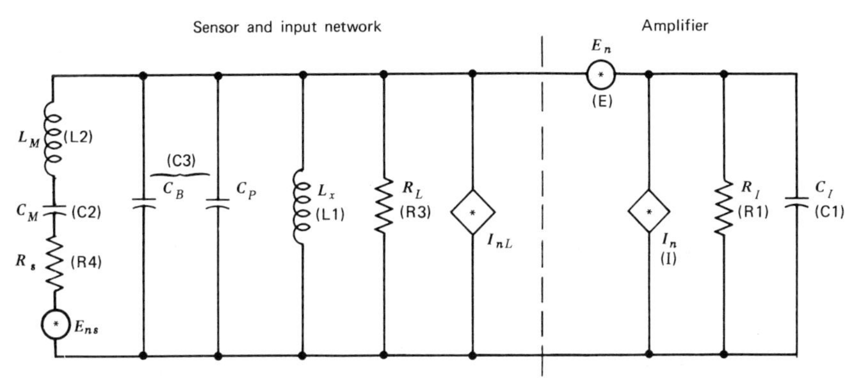

Fig. 8-18. Noise model for computer.

Symbol equivalents are

L_M = L2 = mechanical inductance
C_M = C2 = mechanical capacitance
R_s = R4 = loss resistance
C_B = block or bulk capacitance
C_P = cable capacitance
$C_B + C_P$ = C3
L_x = L1 = external inductance
R_L = R3 = load resistance
R_I = R1 = amplifier input resistance
C_I = C1 = amplifier input capacitance
E_n and I_n = E and I = amplifier equivalent noise generators
$I_{nL} = (4kT/R_L)^{1/2}$

The user must supply values for C1, C2, C3, L1, L2, R1, R3, R4, E, and I.

The system equivalent input noise parameter is

$$E_{ni}^2 = 4kTR_s + E_n^2\left(\frac{Z_s + Z_L}{Z_L}\right)^2 + \left(I_n^2 + \frac{4kT}{R_L}\right)Z_L^2 \qquad (8\text{-}24)$$

Gain between V_s terminal and amplifier input is

$$K_t = \frac{Z_I Z_L}{Z_s(Z_I + Z_L) + Z_I Z_L} \qquad (8\text{-}25)$$

where

$$Z_s = R_s - j\left(\frac{1 - \omega^2 L_M C_M}{\omega C_M}\right)$$

$$Z_L = \frac{j\omega L_x R_L}{j\omega L_x - \omega^2 L_x(C_P + C_B)R_L + R_L}$$

and

$$Z_I = \frac{R_I}{1 + j\omega C_I R_I}$$

The computer program calculates noise from Eq. 8-24. The first term, the thermal noise of R_s, is printed as SENSOR EN. The second term, the E_n noise voltage term, is printed as AMPL EN. The third term, $I_n Z_L$, is the AMPL IN*ZS term. Finally, the last term, the noise current of R_L times Z_L, is given as the LOAD EN term in the computer printout.

Equivalent input noise has four terms. The noise current terms dominate. At low frequencies, Z_L is extremely large since it is primarily the impedance of C_B and C_P. To keep the I_n term small, R_L should be large and I_n small. This implies a transistor biased at a very low collector current (in the μA range or less), or a FET with low gate leakage current. The $I_n Z_L$ term is most prominent at low frequencies where 1/f noise is likely to manifest itself. This constitutes a second vote for the FET since its I_n has very little 1/f noise. Also, the FET requires little or no biasing so R_L can be extremely large. Resonating the sensor decreases the equivalent input noise, that is, increases the signal-to-noise ratio at the resonant frequency.

8-12 A SAMPLE PROBLEM

The NOISE program is used to analyze the system shown in Fig. 8-19. The mechanics of assembly of the program card deck, including required subroutines, program and subroutine statements, and so on, are discussed in this

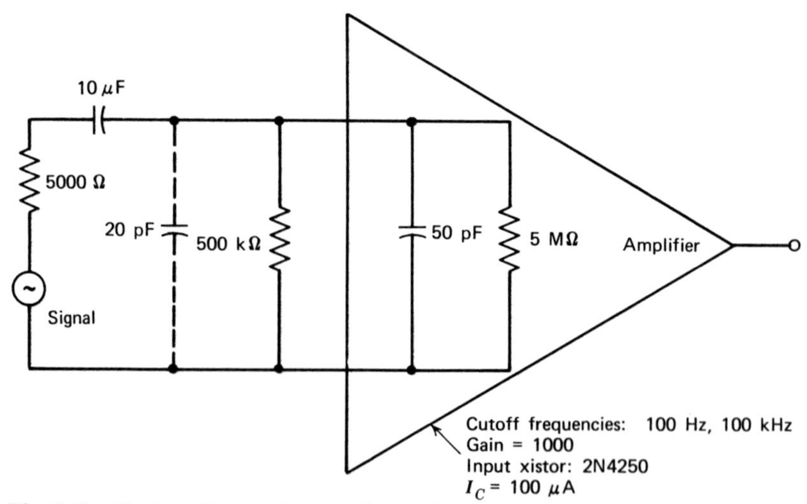

Fig. 8-19. System diagram for sample problem.

section, along with information necessary for setting up the analysis; the results obtained from running the program are discussed in the next section.

We shall run all six calculations listed in Section 8-1 on the Fig. 8-19 system. The particular version of NOISE described here requires that the user assemble seven data cards. Five of the cards are considered in Table 8-1.

Source data are contained in cards 1 and 2. *A value for each element must be provided even if that element is not present in the noise model of the system being studied.* For example, a resistive sensor as discussed in Section 8-6 does not have elements C3, R2, R6, L1, or L2, nor does it use values for I2, T0, or F6. *Harmless constants must be provided for those locations on the card.* A set of such constants is shown in Table 8-1.

For this example, data values from Fig. 8-19 are

C1 = 5E–11 (50 pF)
C2 = 2E–11 (20 pF)
C6 = 1E–5 (10 μF)
R1 = 5E6 (5 MΩ)
R3 = 5E5 (500 kΩ)
R4 = 5000 (5000 Ω)

Values for the *other* entries on cards 1 and 2 are obtained from the harmless constants shown in the figure. If we use T = 3.0E2, we are assuming the temperature to be 300°K.

Card 4N is used when the NOSMOD representation for the system electronics is used. When using XISTOR, we present the data according to card 4X. Either card may be used, but not both.

Entries on the NOSMOD card includes values for E and I, corresponding to midband values of E_n and I_n, and frequencies F1 through F4, corresponding to the break frequencies of those noise generators for the 2N4250 transistor. Harmless constants are given in the Table. For this example, the numbers used are

E = 2.0E–9 (2×10^{-9} V)
I = 3.0E–13 (3×10^{-13} A)
F1 = 3.0 (3 Hz)
F2 = 1.0E7 (10 MHz)
F3 = 90. (90 Hz)
F4 = 1.0E6 (1 MHz)

No harmless constants are listed in Table 8-1 for the XISTOR subroutine. If the user selects XISTOR, he must supply meaningful values for all nine locations. Symbol equivalents were given in Section 8-4.

Cards 5 and 6 are used to enter data for AMPLGN. Harmless constants are given in the figure. This example requires that we supply values for the break frequencies F0 and F5, and the midband gain K2. We are using the

Table 8-1. Contents of Data Cards 1, 2, 4, 5, and 6. Values Must Appear in Every Space as Indicated. (Values must be correctly aligned in card field.)

	SOURCE DATA				NOSMOD DATA		XISTOR DATA		AMPLGN DATA			
	Card 1		Card 2		Card 4N		Card 4X		Card 5		Card 6	
Card Columns	Qnty.	Value	Qnty.	Value	Qnty.	Value	Qnty.	Value	Qnty.	Value	Qnty.	Value
1–8	C1	2.0E–12	I2	1.0E–12	E	2.0E–9	I4		F0	1.0E–6	S1	1.0E9
9–16	C2	1.0E–12	L1	1.0E4	I	3.0E–13	B		F5	1.0E10	S2	1.0E9
17–24	C3	2.0E–12	L2	0.0E1	F1	3.0	F8		N8	1.0E0	S3	1.0E9
25–32	C6	1.0E–3	T	3.0E2	F2	1.0E7	N5		N9	1.0E0	S4	1.0E9
33–40	R1	1.0E8	T0	1.0E0	F3	90.	N6		K2	1.0E0	S5	1.0E10
41–48	R2	1.0E8	F6	1.0E–3	F4	1.0E6	F7				Q0	1.0E–3
49–56	R3	1.0E8					Q1					
57–64	R4	1.0E0					Q2					
65–72	R6	1.0E0					I7					

A7 and A8 representations discussed in Section 8-5, with N8 = N9 = 1. Specific entries to card 5 for this example are

F0 = 100 (100 Hz)
F5 = 1E5 (100 kHz)
K2 = 1000

Card 6 must be included; all values can be as shown in Table 8-1.

We must tell the computer what calculation to make. This is done on data card 3. The version of NOISE considered here consists of a main program and 20 subroutines. The card deck is assembled with the main program followed by the subroutines used for a particular calculation and then the seven data cards. The main program and all subroutines could be included for every calculation, but it is more efficient if only those subroutines actually used for a calculation are included.

Names of the subroutines used for each calculation are shown in Table 8-2. Main program and subroutines statements are given in Appendix IV.

The contents of data card 3 are also shown in Table 8-2. That card contains (in integer format 3I2) the following

1. In column 2: the desired calculation (a number from 1 to 6).
2. In column 4: the source used (a number from 1 to 6).
3. In column 6: the model used (1 for NOSMOD, 2 for XISTOR).

Data card 7 contains information regarding the frequency or frequencies to be used in each calculation. These data are shown in Table 8-2 for the example being considered here.

8-13 SAMPLE RESULTS

The results of *calculation 1* for the example of the preceding section are

FREQUENCY SENSOR EN	LOAD EN	AMPL EN	AMPL IN*ZS	SUM NOISE	NANOVOLTS
100 TO	1000				
272.983	27.298	60.832	49.913	285.406	
1000 TO	10000				
863.249	86.325	191.709	143.934	900.067	
10000 TO	100000				
2729.830	272.983	606.467	451.348	2845.700	

TOTAL NOISE VOLTAGE 2998.27 NANOVOLTS

Table 8-2. List of Subroutines and Card Contents

Calculation	Subroutines Used	Data Card 3	Data Card 7**
No. 1. Total Equivalent Input Noise	PRNT2, PRNT3, INTEG, KVAL, MODEL, NOSMOD or XISTOR, SOURCE, SORCE1*	1 1* 1	100. 1000. 10000. 100000.
No. 2. Input Network Frequency Response	READ1, SOURCE, SORCE1*, PRNT4, PRNT5	2 1* 1	100000. 1000000.
No. 3. Input Noise at One Frequency	PRNT3, KVAL, MODEL, NOSMOD or XISTOR, SOURCE, SORCE1*	3 1* 1	1000.
No. 4. Input Noise versus Frequency	READ1, PRNT2, PRNT3, KVAL, MODEL, NOSMOD or XISTOR, SOURCE, SORCE1*	4 1* 1	100. 300.
No. 5. Total Noise at the Output	PRNT2, PRNT3, INTEG, KVAL, KVAL2, MODEL, NOSMOD or XISTOR, SOURCE, SORCE1*, AMPLGN	5 1* 1	10. 100. 1000. 10000. 100000. 1000000.
No. 6. Total System Gain	READ1, SOURCE, SORCE1*, AMPLGN, PRNT4, PRNT5	6 1* 1	100. 1000.

* Insert appropriate SORCE number, 1 through 6.

** Frequencies and frequency ranges for the sample calculations.

The computer calculates and integrates the noise in each frequency decade. It prints the rms value of the noise contribution of each of the noise generators in each frequency interval, as well as the sum which is the *total* equivalent input noise.

The load resistance is 100 times the source; because of voltage division in the network its noise effect is only $\frac{1}{10}$ of the sensor noise. Note that the E_n and I_n contributions are about equal. This shows that the amplifier is operating near its optimum source resistance. The noise of the source is the dominant noise mechanism, which indicates that the amplifier is well designed. The NF of this amplifier would be 0.35 dB.

For the input network transfer function, that is, the gain of the input network including the amplifier input impedance, we select *calculation 2* and rerun the program:

FREQUENCY	GAIN	DB
100000	.967	−0.296
125892	.954	−0.409
158489	.935	−0.583
199526	.907	−0.844
251188	.868	−1.230
316227	.815	−1.777
398106	.748	−2.525
501186	.669	−3.496
630956	.582	−4.694
794327	.495	−6.100
1000000	.413	−7.678

The frequency range chosen was 100 kHz to 1 MHz. The gain is near unity, rolling-off 3 dB at about 400 kHz.

Calculation 3 determines the equivalent input noise at a single frequency. The choice is 1000 Hz. The computer printout yields

SENSOR EN	LOAD EN	AMPL EN	AMPL IN*ZS	SUM NOISE NANOVOLTS
9.099	.910	2.023	1.566	9.496

The noise given is in a one-cycle bandwidth at 1000 Hz, and the noise contribution of each of the principal noise mechanisms is indicated. As before, the dominant noise source is the sensor thermal noise.

We select *calculation 4* to determine the equivalent input noise versus frequency. The program calculates and prints the noise at each of 10 frequencies/decade over the range requested. The 100 to 300 Hz band was selected.

```
FREQUENCY
 SENSOR EN  LOAD EN  AMPL EN  AMPL IN*ZS  SUM NOISE  NANOVOLTS

  100
    9.099      .910     2.050      2.068       9.554
  126
    9.099      .910     2.044      1.964       9.531
  158
    9.099      .910     2.039      1.878       9.512
  200
    9.099      .910     2.035      1.807       9.498
  251
    9.099      .910     2.032      1.748       9.486
  316
    9.099      .910     2.030      1.700       9.477
```

Using *calculation 5* the total noise at the output of the amplifier is obtained:

```
FREQUENCY
 SENSOR EN  LOAD EN  AMPL EN  AMPL IN*ZS  SUM NOISE  NANOVOLTS

      10 TO      100
     41662.      4166.     9467.     10717.      44245.
     100 TO     1000
    259517.     25952.    57814.     47102.     271263.
    1000 TO    10000
    851784.     85178.   189163.    142023.     888113.
   10000 TO   100000
   2339835.    233983.   519763.    386805.    2439128.
  100000 TO  1000000
   2029450.    202945.   457083.    346735.    2118727.
TOTAL NOISE VOLTAGE =3361939.   NANOVOLTS
```

Although the frequency band of interest, as previously stated, is 100 Hz to 100 kHz, this calculation includes a decade beyond each limit to determine if any noise is coming through the roll-off region. Calculation 5 takes into account the input network transfer function and the gain as stored in AMPLGN. Total noise at the output is found to be 3.4 mV. There is little additional noise in the 10- to 100-Hz region, but there is a significant contribution in the 100 kHz to 1 MHz band.

The total equalized system gain is determined using *calculation 6*:

FREQUENCY	GAIN	DB	FREQUENCY	GAIN	DB
100	699.414	56.894	1258	985.933	59.876
126	774.511	57.780	1584	987.028	59.886
158	836.524	58.449	1995	987.675	59.892
199	884.274	58.931	2511	988.011	59.895
251	918.977	59.265	3162	988.110	59.895
316	943.084	59.490	3981	987.989	59.894
398	959.311	59.639	5011	987.626	59.891
501	969.988	59.735	6309	986.941	59.885
630	976.907	59.796	7943	985.790	59.875
794	981.342	59.836	9999	983.930	59.859
1000	984.160	59.861			

The response peaks near 3162 Hz. Gain at 100 kHz is 683.434, and at 1 MHz it is 41.111.

8-14 INTEGRATION

A simple method of summing the noise components over a wide bandwidth is to use Simpson's rule for approximate integration. This is an important part of the noise analysis program. The amplifier and equalizer response curve in Fig. 8-20*a* is divided into intervals. Where the function is changing rapidly, the divisions are many; they are few where it is varying slowly. In the slowly varying region from f4 to f5 there is one division; there are four divisions in the critical region from f6 to f10. The numerical integration uses more data points in the highly variable region.

Simpson's rule for approximate integration is based on fitting a parabola to every three adjoining values of the function. The band for integration is divided into an even number of equal intervals; in this instance the intervals are ΔF. As shown in Fig. 8-20*b*, the ordinate values have been given the symbols $a, b, \ldots, g$; these values are known.

It follows that the integral of this function is

$$\text{Area} = \frac{\Delta F}{3}(a + 4b + 2c + 4d + 2e + 4f + g) \qquad (8\text{-}26)$$

The coefficients of the terms in Eq. 8-26 follow an easily recognized pattern.

Simpson's rule is used in NOISE, and is located in the INTEG subroutine.

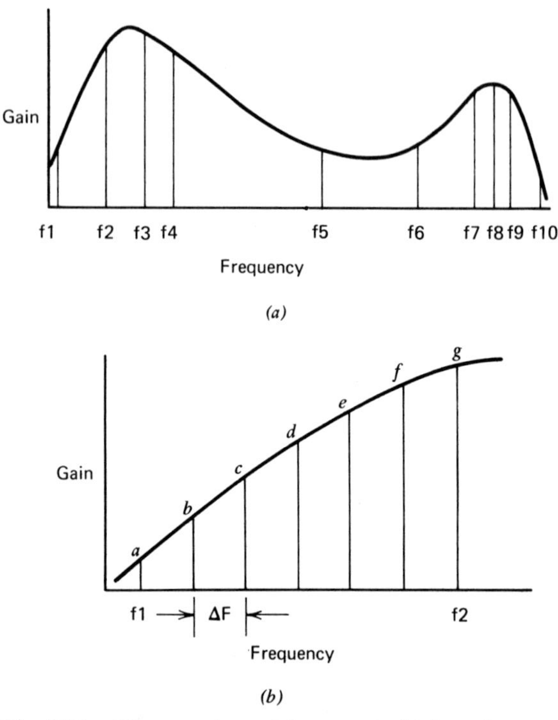

Fig. 8-20. Diagrams to explain numerical integration.

8-15 USE OF CIRCUIT ANALYSIS PROGRAMS FOR NOISE PROBLEMS

Various circuit analysis programs are available for computer-aided design. Programs for ac analysis, such as ECAP and HICAP, can be used for noise problems [2]. We insert the noise mechanisms in the computer model as noise voltage and current generators. For transistor noise, a voltage generator E_n is placed in series with the base and a noise current generator I_n in parallel with the emitter-base junction. For the thermal noise of resistors, a noise voltage or current generator is placed in series or parallel with each of the resistors. Calculate the output resulting from each generator acting alone. To obtain the complete noise voltage output, the contributions from each generator must be added in mean square terms. Taking the square root of this sum yields rms noise output voltage.

ECAP is capable of analyzing the noise of cascaded stages. Also, it is useful for a worst case analysis. In general, ECAP is not as applicable to noise problems as the NOISE program. Universal circuit analysis programs cannot integrate the noise over a bandwidth. They only calculate the noise as a function of frequency.

Although it is possible to use a general circuit analysis program for noise problems, it is probably faster and more convenient to use NOISE or to write your own program.

A program specifically written for noise calculations has been reported [3]. It would appear to be faster than general circuit analysis programs, but does not calculate E_{ni}, does not integrate, and does not provide for excess noise. It does consider all noise sources, including those in cascaded stages that follow the first stage.

SUMMARY

a. Computer-aided design is useful for low-noise systems to calculate E_{ni} at one frequency or over a band of frequencies, response curves for the sensor-network and for the amplifier, and spot and broadband values of output noise.

b. The NOISE program includes provision for six different sensor models. The input amplifier stage can be represented by $E_n(f)$ and $I_n(f)$, or if bipolar, by hybrid-π parameters.

c. NOISE includes a universal equalization curve to represent amplifier response.

PROBLEMS

1. Derive Eq. 8-12 for the transfer voltage gain of the SORCE1 network.
2. Determine the total noise at the amplifier input terminals from Fig. 8-8, and divide by $|K_t|^2$ to verify Eq. 8-11.
3. Derive Eq. 8-25 for the transfer voltage gain of the SORCE6 network.
4. Determine the total noise at the amplifier input terminals from Fig. 8-18, and divide by $|K_t|^2$ to verify Eq. 8-24.
5. Consider the results of the sample computer runs given in Section 8-13. Use calculation 5 for 1000 to 10,000 Hz to check calculations 1 and 6. Approximate where necessary.

REFERENCES

1. Motchenbacher, C. D., Research Engineer, Honeywell, Inc., Hopkins, Minnesota 55343.
2. "1620 Electronic Circuit Analysis Program (ECAP) User's Manual," Application Program No. 1620-EE-02X, IBM, White Plains, N.Y., 1970 ed.
3. Rohrer, R., L. Nagel, R. Meyer, and L. Weber, "Computationaly Efficient Electronic-Circuit Noise Calculations," *IEEE J. Solid-State Circuits*, **SC-6**, 4 (August 1971) 204–212.

Chapter 9

NOISE IN PASSIVE COMPONENTS

The input transistor and the source resistance should be the dominant noise sources in an amplifier. Unfortunately, as we have seen, transistors are not the only circuit elements that generate noise. Passive components located in the low-signal level portions of the electronics also can be major contributors. This chapter discusses noise in passive components: resistors, capacitors, diodes, batteries, and transformers. The noise mechanisms of these components are studied, and methods for minimizing their effects are considered.

9-1 RESISTOR NOISE

The total noise of a resistor is made up of thermal noise and excess noise as discussed in Chapter 1. All resistors have a basic noise mechanism, thermal noise, caused by the motion of charge carriers. This noise voltage is dependent on the temperature of the resistor, the value of the resistance, and the bandwidth of the measurement. In every conductor above absolute zero there are charge carriers excited by thermal energy, jumping about within the conductor. Each jump is equivalent to a small burst of current. The rms value of the thermal noise voltage at the terminals of a resistance R is

$$E_t = (4kTR\,\Delta f)^{1/2} \tag{1-3}$$

There are four commonly used types of fixed discrete resistors: carbon composition, deposited carbon, metal film, and wirewound. Each has particular characteristics suitable for specific applications in electronic circuitry.

Often an excess noise is generated when dc flows through a resistor. Excess noise gets its name because it is found to exist *in addition to* the fundamental thermal noise of the resistor. It is also referred to as *current noise*. Excess noise usually occurs when a current flows in any discontinuous conductor such as a manufactured resistor. A carbon composition resistor is made of carbon granules mixed with a binder and squeezed together. Dc flows unevenly through the resistor because of variations in conductivity. There is something like microarcs between the carbon granules, resulting in spurts or bursts of current flow. These spurts of current can be the excess noise. The more uniform the resistor, the less the excess noise. Composition carbon resistors usually generate the greatest noise, whereas tin oxide, metal film, and wirewound are all lower noise devices.

Excess noise usually has a l/f noise power spectrum. Noise power varies inversely with frequency, and therefore noise voltage increases as the *square root* of the decreasing frequency. There is equal noise power in each frequency decade. Since the noise in each of these intervals is uncorrelated, the noise contained in a wide frequency band increases as the square root of the decades of frequency in that band.

The total noise generated in a resistor is illustrated in the graph of Fig. 9-1.

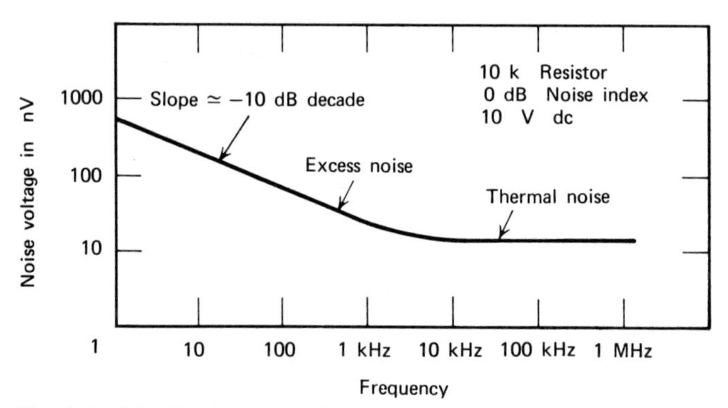

Fig. 9-1. Total noise of a resistor.

The thermal noise component dominates at high frequencies, and the excess l/f noise component dominates at low frequencies. This example is a fairly noisy carbon composition resistor.

Noise Index

Excess noise power is proportional to l/f, and excess noise voltage is proportional to the level of dc flowing through a resistor. An experimentally determined equation for the excess noise voltage E_{ex} in a resistor is

$$E_{ex}^2 = \frac{KI_{DC}^2R^2}{f} \tag{9-1}$$

where K = a constant, dependent on manufacturing methods
I_{DC} = direct current flowing through the resistor

Using these measured properties we define a *noise index* (NI) to express the amount of excess current noise in a resistor. *The noise index is the rms value in μV of noise in the resistor per V of dc drop across the resistor in a decade of frequency.* Thus, even though the noise is caused by dc, using the definition of NI results in an expression that is independent of I_{dc} and R. We obtain

$$\text{NI} = \frac{E_{ex}}{V_{DC}} \times 10^6 = \frac{K'}{f^{1/2}} \quad \text{in} \quad \mu\text{V/V}$$

NI is almost always expressed in dB:

$$\text{NI} = 20 \log \left(\frac{E_{ex}}{V_{DC}}\right) \quad \text{in dB (in a decade)} \tag{9-2}$$

To use this equation, E_{ex} *must be in μV per frequency decade.* The equation gives NI = 0 dB when $E_{ex}/V_{DC} = 1$. For most practical resistors the ratio E_{ex}/V_{DC} is less than unity. Resistors with little excess noise have noise indexes of -20 dB or lower. The conversion from μV/(V)/(decade) to dB is possible using scale 7 of the Quan-Tech noise rule. (See also Fig. 9-3.)

At one time, a commonly used figure of merit was the ratio of the mean rectified noise in μV to the direct voltage in V. The chief disadvantage of this method lies in the fact that the noise voltage is proportional to the bandwidth of the measuring system and this was not always specified. Several years ago, the National Bureau of Standards proposed a current-noise index called conversion gain, which is defined as the ratio of the available mean current-noise power-spectral density at 1000 cps to the power supplied by the dc supply, expressed in dB [1]. More recently, the same organization recommended the use of NI as given in the preceding paragraph [2]. This index is called the μV/V index and it can be shown that it differs from conversion gain by 159.6 dB.

Resistor Noise Calculation

To use the μV/(V)/(decade) figure both the NI and applied dc voltage must be known. If the NI is not specified by the manufacturer, it must be measured since it cannot be calculated. A method of measurement is outlined in the literature [2,3]. A resistor-noise test set made by Quan-Tech Laboratories can be used to measure the NI of resistors.

To illustrate the determination of resistor noise consider this example. We are given a 10-kΩ composition resistor with a NI of 0 dB or 1 μV/V. It is used in a circuit with a frequency response from 10 Hz to 10 kHz, and the dc voltage drop across the resistor is 10 V. Consequently, the resistor has 1 μV of noise/V of dc in each decade of frequency. Since excess noise increases linearly with supply voltage, there is 10 μV/decade for a 10-V dc drop. In the 3 decades from 10 Hz to 10 kHz, there is equal noise in each decade of frequency. These noise components add in an rms manner. The rms noise in 3 decades is 1.732 times the noise in 1 decade. The total excess noise voltage across the 10-kΩ resistor is, therefore, 1 μV/(volt)/(decade) $\times$ 10 V $\times$ 1.732 for a total noise of 17.3 μV. The *thermal noise* in a 10-kHz bandwidth is only 1.25 μV. You can see that the excess current noise is considerably greater than the limiting thermal noise. The total noise of this 10,000-Ω resistor over the 10 Hz to 10 kHz bandwidth is the root of the mean square sum of the excess noise and thermal noise, or about 17.4 μV.

Bandwidth Correction

Since the NI is noise/decade, it is necessary to multiply by the square root of the number of decades for wideband applications. A chart of bandwidth multiplying factors for various frequency ratios is shown in Fig. 9-2.

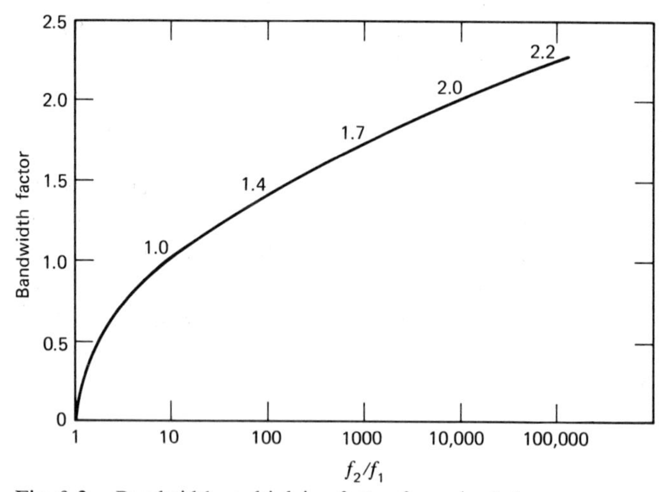

Fig. 9-2. Bandwidth multiplying factor for noise index.

When analyzing the noise of a circuit at a single frequency, it is necessary to know the spectral density of the excess noise.* Spectral density is the noise

* See the discussion of spectral density in Section 1-4.

voltage in a 1-Hz bandwidth at a specific frequency. The NI is the integrated noise in a decade; we work backwards to find the noise spectral density at 1 kHz. The noise voltage $E^2(\Delta f)$ in a bandwidth Δf is the integral of the spot noise voltage or spectral density:

$$E^2(\Delta f) = \int_{f_1}^{f_2} E_n^2(f)\, df \tag{9-3}$$

where f_1 and f_2 are the frequency limits of the band, and $E_n^2(f)$ is the mean-square noise voltage as a function of frequency. For 1/f noise power spectrum, the noise at any frequency is

$$E_n^2(f) = \frac{K}{f} \tag{9-4}$$

Substituting Eq. 9-4 into 9-3 yields

$$E^2(\Delta f) = K \ln (f_2/f_1) \tag{9-5}$$

For a 1-decade frequency range, $f_2/f_1 = 10$. Then Eq. 9-5 becomes

$$E^2(\Delta f) = K(2.303) \tag{9-6}$$

Since a decade was assumed to arrive at Eq. 9-6, $E^2(\Delta f)$ is the square of the NI for $V_{DC} = 1$ V $[E^2(\Delta f) = (\text{NI})^2]$. It follows that

$$K = \frac{(\text{NI})^2}{2.303} \tag{9-7}$$

This value of K can be substituted into Eq. 9-4. Since $E_n(f)$ is the spectral density S, a general expression results:

$$S = \frac{0.66(\text{NI})}{f^{1/2}} \qquad \text{in V/Hz}^{1/2} \tag{9-8}$$

Specifically at $f = 1000$ Hz the spectral density becomes

$$S_{1000} = 2.0208(\text{NI}) \tag{9-9}$$

The noise spectral density of excess noise at 1 kHz is 0.0208 times the NI in μV/(V)/(decade). The spectral density at any other frequency is inversely proportional to the square root of the frequency ratio. For instance, from Eq. 9-9, the value of S at 1/100 the frequency or 10 Hz is 0.208 times the NI. The conversion from NI in dB to noise spectral density at 10 Hz, 100 Hz, and 1 kHz is shown in Fig. 9-3. Scale 7 of the Quan-Tech Noise Calculator can be used for conversion of NI to S at 1 kHz.

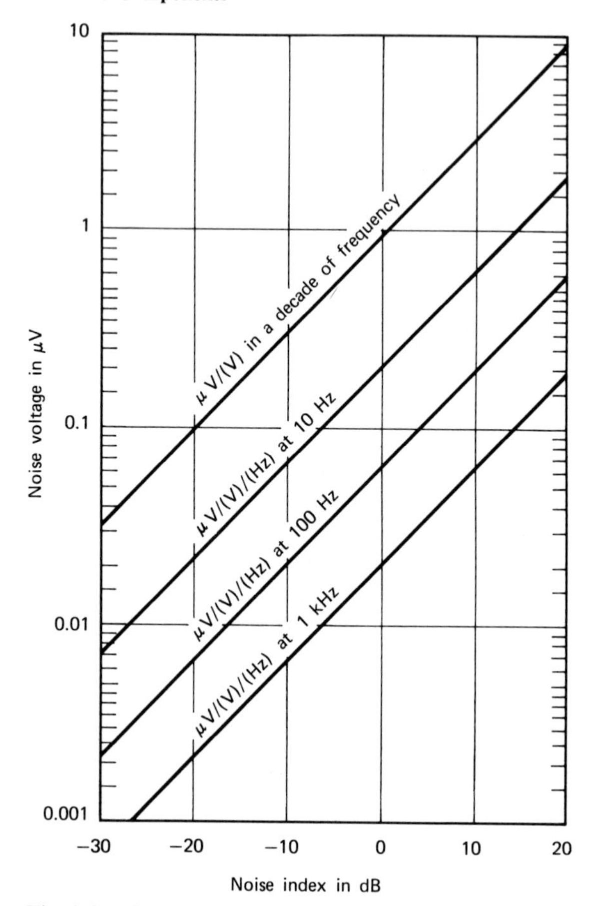

Fig. 9-3. Conversion of noise index in dB to μV/V.

Noise Index of Commercial Resistors

It is difficult to obtain exact noise information from resistor manufacturers. One reason for this is economic: a noise test is an extra measurement not usually made by the manufacturer. Noise variations depend on manufacturing processes and process changes. It has been shown that the greater the noise variation, the lower the reliability [4]. Resistors with high values of NI tend to be less stable. Figure 9-4 is a chart of the range of noise indexes based on data taken on different types of resistors. Typically, there is a 20-dB spread from minimum to maximum within a type and value. If the excess noise is critical, it is best to place a maximum noise spec on the resistor. An additional charge of two cents per resistor may be normal.

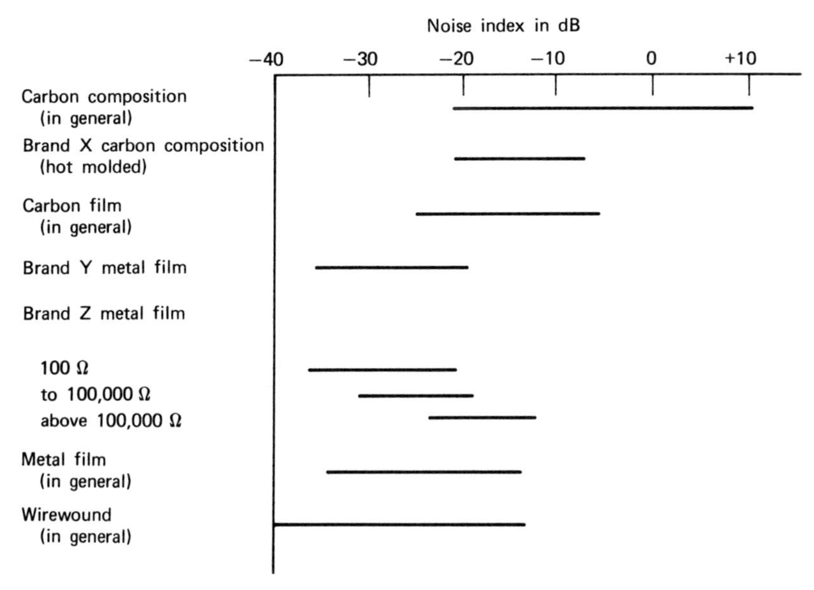

Fig. 9-4. Noise indexes of resistors.

It is to be remembered that excess noise is dependent on the voltage drop across the resistor. If there is no voltage drop, you do not need a low-noise resistor. Excess noise in a resistor is only significant at low frequencies when there is a voltage drop across the resistor.

As pointed out earlier, excess noise is caused by current flowing through a discontinuous conductor. From the figure we conclude that the carbon composition resistor, made of compacted carbon grains, has the highest noise; the wirewound resistor can have the lowest noise. In general, the larger the resistor, the higher the wattage rating, or the thicker the film element; the lower the noise. High-wattage resistors have less noise than low-wattage resistors. Also, high-resistance values have a higher NI than low-resistance values. For example, a high-resistance metal film resistor has a very thin film of metal on a substrate and so it is more affected by surface variations in the substrate. Similarly, a high-resistance composition carbon resistor has less carbon and more binder so it is less uniform.

This variation of NI with resistance and process type is graphically shown in Fig. 9-5 for a type of carbon film resistor. Note that the helixed resistors have a lower NI than the nonhelixed. This is because the resistance per square of a nonhelixed resistor is much higher, since it is a thinner film. Note also the increasing NI with increasing resistance value.

Carbon composition resistors are generally made by mixing a carbon powder with a binder and molding it around lead wires. A second method is to put a

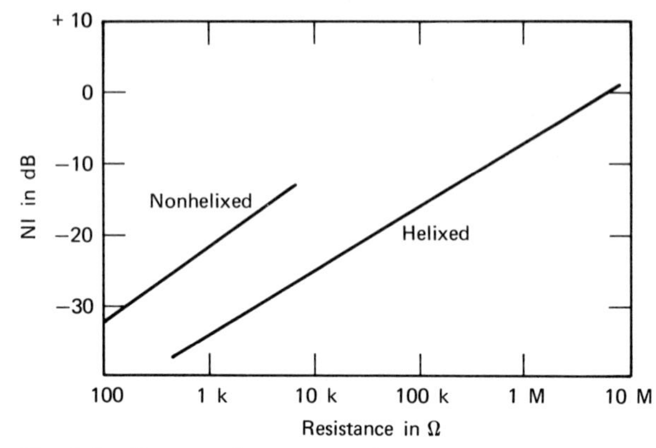

Fig. 9-5. Excess noise versus resistance.

film of resistive material on a tube and insert lead wires into the tube. The second method is generally noisier. Since both of these methods can result in relatively discontinuous material and poor end contacts, the excess current noise is highest. They have the advantage of being inexpensive and are the more widely used.

Deposited carbon resistors are made by depositing a carbon film on a ceramic rod and then cutting spiral grooves in the ceramic and carbon. With deposited carbon resistors it is important to have a good end contact to avoid generating excess noise at the terminations. Carbon film resistors formerly were lower noise, but the ordinary composition carbon resistors have now improved to the point where they are becoming competitive with deposited carbon. Further, deposited carbon resistors are declining in popularity and are probably a poor choice for low-noise design. In general, carbon film resistors are about 20-dB noisier than metal film resistors.

Metal film resistors are the best choice for low-noise applications. They are made by evaporating a thin film of metal on a ceramic core. The resistive path is frequently helixed to increase the resistance. Not all metal film resistors are low noise. For instance, those with poor end terminations generate significant excess current noise at the end contacts. Poor helixing also makes them noisy. Care is necessary in the handling of the glazed-type metal film resistors. With these units, if the leads are pulled too hard or bent very close to the body, the seal will crack and the resistor will become noisy. The molded metal film resistors are not as susceptible to this problem; they are slightly more expensive.

Wirewound resistors can have the lowest excess noise of all. Their principal noise source is at the end termination. Those with crimped terminations can be noisy; the ones with solder or welded terminations will probably be low noise. There are disadvantages to wirewound resistors, however. They tend to be expensive, they are limited to the lower values of resistance, and they are more frequency sensitive. For general low-noise design, metal film resistors turn out to be a better choice.

IC Resistors

Integrated circuit resistors can be classified according to the method of manufacture as *thin film, thick film*, or *monolithic*. Thin-film and thick-film devices differ in the method of deposition of the resistor material on the substate. Thin-film resistors are formed by a type of atom-by-atom process, such as vacuum deposition, cathode sputtering, and vapor plating. Thick-film devices, on the other hand, are assembled by screen deposition of an ink, and the curing or firing of this ink in a furnace.

Thin films are depositions on an insulating substrate. The film is usually protected by an insulating coating. Terminals are formed at ohmic contacts with the film made through etched apertures in the coating.

The more common thin-film materials are nickel-chromium alloys (such as Nichrome®), tantalum, and cermets. Cermets are metal-ceramic combinations such as chromium and silicon monoxide. Tin oxides are occasionally used. The decision of material type to be used for a given application depends on factors such as temperature coefficient, resistivity, and stability with time. Materials used for thin-film resistors have NIs comparable to those of discrete metal-film resistors [5]. Titanium nitride, for example, has a NI of less than 0.1 μV/V/decade or -20 dB.

Thick-film resistors are fired in an atmosphere usually between 500 and 1100°C. At these temperatures, the solids oxidize, melt, and sinter together to solidify the film, level the surface, and bond the film to the substate. Organic binders and vehicle solvents are evaporated off.

Precious metal thick films are known to exhibit NIs of less than 0.1 μV/V/decade. Cermet thick films, the most common of which is a mixture of palladium, silver, palladium oxide, and glass, have been reported to have NIs more than ten times that of the precious metal film, 1 to 10 μV/V/decade [6]. Apparently, the NI of cermet thick-film resistors is highly dependent on resistance value. Noise indexes have been reported as -35 dB for 100 Ω, -10 dB for 10 kΩ, and $+17$ dB for 200 kΩ units [7].

Monolithic resistances fabricated during the base diffusion or the emitter diffusion operations are normally isolated from other IC components by junction isolation. These reverse-biased *pn*-junctions exhibit some shot noise. Excess noise is possible at the resistor contacts. Resistances made from FET-like structures can have noise sources of the types discussed in Chapter 5.

9-2 NOISE IN CAPACITORS

In general, capacitor noise is not a problem in circuit design. An ideal capacitor is noiseless. It is not a thermal noise source. A real capacitor, however, is not perfectly lossless. It has a certain amount of series resistance and shunt leakage resistance. These real components of the impedance contribute thermal noise, but they are usually negligible.

The noise model of a capacitor is a noise current generator shunted by the capacitor's impedance. The main noise contribution is at low frequencies, where the reactance of the capacitor does not effectively shunt the internal noise. Low-frequency excess noise is the dominant noise mechanism.

For low-frequency circuit applications, large values of capacitance are usually required. Electrolytic capacitors are often used because of their smaller physical size. Tantalum electrolytics have been used successfully in low-noise circuits. Aluminum electrolytics with their higher leakage and higher forming currents are less desirable.

Electrolytic capacitors are used in two principal ways: bypass capacitors and coupling capacitors. No instances have been observed when a bypass capacitor added noise to the circuit. Its own impedance effectively shunts the noise generators.

Coupling capacitors occasionally add noise. After an electrolytic capacitor has been reverse biased, it generates bursts of noise for a period of time, from a few minutes to several hours. Usually a transient causes the reverse bias. Occasionally, during turn-on, the coupling capacitors are reverse biased as one stage carries "on" faster than the next. After this happens, the capacitor may be noisy for a while. There are three ways of avoiding this problem: one is to design the circuit so that there is no reverse bias at any time and no turn-on transient. A second solution is to place a low-leakage silicon diode in parallel with the capacitor. Any reverse bias forward biases the diode and prevents the capacitor from breaking down. A third method is to use two electrolytic capacitors of twice the size connected back to back (in series). This forms a nonpolar electrolytic capacitor. Regardless of the polarity of applied voltage, the proper capacitor takes the voltage drop.

9-3 NOISE OF REFERENCE AND REGULATOR DIODES

In a forward-biased diode the principal noise is the shot noise attributed to the dc. There is also excess noise in the forward-biased diode; as discussed in Chapter 4 the base-emitter diode of a transistor exhibits excess noise. *A reverse-biased diode has two breakdown mechanisms that exhibit different noise properties. These are the Zener and avalanche mechanisms.*

Internal field emission, the Zener effect, is the predominant mechanism in diodes with low (7 V or less) reverse-breakdown voltages. At a very thin junction, the electric field can become large enough for electrons to jump the energy gap. The Zener mechanism exhibits shot noise. There is little excess noise. Some Zener diodes have l/f noise corners lower than 10 Hz. Being shot-noise limited, the rms noise voltage can be calculated from the direct current by using $I_{sh} = (2qI\,\Delta f)^{1/2}$. The Zener diode is a low-noise regulator or reference diode.

In a diode with a larger junction width breakdown occurs at a lower voltage than can be explained by the Zener effect. This is avalanche breakdown. As the reverse bias on the diode is increased, carriers are accelerated to an energy level great enough to generate new hole-electron pairs on collision. These new pairs are accelerated by the electric field of the bias voltage and generate more hole-electron pairs. The process cascades and is self-sustaining as long as the field is maintained.

The noise of an avalanche breakdown is larger and more complicated than Zener noise. There is noise due to the avalanche mechanism and also a multistate noise mechanism. As with forward-bias shot noise, the avalanche mechanism results in random arrival of carriers after crossing the junction. In this case, there are bundles of carriers that give an amplified shot noise. Avalanche noise has a flat frequency spectrum, hence it is white.

The more troublesome noise in a reverse-biased avalanche diode is the multistate noise. It gets its name from the fact that the noise voltage appears to switch randomly between two or more distinct levels. These levels may differ by many mV. As the reverse current of the diode is increased, the higher level is favored until finally it predominates. Although the average noise current is constant, the period is completely random ranging from μsec to msec. The switching time between levels is extremely fast. Multiple-level operation is apparently due to defects and localized inhomogeneities in the junction region. These can create a local negative-resistance area with either switching or microplasma oscillation.

Multistate noise has a wide frequency range, but it is predominantly l/f noise. It is definitely an excess noise mechanism that does not need to exist. It is process dependent and varies from manufacturer to manufacturer. Lower noise units can be selected from a production run. High multistate noise generally indicates lower device reliability. If a low-noise reference diode is needed, the most acceptable solution is to use the lower voltage Zener diode. When the need exists for a reference in the avalanche region, greater than 5 to 50 V, select units for low noise at the proper circuit current level.

Noise behavior of breakdown diodes is depicted in Fig. 9-6. For this diode series, units with breakdowns below 3 V exhibit very little noise; units breaking down above 5 V are quite noisy. For these data a reverse current of 250 μA was passed through all units when measuring noise.

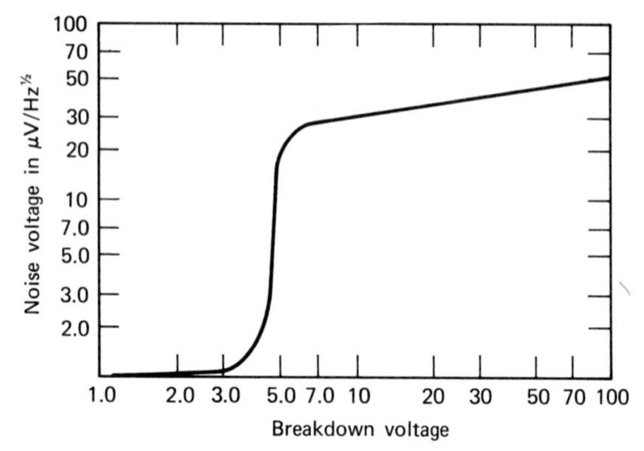

Fig. 9-6. Noise voltage versus breakdown voltage for a family of breakdown diodes.

Breakdown diodes are not recommended as coupling or biasing elements in a low-noise amplifier because of their noise behavior. When a regulator or reference diode is used in the power supply of a low-noise amplifier, it should be decoupled by an *RC* network or a capacity multiplier as described in the section on power supplies in Chapter 11. If a reference diode is used as a calibration source, select a low-voltage Zener type and shunt it with a large bypass capacitor.

9-4 BATTERIES

Batteries are desirable power supplies for low-noise amplifiers for several reasons. The battery can be located with the amplifier in a shielded case to avoid pickup. An isolated battery power supply is less likely to have ground loops and 60-cycle pickup.

In general, batteries are not noise sources. Since a battery has current flowing and an internal impedance, there is some noise. A battery acts as a large capacitor and therefore shunts its own internal noise. Only when nearly exhausted does the noise rise. If the noise of a battery supply is a problem, bypass the supply or decouple the noise with a capacity multiplier. This also reduces the series impedance of the power supply and helps to prevent regeneration and motorboating in the amplifier.

9-5 NOISE EFFECTS OF COUPLING TRANSFORMERS

Transformers are used in several types of low-level applications: data acquisition, data transmission, geophysical measurements, dc chopper amplifiers, bridges, and so forth. They perform one or more of the following: isolation, impedance matching, noise matching, and common-mode rejection. Low-level transformers are used to discriminate against interference that often accompanies signal voltages. These interfering signals come from stray magnetic fields, ground loops, common-mode signals, and machine-made noise.

Types of magnetic components used in low-level systems include input transformers, chopper input transformers, interstage transformers, output transformers, filter reactors, and low-pass filters. Our particular interest in this section is with input transformers and chopper transformers. An input transformer is used to match the amplifier noise characteristics to the sensor impedance. Thus it is possible to design an amplifier for operation at its optimum noise factor F_{opt} and then transform the sensor impedance to look like R_o. This transformer noise matching is discussed in Chapter 7.

There are several second-order effects that can add noise to a low-level transformer circuit. Some of the considerations necessary in the proper application of transformers are common mode rejection, magnetic shielding, primary inductance, frequency response, and microphonism. The following sections discuss each of these effects.

Electrostatic Interwinding Shielding

All instrumentation transformers incorporate electrostatic Faraday shields between the windings for isolation and common-mode rejection. This is illustrated in Fig. 9-7. This interwinding shield may consist of a copper foil interleaved between the layers, or it may be a complete box-type construction as shown in Fig. 9-7*a*. Consider that a common-mode voltage between the sensor and ground is present at the primary terminals of the input transformer. Any capacitive coupling between the primary and secondary windings couples part of the common-mode voltage to the secondary. The interwinding Faraday shield terminates the electrostatic field from the primary. The more the shield encloses the primary winding, the better the common-mode rejection.

The effectiveness of the electrostatic shielding can be tested. Refer to Fig. 9-7*b*. A common-mode voltage V_{cm} is connected between primary and ground. The secondary voltage V_s is measured. Any coupling impedance, X_{CM}, is considered to be an equivalent primary-to-secondary capacitance. Therefore, the equation $V_s/V_{cm} \simeq R_L/X_{\mathrm{CM}}$ can be solved for the interwinding capacitance. Typical capacitances can be as high as 10 pF and as low as 5×10^{-7} pF.

In addition to low-interwinding capacity, common-mode rejection requires a balanced primary with symmetrical capacitances. If there is unbalanced shunt capacity between the center tap and the two ends of the primary, a common-mode voltage injected in the center tap causes unbalanced voltages

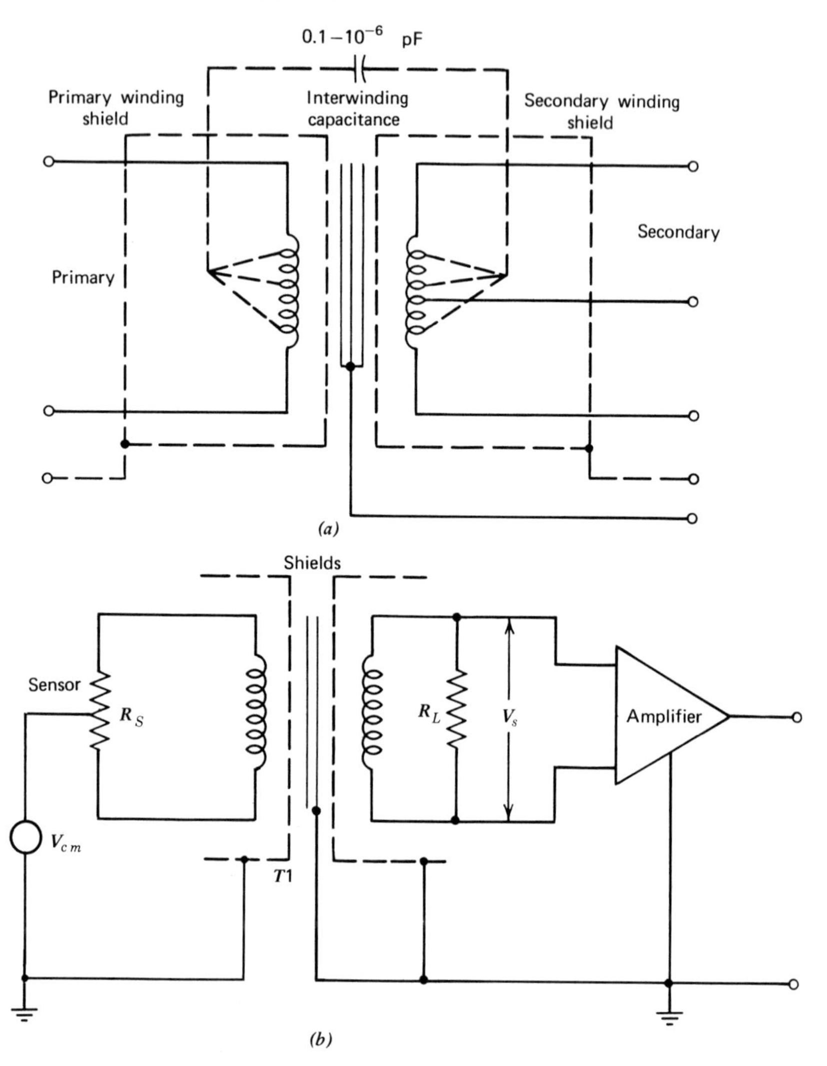

Fig. 9-7. Winding shields in instrumentation transformers.

in the primary winding. This, then, is a true differential mode signal. The common-mode rejection of a good transformer is difficult to measure and even more difficult to specify. There are several ways in which it can be measured. A circuit for measuring common-mode rejection is shown in the James Electronics Instrument Transformer catalog, and is discussed in the literature [8].

Magnetic Shielding

For low-level use, a transformer must be shielded from externally caused magnetic fields. Magnetic pickup also can be reduced by coil construction techniques. External fields are rejected by the "hum-bucking" winding technique. "Hum-bucking" implies that the transformer has two secondary windings oppositely wound on the core. An externally generated magnetic field induces equal and opposite voltages in the windings. This is particularly effective when the field is distant and the transformer can be spacially oriented.

Magnetic shielding is accomplished by surrounding the transformer with multiple layers of mumetal interleaved with heavy copper layers. Multiple layers provide higher attenuation than a single layer of equal thickness. Magnetic shielding provides a low reluctance path around the transformer core for the interfering signal. Any fields that penetrate the first magnetic layer induce eddy currents in the copper layer beneath it. These currents generate a reverse magnetic field to oppose the interfering field. Multiple layers can be noted in the expanded view of a transformer given in Fig. 9-8.

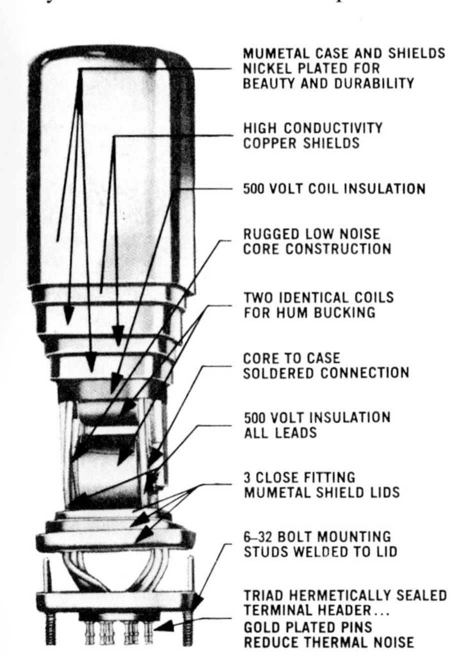

Fig. 9-8. Exploded view of well-shielded instrumentation transformer (courtesy of Triad, a division of Litton Industries).

Magnetic-shielding effectiveness is the ratio of the pickup in a transformer with and without shielding. It can be measured by placing the transformer between two large Helmholtz coils. The coils are driven with an ac that sets up a known magnetic field intensity between them. This test is described in MIL-T-27 [9].

Transformer Primary Inductance

Transformer primary winding inductance is important because it determines the low-frequency response of the transformer, and it increases the effect of the equivalent noise voltage of the amplifier. The low-frequency cutoff occurs when the primary impedance equals the parallel resistance of the source and reflected load resistance. Below this frequency the equivalent noise voltage generator of the amplifier is increased as shown in Eq. 8-22.

Primary inductance varies with frequency and measuring voltage. At very low levels of flux density (low-measuring voltage or high frequency), core permeability and therefore inductance are low valued. In low-noise applications the signal levels are extremely small, and the core has its lowest permeability. Since inductance is normally measured at a higher level, the *actual* primary inductance is smaller than indicated on the spec sheets. This effect is illustrated by the data on three James Electronics transformers given in Table 9-1 [10].

Table 9-1. Inductance at Three Levels

Transformer Series	2000-Gauss Relative Inductance	200-Gauss Relative Inductance	40-Gauss Relative Inductance
"8100"	1	0.83	0.72
"8300"	1	0.81	0.69
"8700"	1	0.75	0.62

To measure the inductance of a low level transformer accurately, use a low-frequency impedance bridge with very small drive signals. Measure the inductance at the lowest frequency used. Inductance measurement is discussed in the referenced article [8].

It is difficult to control the primary inductance of a transformer. The number of turns can be held constant, but the core permeability and stacking efficiency of the laminations vary from unit to unit. Permeability can vary over a 4:1 range. The permeability of the core is more constant if an air gap is introduced, but this increases the size of the transformer and makes it more susceptible to magnetic pickup. Because of permeability tolerances, the minimum acceptable inductance must be determined. One of the differences between transformer manufacturers is how well they handle their stacking factors. If they stack tighter and increase the permeability of the core, the transformer has greater inductance or smaller size. This reduces the cost as well.

Frequency Response

The low-frequency response of a transformer is determined by the shunting effect of its primary inductance. The high-frequency cutoff is determined by the shunting effect of its winding capacitance. The widest frequency response is obtained when the primary and secondary inductances are terminated in their nominal impedances. A transformer can be used, of course, at other than its nominal impedances. Usually, low-frequency response can be improved by lowering the source impedance below nominal. The converse also applies.

In general, the equivalent input noise increases when operating outside the pass band. Beyond roll-off, amplifier noise remains constant while the signal passed by the transformer falls off; therefore the equivalent input noise increases. In addition, core losses and transformer noise increase with frequency.

Microphonics and Shock Sensitivity

High-inductance transformers generate extraneous voltages when shocked or vibrated. This can be caused by change in the permeability of the core in the presence of the internal or an external field, or by a change in coupling with the internal or an external magnetic field. If a transformer is vibrated, the laminations may shift in position or the core may be stressed. In either case, the permeability of the core is modulated. If there is a magnetic field present, the changing inductance generates a voltage. With very high inductance windings, just squeezing the transformer with your fingers generates a voltage.

An external dc magnetic field is also a problem. A power supply choke is one source of such a field. The earth's field is another. If the transformer vibrates in the field, or if the field source vibrates, a voltage can be generated directly in the coil or indirectly by inducing a voltage in the shield and then into the coil. Motion of a magnetized shield also induces a voltage in the transformer windings.

Laminations and shielding cans should be demagnetized by the manufacturer before assembly in order to minimize the residual flux in those parts of the transformer assembly. Once it is assembled it is difficult for the user to degauss a transformer adequately. Inadvertent magnetization can result from the passing of a dc bias current through a winding or from measuring coil continuity with an ohmmeter. A small dc in the windings, such as a transistor base current, can make the transformer more microphonic. If a transformer is inadvertently magnetized, it can be degaussed by passing low-frequency ac through the windings and slowly decreasing the amplitude of this wave to zero.

To minimize shock sensitivity the following can be done:

1. Have the core vacuum-impregnated so that the laminations cannot move with respect to one another.
2. If space permits, acoustically isolate the coil and laminations from the shields and the shields from each other.
3. Specify that the core and shields are to be demagnetized before assembly.
4. Shock mount the transformer on the chassis.
5. Shock mount any other magnetic field sources such as power transformers or chokes.
6. DO NOT check the dc continuity of any of the transformer windings unless they can be adequately demagnetized.

One last thought on transformers. In a circuit formed of two wires of different metals, if a temperature difference exists between junctions, a direct voltage is generated. This thermoelectric effect is a problem in chopper transformers. The thermally generated voltage (thermal EMF) can add or subtract from the dc signal being modulated and cause an error. Therefore locate the transformers away from any heat sources.

SUMMARY

a. In addition to thermal noise, resistors exhibit excess noise when dc is present.

b. Excess noise can be minimized by proper selection of the resistor manufacturing process. Metal film units are usually superior.

c. Noise index is given in units of μV/V/decade, or

$$NI = 20 \log \left(\frac{E_{ex}}{V_{DC}}\right) \quad \text{dB}$$

over the frequency decade used to determine E_{ex}.

d. Capacitors usually do not present a noise problem because their own C shunts the internal noise. Electrolytics with high leakage can be troublesome.

e. Zener diodes are low-noise devices. Avalanche-breakdown devices can be noisy, and selected units may have to be used if the application is critical.

f. Breakdown diodes do not make good coupling or biasing elements in a low-noise system.

g. Batteries are not a source of noise except when nearly exhausted.

h. Coupling transformers should have electrostatic interwinding shielding and magnetic shielding.

i. To control microphonics in interstage transformers, manufacturing techniques and handling and inspection must be tightly controlled.

PROBLEMS

1. Verify the five bandwidth correction factors shown in Fig. 9-2.

2. Determine the μV of resistor excess noise in a frequency decade for NI = 0 dB, −10 dB, and −20 dB. Consider that V_{DC} = 6 V, and R = 10 kΩ. How much thermal noise is generated in this element at 300°K over the 1 to 10 kHz decade? Determine the total noise voltage, excess and thermal, in this element for NI = −20 dB.

3. A noise test on a resistor results in the following data: R = 100 kΩ; total noise voltage, including thermal, is 600 nV for a noise bandwidth of 1 to 100 Hz; V_{DC} = 4 V. Determine the noise index in dB, and determine the noise index in μV/V/Hz at 100 Hz.

4. A low-noise wave analyzer is being designed for operation between 10 Hz and 50 kHz. A filter within the analyzer has a noise bandwidth of 3 Hz. The filter contains a 1-MΩ metal-film resistor. Estimate the total noise voltage (thermal plus excess) caused by the resistor. Use worst case philosophy and V_{DC} = 6 V.

5. A coupling transformer has the following characteristics: primary inductance 10 H, primary winding resistance 100 Ω, secondary winding resistance 10 Ω, and nominal impedance ratio 10,000:1000. The transformer is to be used to couple a 10,000-Ω resistive sensor to an amplifier with input resistance of 1200 Ω.

(a) At midfrequency determine the voltage gain between sensor (V_s) and the amplifier input terminal.

(b) Calculate the lower cutoff frequency of the sensor-amplifier system caused by transformer primary inductance.

(c) At f = 200 Hz determine the impedance seen through looking into the primary terminals of the transformer when connected in the system.

(d) The primary winding resistance is 100 Ω and the resistance of the secondary winding is 10 Ω. Estimate the ratio of power lost in these windings to the total signal power delivered to the amplifier. What other effect causes loss in a transformer?

REFERENCES

1. Conrad, Jr., G. T., "A Proposed Current-Noise Index for Composition Resistors," *IRE Trans. Component Parts*, **CP-3**, 1 (March 1956), 14–20.

2. Conrad, Jr., G. T., N. Newman, and A. P. Stansbury, "A Recommended Standard Resistor-Noise Test System," *IRE Trans. Component Parts*, **CP-7**, 3 (September 1960), 1–4.
3. Stansbury, Alan, "Measuring Resistor Current Noise," *Elec. Equipment Eng.*, **9** 6 (June 1961), 11–14.
4. Curtis, J. G., "Current Noise Tests Indicate Resistor Quality," *Int. Elec.*, **7**, 2 (May 1962).
5. Berry, R. W., P. M. Hall, and M. J. Harris, *Thin Film Technology*, Van Nostrand Reinhold, New York, 1968, p. 336.
6. Burks, D. P., "Thick Film Materials and Their Capabilities," *1969 IEEE International Convention Digest*, pp. 180–181.
7. Thomas, H. E., *Handbook of Integrated Circuits*, Prentice-Hall, Englewood Cliffs, N.J., 1971, p. 125.
8. Sommer, B. I., and G. W. Plice, "Specification and Testing of Shielded Transformers," *Electro-Technology*, **71**, 5 (May 1963), 102–105.
9. U.S. Naval Supply Depot, 5801 Tabor Avenue, Philadelphia, Pa. 19120.
10. "Instrument Transformers by James," James Electronics, Inc., Chicago, Ill.

Chapter 10

BIASING WITHOUT NOISE

From the discussion in Chapters 3 and 8 we are aware of the noise behavior of sensors. The noise behavior of transistors and other components was discussed in Chapters 4, 5, 6, and 9. When these devices are assembled to form a low-noise system, we strive to minimize the noise contribution of each element. Resistors used for biasing the low-signal-level transistor stage are in a particularly sensitive position in the system. The thermal noise and excess noise that they add to the system must be carefully observed and controlled.

We desire that biasing contribute no noise. This is asking a great deal from any electronic circuit. However, we find that the optimum can be approximated if we are willing to add a few additional passive components to the first amplifying stage.

This chapter includes an element-by-element discussion of low-noise biasing.

10-1 THE BIPOLAR TRANSISTOR STAGE

A transistor can be connected to operate in one of the three useful configurations: common emitter (CE), common base (CB), and common collector (CC). If electrical noise were the sole consideration, all signal input stages would be connected CE, for that configuration provides the maximum power gain. As discussed in Chapter 2, noise contributions of stages beyond the first have virtually no effect on system NF when the input stage provides high gain.

Considerations other than noise sometimes predominate, and therefore CC and CB stages are used for input stages when impedance, bandwidth, and stability are of paramount importance. It is shown in Chapter 12 that NF is independent of configuration.

It is also true that biasing of the bipolar transistor is independent of configuration. The requirements are that the emitter-base diode be forward biased, and the collector-base diode be reverse biased. Consequently, we consider only the more widely used CE connection for our discussions here. The biasing of a stage and the amplification it provides are interrelated; let us take a closer look at these two parts of the transistor picture before returning to noise considerations.

The voltage amplification provided by any device can be expressed in terms of its current gain A_i, its input resistance R_i (or input impedance), and the effective load R_L that the device feeds. Thus

$$A_v = \frac{V_o}{V_i} = -\frac{A_i R_L}{R_i} \tag{10-1a}$$

The minus sign appears in Eq. 10-1 to represent phase reversal present in the CE connection. Equation 10-1 is not practical for tube and FET amplifiers because they are not current amplifiers and therefore do not have a relevant A_i. For these devices, a widely used approximation is

$$A_v \simeq -g_m R_L \tag{10-1b}$$

Base-biasing elements are not shown in the simple CE stage of Fig. 10-1*a*. They are considered later. The low-frequency hybrid-π equivalent circuit for this stage is given in Fig. 10-1*b*. Note the load resistance R_C and the local feedback element R_e. When used, R_e provides negative feedback that stabilizes

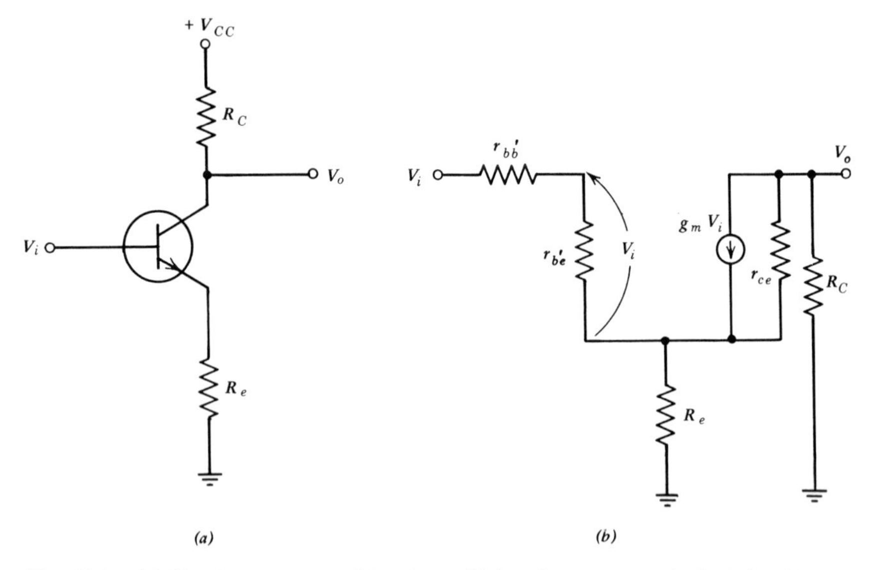

Fig. 10-1. (*a*) Simple common-emitter stage; (*b*) low-frequency equivalent circuit.

the gain of the stage against variations in parameters, and that element raises the input resistance of the stage well above the value of R_i obtained in the absence of feedback.

As long as R_C is smaller in value than r_{ce}, the current gain of the transistor CE stage is

$$A_i \simeq \beta = g_m r_{b'e} \tag{10-2}$$

The input resistance is

$$R_i = r_{bb'} + r_{b'e} + (1 + \beta)R_e \tag{10-3}$$

The $(1 + \beta)$ multiplier occurs because the current through R_e is $I_b + I_c$, and since $I_c = \beta I_b$, that current is $(1 + \beta)I_b$.

For simplicity in these equations we consider that r_{ce} is connected in parallel with R_C. The effective load on the stage is

$$R_L = R_C \| r_{ce} \tag{10-4}$$

Therefore, Eq. 10-1 can be expressed in terms of transistor parameters and fixed resistances:

$$A_v \simeq -\frac{g_m r_{b'e} R_L}{r_{bb'} + R_e + r_{b'e}(1 + g_m R_e)} \tag{10-5}$$

Transconductance g_m is considered to vary linearly with direct collector current; as noted in Section 4-4, $g_m = \Lambda I_C$. Element $r_{b'e}$ is commonly represented in terms of β and g_m: $r_{b'e} = \beta/g_m$.

As an example consider a stage of the Fig. 10-1 variety with $I_C = 100\ \mu\text{A}$, $\beta = 100$, $r_{bb'} \simeq 0 = R_e$, $R_C = 10\ \text{k}\Omega$, and $r_{ce} = 40\ \text{k}\Omega$. We wish to calculate the voltage gain. From Eq. 10-4 $R_L = 8\ \text{k}\Omega$. We find that $g_m = 0.004$ mho and $r_{b'e} = 25\ \text{k}\Omega$. The voltage gain from Eq. 10-5 is 32, and $R_i = 25\ \text{k}\Omega$ from Eq. 10-3.

10-2 A PRACTICAL CE STAGE

The addition of base-biasing elements R_A and R_B in Fig. 10-2 can reduce the gain of the CE stage and decrease R_i from the values predicted in the preceding section. We shall investigate the effect of those resistors, and of coupling capacitor C_C, source resistance R_s, and bypassed emitter resistance R_E in this section.

For stabilization of the operating point in the context of unit-to-unit variations in β_{dc} and in the voltage drop V_{BE}, dc feedback is often employed in CE stages. Emitter resistance R_E provides for such stability, and helps to

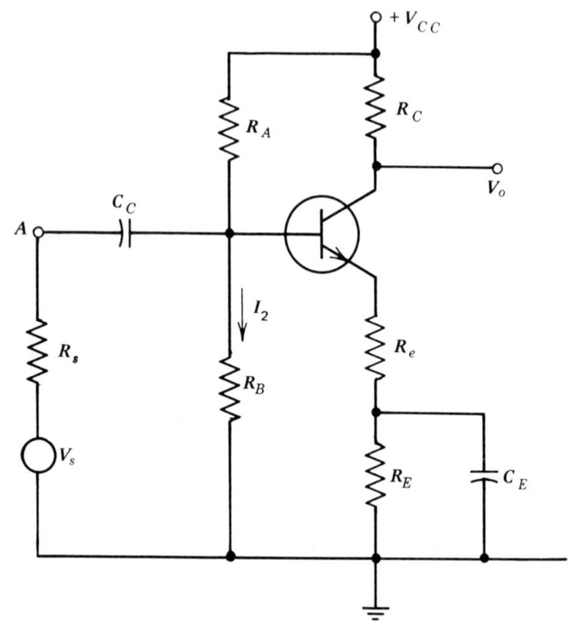

Fig. 10-2. Biased CE stage.

overcome Q-point drifting caused by temperature effects on the transistor, as well as by manufacturing tolerances. If, for any reason, the direct emitter current were to increase, the $I_E R_E$ drop is of such a polarity that it would tend to lower the available base-emitter direct voltage and consequently reduce the emitter current. (This is most easily understood by considering the voltages $I_E R_E$, $I_2 R_B$, and V_{BE} around the loop in Fig. 10-2.) We consider that the reactance of C_E is so small that R_E is successfully bypassed and no ac feedback results from the presence of that element.

Refer again to Fig. 10-2. To the right of point A, and for the moment neglecting C_C, the input resistance of the transistor stage is no longer as given by Eq. 10-3. Now it is

$$R_i' = R_A \| R_B \| R_i \tag{10-6}$$

If a large input resistance is required, elements R_A and R_B must be high-valued resistors. However, R_i' can never be greater than R_i.

When discussing circuit gain it is imperative that a clear description be given of the two points whose voltages are being compared. The addition of R_A and R_B does not change the collector-to-base voltage gain as calculated earlier, but it can seriously affect the gain between output and signal source,

V_o/V_s. That ratio is

$$K_t = \frac{V_o}{V_s} = A_v\left[\frac{R_i'}{R_s + R_i'}\right] \tag{10-7}$$

Note that the multiplier in brackets is always less than unity.

When the reactance of C_C is not negligible, at low frequencies,

$$K_t = A_v\left[\frac{R_i'}{R_s + R_i' - jX_C}\right] \tag{10-8}$$

where $X_C = 1/\omega C_C$. The lower cutoff frequency f_l is the value of f that results in equal real and imaginary parts of the denominator. From Eq. 10-8 we find

$$f_l = \frac{1}{2\pi(R_s + R_i')C_C} \tag{10-9}$$

The lower cutoff frequency is also dependent on C_E. The reader is referred to Section 10-7.

10-3 DESIGNING THE CE STAGE

The process of designing a low-noise transistor amplifying stage involves selection of the transistor type, determination of the operating point, and calculation of the circuit resistances. The transistor type and operating point follow from the discussion given in Chapter 4. Here we concentrate on the other decisions to be made during a design.

Input resistance can be an important specification; we note that there are four contributors to its value. As mentioned, R_A and R_B parallel R_i of the transistor. Local feedback element R_e and external or overall feedback can be used to increase input resistance. Without using feedback, the major factor influencing input resistance is direct collector current. We have seen that $r_{b'e} = \beta/g_m = \beta/\Lambda I_C$. For $\beta = 100$, the expected values of $r_{b'e}$ are 2.5, 25, and 250 kΩ for $I_C = 1$ mA, 100 μA, 10 μA, respectively.

The contribution of R_e to input resistance is $(1 + \beta)R_e$, but *in order to double input resistance using local feedback, one-half of the voltage gain of the stage is sacrificed. Unbypassed emitter resistance is a significant source of noise.* The use of R_e cannot be justified in most cases because it becomes the dominant source of noise. The increase in input resistance or gain stability that results from the local feedback is not of sufficient value to warrant the inclusion of this additional noise source. Where higher input resistance is mandatory, overall negative feedback returned to a very small R_e is possible. A FET may be the best solution.

The voltage gain provided by a transistor stage is directly proportional to the value of the total effective load resistance R_L. That load is the result of three parallel effects: the output resistance of the transistor r_{ce}, the load resistance R_C, and the input resistance of the second stage. One design problem is to determine R_C in order to maximize gain. As a start, we could allot two-thirds of V_{CC} to the dc drop across R_C. Since I_C is known, it follows that $R_C = 2V_{CC}/3I_C$. The remaining one third of V_{CC} is available for $V_{CE} + I_C(R_e + R_E)$. If R_e and R_E are not used, the value selected for R_C may be larger.

To provide a stable operating point it is necessary for R_E to be as large as possible and the parallel combination of R_A and R_B to be as small as possible. Elements R_B and R_E can be eliminated from a design if the resulting operating point excursions caused by temperature and manufacturing tolerances are acceptable.

A powerful worst case design method is considered in the literature [1]. Our calculations in this discussion are somewhat simplified from that presentation; yet the results are acceptable for many designs. Refer to Fig. 10-3.

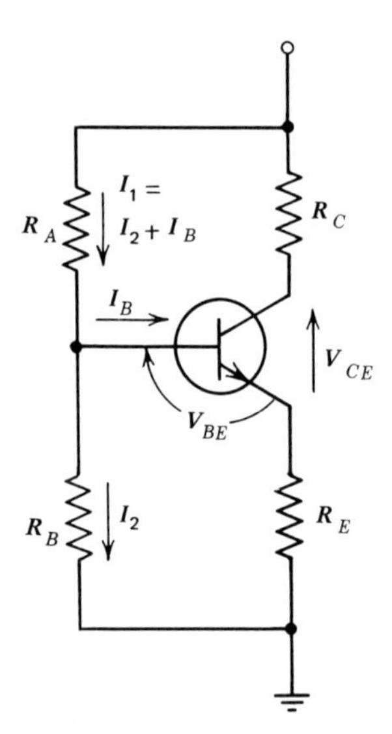

Fig. 10-3. Dc equivalent circuit for biased CE stage.

We use the criterion that the current I_2 that flows through R_B is four or more times the nominal base current of the transistor. In this manner, changes in the base current do not seriously affect the voltage divider function of R_A and R_B. For the drop from base to emitter, 0.6 or 0.7 V is common for silicon transistors and 0.2 V is typical for germanium devices.

The Kirchhoff equations for the circuit of Fig. 10-3 are

$$\begin{aligned} V_{CC} &= (I_2 + I_B)R_A + I_2R_B \\ I_2R_B &= V_{BE} + I_ER_E \end{aligned} \tag{10-10}$$

It has been noted that two-thirds of V_{CC} will be alloted to the drop across R_C. This leaves one third of V_{CC} for $V_{CE} + I_ER_E$. After selection of a value for the V_{CE} coordinate of the Q point, R_E is determined from

$$R_E = \frac{V_{CC}/3 - V_{CE}}{I_C}$$

Note that $I_E \simeq I_C$, and I_C is determined from noise data. Since I_C and β_{dc} are known, $I_B = I_C/\beta_{dc}$. Also, we make $I_2 = KI_B$ with $K \geq 4$. The second of Eqs. 10-10 can be solved for R_B. With this value for R_B we can enter the first of Eqs. 10-10 to calculate R_A.

Example

SPECIFICATIONS. A stage of the Fig. 10-2 type is to be designed. The following specifications apply:

Transistor: low-noise *npn* silicon, $\beta = 150 = \beta_{dc}$ at the selected operating point.

Power supply: $V_{CC} = 10$ V.

Operating point: $I_C = 100\ \mu$A, $V_{CE} = 2$ V.

SOLUTION. If we use the two-thirds rule, R_C is

$$R_C = \frac{(2/3)10}{100 \times 10^{-6}} = 67\ \text{k}\Omega$$

The voltage across R_E is $10 - 6.7 - 2 = 1.3$ V. Therefore

$$R_E = \frac{1.3}{100 \times 10^{-6}} = 13\ \text{k}\Omega$$

This must be bypassed. From Eq. 10-10 we find $I_2R_B = 1.9$ V. The nominal $I_B = 100 \times 10^{-6}/150 = 0.7\ \mu$A. We make $I_2 = 7\ \mu$A. Then

$$R_B = \frac{1.9}{7 \times 10^{-6}} = 271{,}000\ \Omega$$

and from Eq. 10-10 we obtain R_A:

$$R_A = \frac{8.1}{7.7 \times 10^{-5}} = 1{,}050{,}000\ \Omega$$

The maximum collector-to-base voltage gain of this stage can be calculated from Eq. 10-1:

$$A_v = \frac{-150(67000)}{37500} = -268$$

Laboratory measurements of the gain of this stage may yield as little as one-half of the predicted value. The major reason for the difference lies in the fact that the calculations omitted element r_{ce}. If that element is 67 kΩ, the gain is cut in half.

10-4 NOISE IN THE CE STAGE

The thermal noise contribution of the collector return resistor R_C is not significant if the stage has high gain. Low-frequency excess noise is more likely to be troublesome. A metal-film resistor can be used if the noise problem is considered serious.

When an unbypassed emitter resistor R_e is used to increase amplifier input resistance and gain stabilize the stage, the noise contribution of that resistor must be considered. The noise of R_e is effectively in series with the signal source and base resistance. In circuits that use an emitter resistor, *the value of the resistor should be low compared to the resistance of the source*. When R_s is very small or zero, $R_e \ll r_{bb'}$ for best results. If excess noise is a problem, a metal-film unit can be used.

Resistors R_A and R_B shunt the signal source. *For low noise these elements should be much larger than the source resistance*. On the other hand, they should be low valued for good bias stability. A compromise may be necessary. Low-noise metal film resistors are recommended for R_A and R_B.

Let us calculate the excess noise contributions of R_A and R_B. Consider the following data:

Worse case NI $= -20$ dB at 1 kHz; $R_s = 10^4\ \Omega$.

Evaluate noise at 10 Hz for $R_A = 1.05$ MΩ and $R_B = 271$ kΩ. The thermal noise generated in each element in a 1-Hz bandwidth is

$$E_t = 12.6 \text{ nV} \qquad \text{from } R_s$$

$$E_t = 130 \text{ nV} \qquad \text{from } R_A$$

$$E_t = 66 \text{ nV} \qquad \text{from } R_B$$

The thermal noise of R_A and R_B referred to the signal-source location is calculated by multiplying the thermal noise generator by R_s/R_A and R_s/R_B, respectively. This results in

$$E_{ts} = 12.6 \text{ nV} \qquad \text{from } R_s$$

$$E_{tA} = 1.2 \text{ nV} \qquad \text{from } R_A$$

$$E_{tB} = 2.4 \text{ nV} \qquad \text{from } R_B$$

Hence, for this circuit, the thermal noise contributions of R_A and R_B are small.

Excess noise, however, can be a problem. A NI of -20 dB implies low-noise resistors. It is equivalent to 0.0021 μV/V at 1 kHz. We multiply this figure by the square root of the frequency ratio (1000 Hz:10 Hz) to refer the values to 10 Hz. This yields 0.021 μV/V. Then multiply by the direct voltage drops across each element, in this case 8.1 and 1.9 V. The excess noise generated in each resistor is

$$E_x = (0.021)(8.1) = 0.17\ \mu\text{V} \qquad \text{from } R_A$$

$$E_x = (0.021)(1.9) = 0.04\ \mu\text{V} \qquad \text{from } R_B$$

The attenuation multipliers previously used refer these values to the source. Then, at the source,

$$E_{xA} = 1.6 \text{ nV} \qquad \text{from } R_A$$

$$E_{xB} = 1.5 \text{ nV} \qquad \text{from } R_B$$

These can be combined to give the total bias resistor thermal plus excess noise at the source:

$$(E_{xA}^2 + E_{xB}^2 + E_{tA}^2 + E_{tB}^2)^{1/2} = 3.5 \text{ nV}$$

When this figure is compared to the 12.6 nV of thermal source noise, we conclude that in this particular design with the set of resistor values selected, bias circuit noise is not a serious problem. *Can we do better*? This question is answered in the next section.

10-5 NOISELESS BIASING

The network shown in Fig. 10-4 contains an additional resistance R_D connected between the junction of R_A and R_B and the transistor base. It also requires an additional bypass capacitor C_B. This element is chosen so that its reactance at the lowest operating frequency is small compared to R_B. Thus C_B provides an ac ground that eliminates any noise generated in R_A and R_B from getting to the transistor. The only biasing resistor that contributes noise is R_D.

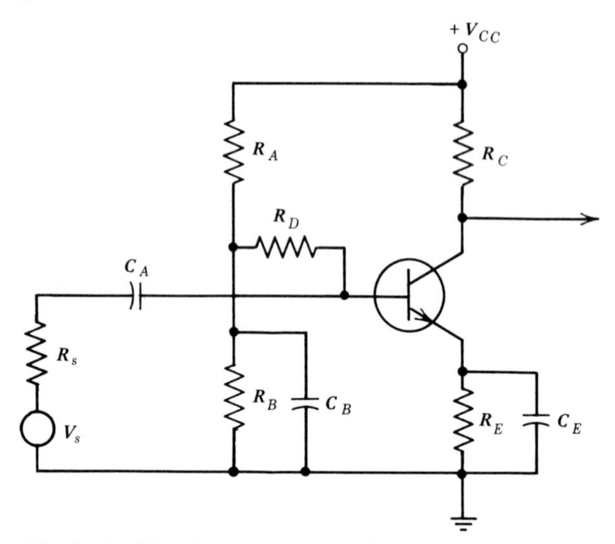

Fig. 10-4. Transistor stage with noiseless biasing.

The dc drop across R_D *need not be large.* Normally it would be less than 1 V, and therefore the excess noise is low. Because R_D parallels the source its thermal noise is important. However, as we have seen, attenuation of that noise is present: R_s/R_D.

In the selection of a value for R_D, we note that the thermal noise voltage generator increases in proportion to the square root of R_D, whereas the passive attenuation behaves as given in the preceding paragraph. It can be shown that *the maximum thermal noise contribution exists when* R_D *is low valued.* This condition should be avoided.

Example

SPECIFICATIONS. Design a stage of the Fig. 10-4 type. The same specifications as given in the Section 10-4 example apply:

Transistor: low-noise *npn* silicon, $\beta = 150 = \beta_{dc}$ at the selected operating point

Power supply: $V_{CC} = 10$ V.

Operating point: $I_C = 100\ \mu$A, $V_{CE} = 2$ V.

Source resistance: $R_s = 10$ kΩ.

SOLUTION. Again we select $R_C = 67$ kΩ and $R_E = 13$ kΩ. The direct voltage at the base terminal is again 1.9 V. To keep the excess noise in R_D low we allot only 0.1 V to the dc drop across that element. Since $I_B \simeq 0.7\ \mu A$, it follows that

$$R_D = \frac{0.1}{0.7 \times 10^{-6}} = 143\ k\Omega$$

The $R_A - R_B$ divider must provide 2 V at the junction of the three resistors. We assume that current drain from V_{CC} is not a problem, and accept 1 mA for the current through the divider. Because I_B is so much smaller than 1 mA it need not figure in the calculations. The result is

$$R_A = 8\ k\Omega, \qquad R_B = 2\ k\Omega$$

The collector-to-base voltage gain can be predicted as before. The measured gain for the noiseless-biased stage is found to be about 165, nearly equal to the result of the previous example. Dc collector current was measured to be 0.105 mA when resistance values as indicated were used.

10-6 EMITTER BIASING

It is clear from the preceding discussions that the most important sources of bias circuit noise are attributable to those elements located in the low-signal level portion of the network to the left of the transistor base terminal. A circuit that requires no specific bias elements in that location is the emitter-bias arrangement shown in Fig. 10-5*a*.

A path for dc base current is always necessary. However, when it is permissible to pass that current through the signal source, the emitter bias circuit has a definite advantage over other biasing methods. Note that dc power supplies of both polarities are required. Element C_E ac grounds the emitter terminal. The direct emitter current is approximately equal to V_{EE}/R_E, and the collector current is, as usual, approximately equal to I_E.

Many IC amplifiers use a form of emitter biasing. In the integrated operational amplifier or op amp shown in Fig. 10-5*b*, $Q3$ is a constant-current source supplying the emitters of $Q1$ and $Q2$ with constant and equal currents. As mentioned, a path for dc base current is required. In the figure the base current of $Q2$ flows through R_R to ground. Element R_R is selected to minimize dc offset in the op amp [2].

Impedances Z_F and Z_I determine the gain of the amplifying circuit. The overall gain is $-Z_F/Z_I$. If Z_F is resistive, a path exists for the $Q1$ base current to flow through that element to the output terminal of the op amp, a location that in normal usage is near dc ground. The base current path of $Q1$ could also be completed through Z_I and the signal source if continuity exists.

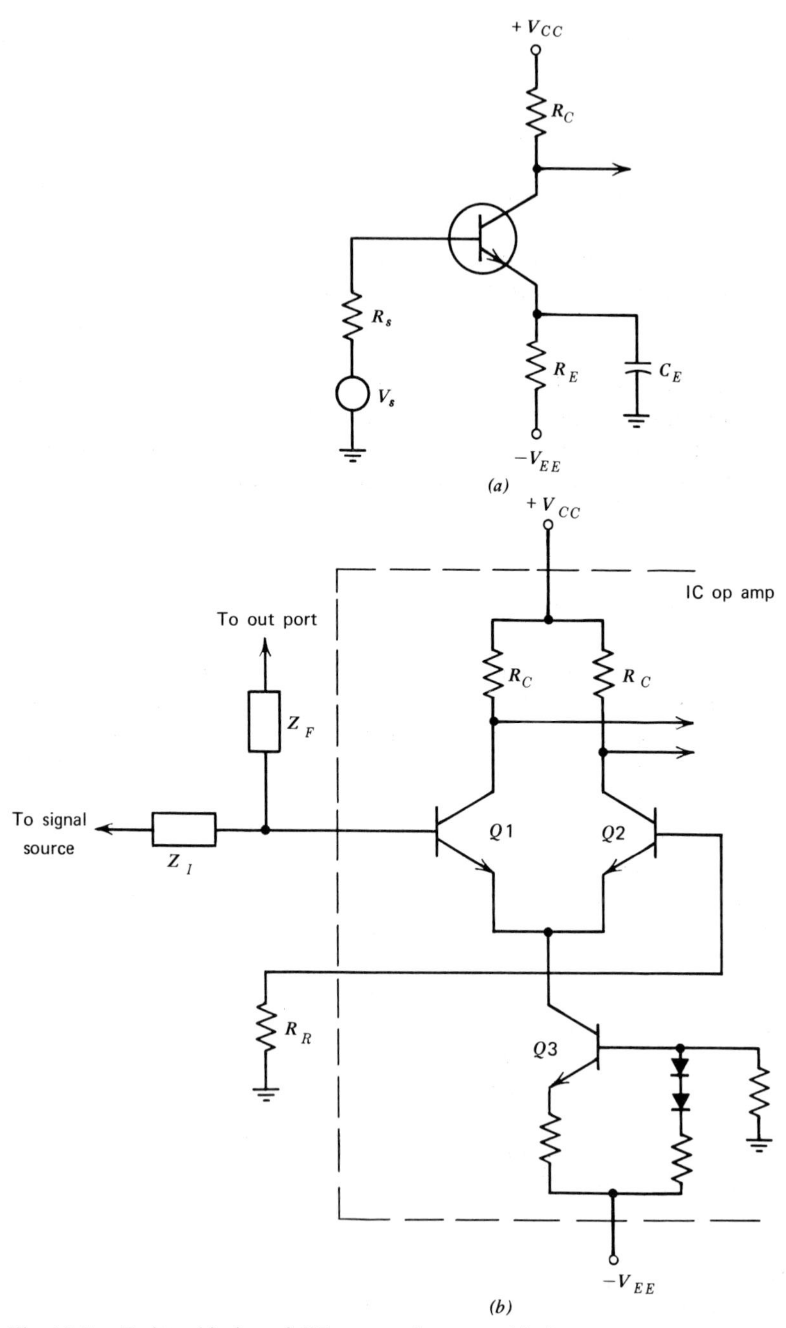

Fig. 10-5. Emitter biasing of CE stage and op amp biasing.

10-7 CAPACITOR SELECTION

The bypass and coupling capacitors shown in the noiseless-biasing network of Fig. 10-4 attenuate low-frequency signals and noise. The selection of values for these elements is the subject of this section.

Coupling or blocking capacitor C_A causes a low-frequency break in the response of the network at the value of f where its reactance is equal to the sum of R_s and the amplifier input resistance (including biasing elements).

The noise requirement for C_A *is more stringent;* X_{CA} *must be much less than* R_s *at the lowest frequency of interest.* Element C_A is in series with R_s, and therefore the noise term I_nR_s becomes $I_n|R_s + jX_{CA}|$. For low noise, $I_nX_{CA} \ll I_nR_s$ is necessary. If C_A sets the gain corner, I_nX_{CA} will be significant. The problem is accentuated by the fact that I_n may be increasing with decreasing frequency.

To remedy the problems noted in the preceding paragraph C_A often is chosen 10 to 100 times larger than the size suggested by the gain criteria. Thus it is recommended that the input coupling capacitance *not* be used as a part of the frequency shaping network.

The purpose of bypass capacitor C_B is to ac ground the node between R_A and R_B, thereby attenuating the noise generated in those resistances. We would select the reactance of C_B to be no greater than 0.1 R_D at the lowest frequency of interest.

The fractions of the noise voltages generated in R_A and R_B that appear across C_B are approximately $1/\omega R_AC_B$ and $1/\omega R_BC_B$. This noise across C_B is in series with resistance R_D. We desire that noise across C_B be as small as possible. It follows that

$$E_{tD}^2 \gg \frac{E_{tA}^2 + E_{xA}^2}{\omega^2 R_A^2 C_B^2} + \frac{E_{tB}^2 + E_{xB}^2}{\omega^2 R_B^2 C_B^2} \tag{10-11}$$

where E_{tD} is the thermal noise in R_D.

Element C_E is used to provide a low-impedance bypass to ac so that resistor R_E is effective only in the dc network. Considerable misunderstanding exists concerning the selection of a value for C_E. In order that the bypass be effective, the impedance of the $C_E - R_E$ network must be low compared to $(R_T + R_i + R_e)/\beta$, where R_T is the Thevenin equivalent of all resistance to the left of the base terminal, and R_i is the input resistance of the transistor alone. Additional discussion of this cause of gain fall-off is given in the literature [1].

To meet the noise specification C_E must effectively bypass the noise of R_E. Since the C_E-R_E network is effectively in series with the signal source, it is desired that

$$E_{ni}^2 \gg \frac{E_{tE}^2 + E_{xE}^2}{\omega^2 R_E^2 C_E^2} \tag{10-12}$$

where E_{tE} and E_{xE} are the thermal and excess noise voltages of R_E.

It is recommended that the low-frequency gain corner be set in a later stage rather than the input stage. In this manner 1/f noise originating in the input stage can be attenuated by the response-shaping network. When following this philosophy, capacitance C_E must be very large and must satisfy Eq. 10-12.

SUMMARY

a. In a conventionally biased transistor stage, base-biasing resistors contribute thermal noise and excess noise to E_{ni}.

b. To refer noise from a resistor R shunting the signal source to E_{ni} multiply the noise of that resistor by R_s/R.

c. Noise in local feedback resistor R_e can be important; use of this kind of feedback is not recommended.

d. The additional resistor and capacitor used in the noiseless bias network can eliminate virtually all noise contributed by the biasing elements.

e. Emitter biasing can be noise-free in certain applications.

f. The capacitor bypassing an emitter resistance should be chosen to minimize the noise contribution of that resistance.

g. Response shaping should usually be accomplished in stages beyond the input stage.

PROBLEMS

1. Use the small-signal equivalent circuit of Fig. 10-1*b* to derive Eq. 10-3 for input resistance.

2. Design a CE stage of the Fig. 10-2 type using a transistor with $\beta = \beta_{dc} = 150$ at $I_C = 50\ \mu A$. The input resistance presented to the sensor must be at least 20 kΩ. Consider $R_s = 20$ kΩ. Calculate the overall voltage gain of your design.

3. As discussed in Section 10-5, resistor noise voltages in R_A and R_B elements can be referred back to the signal source (V_s) terminal by using appropriate multipliers. Consider the circuit with $R_s \| R_A \| R_B \| R_I$, where R_I is the amplifier input resistance. Show that the multiplier for R_A is R_s/R_A, and for R_B it is R_s/R_B.

4. In a noiseless-biasing network, a resistance of 100 kΩ is used for R_D, and $R_A =$

$20\text{k}\Omega$, $R_B = 5\ \text{k}\Omega$. The signal source has thermal noise equivalent to $20\ \text{k}\Omega$. If R_D has a worst case NI $= -10$ dB, determine the thermal and excess noise of R_D referred to the signal source location, and calculate the total noise at that point from R_s and R_D. Consider that $R_E = 0$, $V_{CC} = 10$ V, and $f = 100$ Hz.

5. Design a noiseless-biased CE stage to operate at $I_C = 50\ \mu\text{A}$ and $V_{CE} = 2$ V. The available supply is $V_{CC} = 6$ V and $\beta = \beta_{\text{dc}} = 150$. The input resistance presented to the sensor must be at least $10\ \text{k}\Omega$. Calculate the overall voltage gain of your stage from sensor generator to transistor collector.

6. Select an input coupling capacitor on the basis of (a) frequency response and (b) noise. The input stage is a 2N4250 with $R_s = 100\ \Omega$, $I_C = 1$ mA, and $R_i \simeq r_{b'e} = 5200\ \Omega$. For both calculations consider the important frequency to be 10 Hz. (Use the criteria that $I_n X_C \simeq E_n/3$.)

REFERENCES

1. Fitchen, F. C., *Transistor Circuit Analysis and Design*, 2nd ed., Van Nostrand Reinhold, New York, 1966, pp. 76–80, 163.
2. Fitchen, F. C., *Electronic Integrated Circuits and Systems*, Van Nostrand Reinhold, New York, 1970, p. 196.

Chapter 11

POWER SUPPLIES FOR LOW-NOISE APPLICATIONS

In the design of a dc power source for low-noise electronics consideration must be given to the electrical noise, the ripple generated in the supply, and the common-mode voltages that may be coupled into the supply. If the supply is well filtered, noise is not a serious problem, for the means of successful ripple removal also attenuates noise coming from diodes and resistors in the supply. Common-mode voltages originating in the ac power lines can be isolated by decoupling and shielding.

After discussion of common-mode problems and ripple-attenuating circuits in this chapter a complete practical dc supply for low-noise applications is presented.

11-1 STRAY COMMON-MODE COUPLING

On most power lines there is a lot of noise and other garbage. In addition to these waveforms that can modulate and add to the power frequency, there also can be a common-mode voltage present between the line and ground. This voltage is represented as CMV_1 in the circuit of Fig. 11-1. There may be an additional smaller voltage, CMV_2, from the power line ground to earth ground. The low-voltage power supply transformer is $T1$. The common-mode signal CMV_1 is coupled into the circuit power supply by the unshielded transformer primary to secondary capacitance C_{PS}. Interwinding capacitance

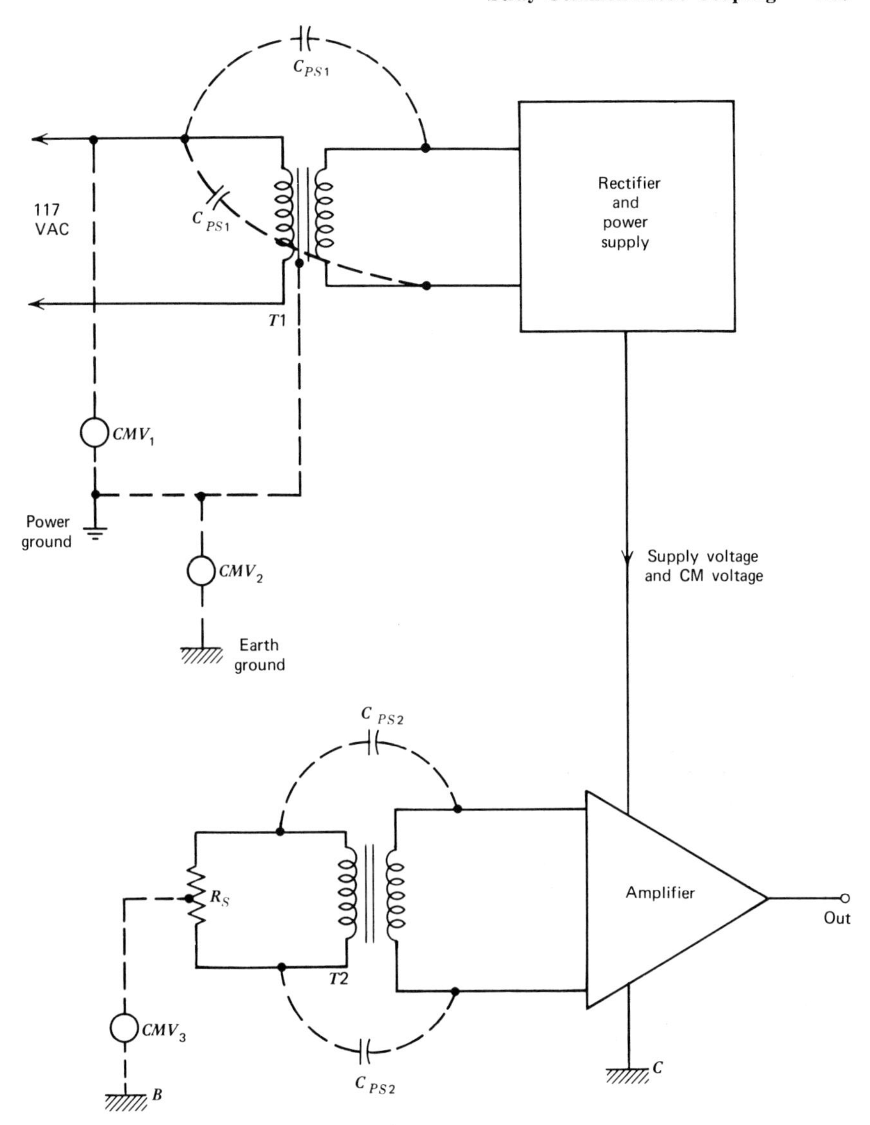

Fig. 11-1. Stray common-mode coupling.

brings the common-mode voltages into the enclosure that houses the circuit power supply and amplifiers. Typically, this noise is picked up by the amplifier and fed to the load as an extra noise mechanism.

There is another way for this common-mode signal CMV_1 to get into the amplifier. At some location the amplifier output is connected to ground or

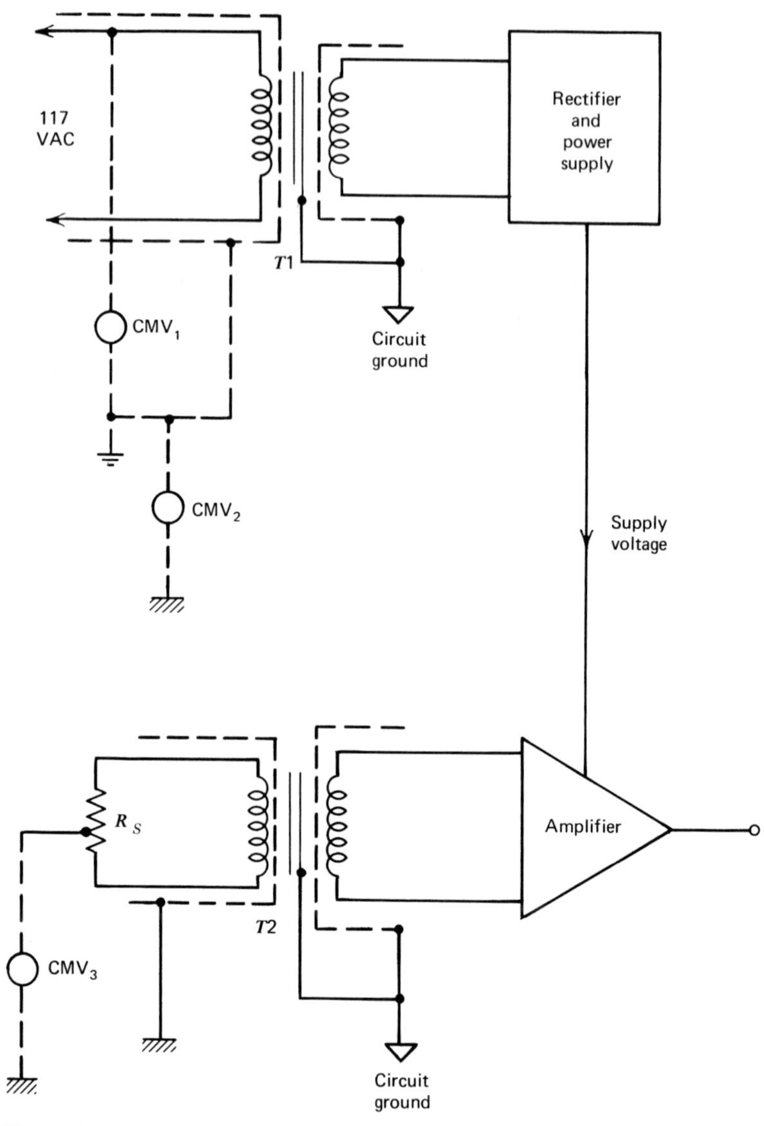

Fig. 11-2. Stray common-mode rejection.

earth. This may be after some additional stages, but eventually the circuit ground is connected to earth or power line ground as at point *C*. This provides two grounds in the system, one from CMV_1 or CMV_2 and the other through the amplifier, and we have a ground loop completed through the output ground connection. This loop provides the potential for a circulating ac through the circuit card and is an additional source of noise and line frequency pickup.

A very small amount of circulating current can cause a lot of pickup. As Ohm pointed out: "One microampere through 4 milliohms is 4 nanovolts," and that is equal to the noise of a 1000-Ω resistor.

A third source of pickup noise, CMV_3, is shown in the figure. This is a common-mode voltage between the signal source and earth. The CMV_3 voltage can be coupled through the interwinding capacity C_{PS2} of transformer $T2$. This network can also include a ground loop.

The question that arises is "how can we eliminate these sources of pickup?" A battery power supply would eliminate the coupling to the line, CMV_1 and CMV_2. To use an ac power supply we wish to break the ground loop. This can be done with a shielded power transformer as shown in Fig. 11-2. Transformers $T1$ and $T2$ now have interwinding Faraday electrostatic shields, as discussed in Section 9-5. The primary shield of transformer $T1$ decouples CMV_1 from the circuit power supply, and the secondary shield decouples CMV_2 from the electronic system. Similarly, in transformer $T2$, shields decouple common-mode voltage CMV_3 from the amplifier input. The level of decoupling is dependent on the effectiveness of the shielding. When using this type of shielded power transformer we do not wish to bring the ac supply line inside the case. The primary transformer leads are shielded and can be brought out to a bulkhead terminal or to a terminal strip external to the shielded enclosure of the dc power supply.

One good source of electrostatically shielded power transformers is the ELECTROGUARD series made by James Electronics of Chicago, Illinois. ELECTROGUARD is offered in standard models from 5 to 100 W with special types up to 2.5 kW. These units are very well shielded with an effective interwinding capacitance of 5×10^{-7} pF. The input transformer $T2$ is a low-level instrument transformer available from companies such as James Electronics, Triad, Southwest Industrial Electronics, and Stevens-Arnold.

11-2 POWER SUPPLY RIPPLE FILTERING

The power supply of a low-noise amplifier should be well filtered. This reduces the ripple and removes noise and interference, as well as harmonics. For low-noise operation in the nV region a ripple of less than 0.1 μV is generally required. A commercial low-ripple regulated power supply may not be the answer. As has been pointed out, the power transformer must have very good common-mode isolation. A highly regulated supply voltage is usually not needed for a low-noise ac amplifier. Theoretically, a well-regulated supply regulates all the ripple out, but this is an expensive way to buy filtering.

There are three principal methods of power supply filtering. The first is the straightforward *L-C* or *R-C* filter as illustrated in Fig. 11-3. A second is the capacity multiplier discussed in the next section and shown in Figs. 11-4,

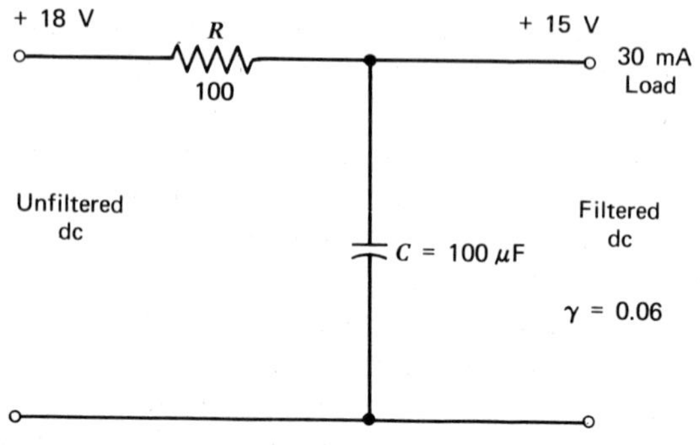

Fig. 11-3. *R-C* filter.

11-5, and 11-6. A third is the ripple clipper circuit shown in Fig. 11-7 and discussed in Section 11-5.

The filters discussed in this Chapter are fed in all cases from a full-wave rectifier. The ripple, then, is primarily a 120-Hz component when the power frequency is 60 Hz. If the peak value of the ac power waveform is designated as V_m, it can be shown that the rms value of the 120-Hz component for an unfiltered full-wave rectified waveform is

$$V_{(120)} = \frac{4V_m}{3\sqrt{2}\pi} \tag{11-1}$$

The average or dc value of that waveform is

$$V_{\rm DC} = \frac{2V_m}{\pi} \tag{11-2}$$

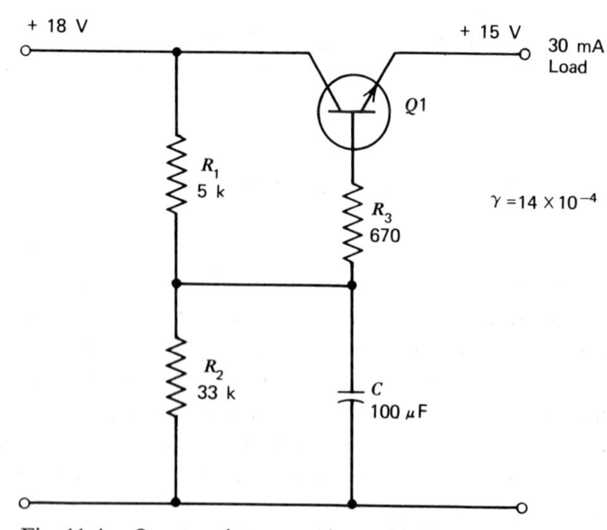

Fig. 11-4. One-transistor capacity multiplier.

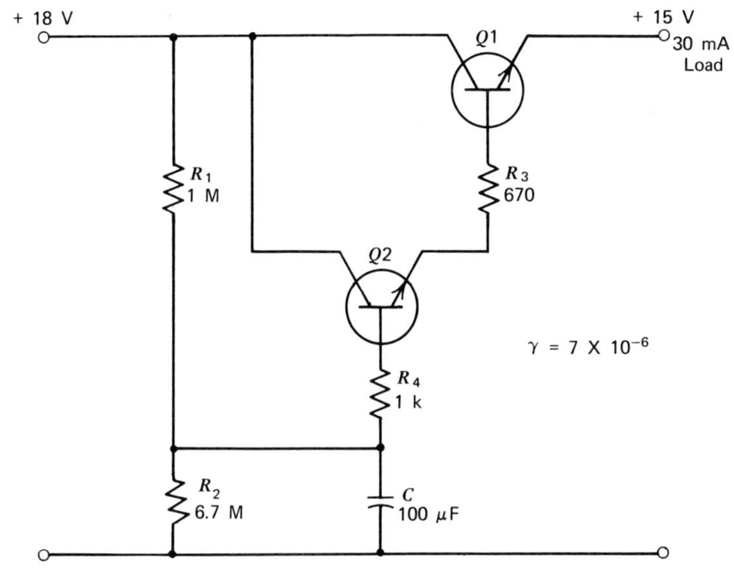

Fig. 11-5. Two-transistor capacity multiplier.

These values for $V_{(120)}$ and V_{DC} are present at the filter input port.

As the reference case consider the *R-C* filter shown in Fig. 11-3. To be effective the filter series resistance must be much larger than the shunt capacitive reactance at the ripple frequency ($R \gg 1/2\omega C$). Then the 120-Hz current is approximately $V_{(120)}/R$ or

$$I_{(120)} = \frac{4V_m}{3\sqrt{2}\pi R} \tag{11-3}$$

The rms value of the ripple voltage across C is $I_{(120)}X_C$ or

$$V_{c(120)} = \frac{\sqrt{2}V_m}{3\pi\omega RC} \tag{11-4}$$

The symbol ω is being used as the frequency of the fundamental or 60-Hz wave.

The ripple factor γ in current use is defined as the ratio of the rms value of the undesired harmonic to the dc level at the load. A perfect filter would have $\gamma = 0$. For the *R-C* filter discussed here,

$$\gamma = \frac{V_{c(120)}}{V_{\mathrm{DC}}} \tag{11-5}$$

Therefore, from Eqs. 11-4 and 11-2 we obtain

$$\gamma = \frac{1}{3\sqrt{2}\omega RC} \tag{11-6a}$$

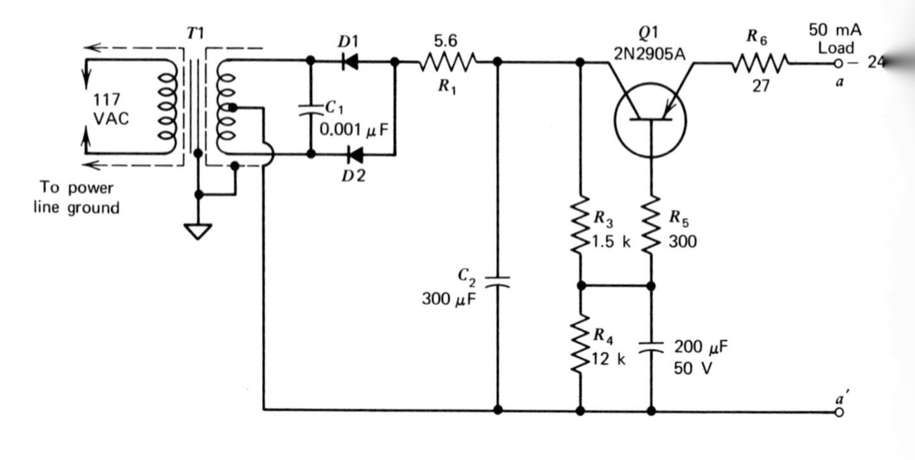

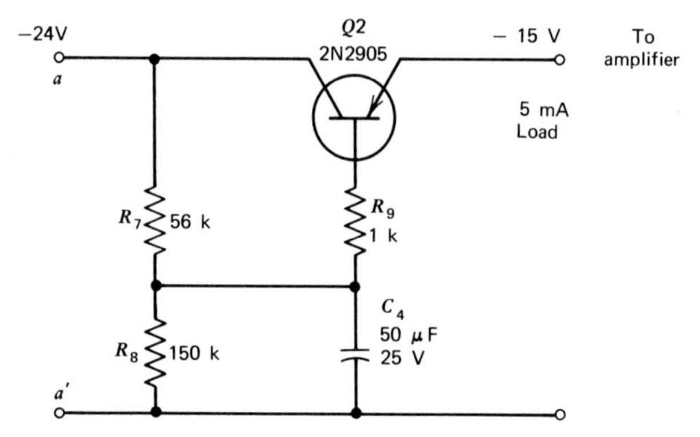

Fig. 11-6. Practical low-noise power supply.

Evaluation of the constant yields

$$\gamma = \frac{0.236}{\omega RC} \tag{11-6b}$$

Thus for second harmonic evaluation, $\gamma = 0.0006/RC$.

To illustrate relative filter effectiveness a 15-V, 30-mA power supply filter is to be developed in three different ways. We start with the *R-C* filter of Fig. 11-3. The ripple factor for $RC = 10^{-2}$ is 0.06 according to Eq. 11-6b. The 100-Ω series resistance in this case is rather high and results in a 3-V drop and poor load regulation. It is used for a direct comparison with the following cases. The 100-μF capacitor is the largest practical size for the application. This value is used in each of the three examples.

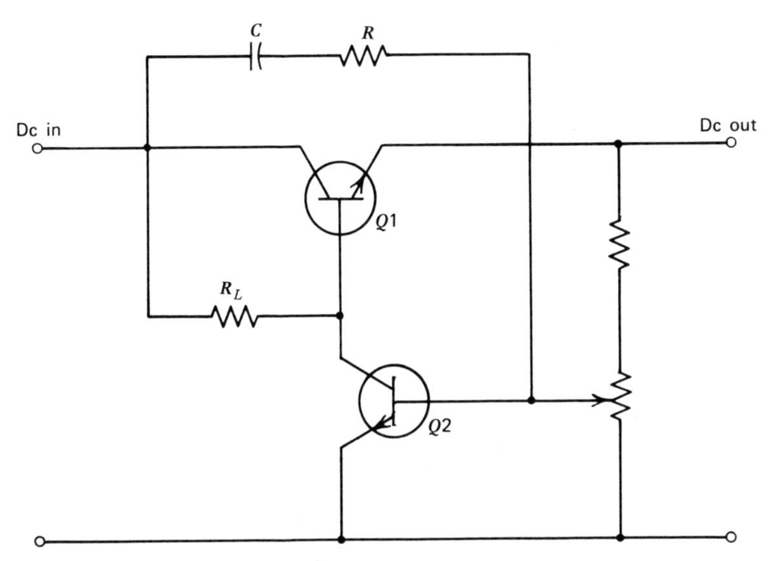

Fig. 11-7. Ripple clipper circuit.

11-3 CAPACITY MULTIPLIER FILTER

The effective capacitance of the 100-μF element can be increased by the capacity multiplier circuit shown in Fig. 11-4. The rectified waveform is filtered by the circuit composed of elements R_1, R_2, and C. It can be shown that the filtering action results from an R-C time constant that is equivalent to the parallel value of R_1 and R_2 times C. In this example $(33k \| 5k)C$ is 43 times larger than the corresponding value in the previously discussed R-C case. Now the value of ripple factor γ is 14×10^{-4}

The capacity multiplier circuit has the potential for increasing the filtering time constant by a value equal to the transistor β. This conclusion follows because the filtering takes place in base current, whereas the load current is the transistor emitter current, and $I_e = \beta I_b$. The presence of ripple at the collector of $Q1$ has little effect since I_e is essentially independent of that voltage.

The capacity multiplier has the potential of oscillating; resistance R_3 is added to spoil the Q and maintain stability.

If lower ripple is needed, a second transistor can be added as shown in Fig. 11-5. This can reduce the ripple by another factor of β. It is possible to use the extra transistor as a trade-off for a smaller capacitor. It may be less expensive to buy the second transistor than a large electrolytic capacitor. This two-transistor circuit is also useful when the supply must deliver a large load current. Transistor $Q1$ handles the current and $Q2$ provides the gain. When

$Q1$ is a low-β power transistor, $Q2$ can be selected to increase the current gain.

The filter factor is the parallel time constant of R_1, R_2, and C. For the circuit shown, $\gamma = 7 \times 10^{-6}$. The current in R_1 and R_2 should be three to four times the base current of $Q2$. Maximum filtering is obtained if $R_1 = R_2$. However, this would drop the output voltage to one-half the input, and the series pass transistor would dissipate more heat. Thus there is usually a compromise in the selected value of R_1. The best choice results in a 1- to 3-V drop across $Q1$ (and also R_1).

11-4 POWER SUPPLY EXAMPLE

The circuit diagram of a successful power supply for a low-noise amplifier is shown in Fig. 11-6. There are essentially three filter sections. They are the *R-C* network R_1-C_2, the capacity multiplier $Q1$, and the capacity multiplier $Q2$. Element C_2 charges to the peak of the secondary voltage and provides some filtering. The two cascaded capacity multipliers perform the main ripple attenuation: $Q1$ provides ripple attenuation at 120 Hz determined by R_3 and C_3; the ripple attenuation of $Q2$ is provided by $R_7 \| R_8$ and C_4.

There are definite advantages to using two separate ripple attenuators. The first attenuator is mounted on the power supply board and handles the current for ten amplifiers. The second ripple filter is mounted on each of the ten amplifier boards. In addition to ripple filtering, it attenuates noise picked up in the leads from the main power supply to the amplifier board. An additional by-product of this type of filtering is the low reverse coupling through the capacity multiplier that minimizes cross talk between the amplifiers and reduces the possibility of oscillation when other stages are connected to the same power supply.

Capacitor C_1 serves two functions. It helps to protect the diodes from the turn-off transient caused by energy stored in the transformer inductance. That capacitor reduces the EMI (electromagnetic interference) resulting from diode switching transients. For best EMI reduction, C_1 should be as large as possible; this condition, however, causes additional current drain on the transformer and thus a compromise is necessary.

11-5 RIPPLE CLIPPER

A third type of filtering circuit is the ripple clipper shown in Fig. 11-7. The ripple is sensed by C and R, amplified by $Q2$, and then fed back to the series pass transistor $Q1$. The ripple signal is fed to the base of $Q1$ out of phase with the incoming ripple and cancels ac fluctuations and ripple. This technique can give any degree of attenuation desired since the feedback signal can be less or greater than the incoming ripple. It is necessary, however, to have the load current remain relatively constant or else the exact cancellation is difficult to maintain. Since the capacity multiplier method is more stable, it is used more often.

11-6 REGULATED POWER SUPPLY

Theoretically, a perfectly regulated supply has no noise or ripple. A typical regulated supply is shown in Fig. 11-8. If the output voltage changes, the voltage across R_2 differs from the Zener voltage across $D1$. This difference is amplified by the op amp and changes the bias, hence the voltage drop across

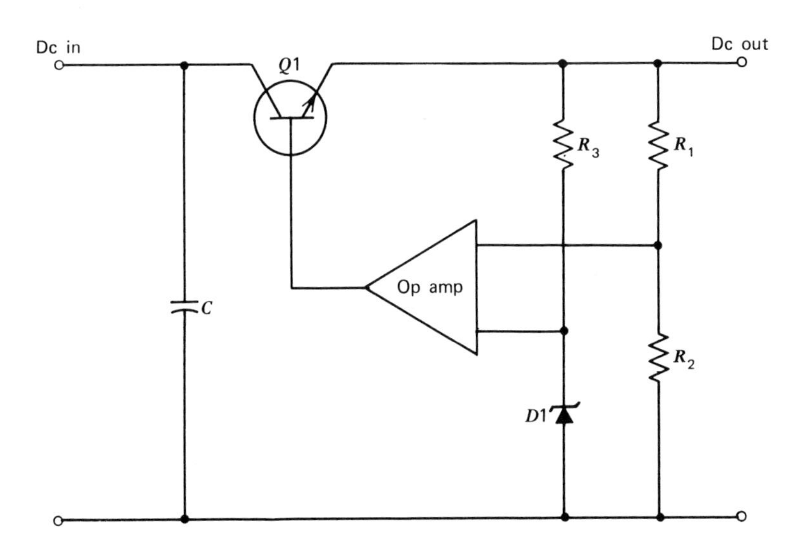

Fig. 11-8. Regulator.

the pass transistor $Q1$. The reference diode $D1$ should be a low-noise, low-voltage Zener diode. Capacitively bypassing $D1$ further reduces the ac noise. Typically, this type of regulator can have an output ripple of 6 to 10 μV.

Some integrated regulators, such as Motorola's MC1461, have a built-in Zener reference. The ripple rejection is stated to be 0.002%/V which is equivalent to 20 μV of ripple per V of supply ripple.

SUMMARY

a. To eliminate stray common-mode coupling, as well as other sources of pickup, transformer shielding is recommended.

b. A power supply that provides good ripple filtering also discriminates against noise reaching the electronic system through the power supply.

c. A ripple factor, the ratio of rms value of undesired harmonic to dc level at the load, is used to compare filter effectiveness.

d. The one-transistor capacity multiplier filter can theoretically reduce the ripple factor by β, the transistor current gain. Practically, the reduction is accomplished by the time constant of a filter in the transistor base circuit.

e. The two-transistor capacity multiplier filter can theoretically reduce the ripple factor by β^2. This circuit can also be used to reduce the size of the required electrolytic capacitor.

f. The ripple clipper circuit uses negative feedback from the load to control conduction in a series-pass transistor.

PROBLEMS

1. Show that the rms value of the 120-Hz voltage component in the output of a full-wave rectifier is given by $V_{(120)} = 4V_m/3\sqrt{2}\pi$.

2. Show that the dc value of the output waveform from a full-wave rectifier is given by $V_{DC} = 2V_m/\pi$.

3. It has been stated that the time constant of the R_1-R_2-C network of Fig. 11-4 is $(R_1 \| R_2)C$. Show that this is true.

4. Calculate the ripple factor of the three sections of the power supply shown in Fig. 11-6.

5. Design a ripple clipper circuit of the Fig. 11-7 type to be used in place of the R-C network of Fig. 11-3. Consider that the transistor $\beta = 200$. Determine values for all passive components.

Chapter 12

CASCADED STAGES AND FEEDBACK

In the preceding text the single common-emitter transistor stage has been used as the prime example of an amplifier input device. Few amplifiers consist of only one stage. Additional stages are almost always required to provide the necessary gain and potential for further signal conditioning. Adding overall feedback to multistage assemblies offers additional performance dimensions.

This chapter discusses noise in the three connections or configurations for the bipolar transistor; bias resistor noise is included. It treats noise in the nine conventional cascading configurations for a transistor pair, and also in FET bipolar pairs. The contribution from the second and later stages to the noise behavior of each pair is analyzed.

The parallel amplifier is introduced as a vehicle for achieving a reduction in the optimum noise resistance of the amplifier in a low-noise system. Because of the widespread application of the differential amplifier, its noise behavior is studied in some detail. The chapter concludes with a discussion of noise in multistage feedback amplifiers.

12-1 TRANSISTOR CONFIGURATIONS

There are three useful connections in which bipolar transistors can be operated: common emitter (CE), common base (CB), and common collector (CC) or emitter follower.

Each configuration offers approximately the same power gain-bandwidth product and the same equivalent input noise; therefore, configuration selection is possible to meet requirements of gain, frequency response, and impedance levels placed on the amplifying system. The CE configuration is used most often because it provides the highest power gain; the other configurations have advantages for certain applications. When a voltage amplifier is desired with high-input impedance and low output impedance, a CC stage can be used. Conversely, when a current amplifier is required with low input impedance and high output impedance, a CB stage is a solution.

Because noise is not dependent on connection, the circuit designer is allowed the option of low-noise operation with the simultaneous freedom to select terminal impedance levels and other nonnoise characteristics. Although parameters such as E_{ni}, E_n, and I_n are usually measured in the CE configuration for convenience, their values apply equally well to the CB and CC orientations, provided that frequencies are limited to those for which the collector-base internal feedback capacitance can be neglected. The equivalent input noise is the same for each configuration; but the output noise is usually not the same. The signal-to-noise ratio, however, does not vary because of the configuration selected.

Before proceeding to study the noise in cascaded stages we investigate the input stage more completely than was done in the noiseless biasing discussion given in Chapter 10. All three orientations are considered, and for each connection a typical design is shown complete with component values. In each instance an expression for the transfer voltage gain and for E_{ni} is derived.

By referring all noise sources to the input, E_{ni} allows us to determine the noise contribution of the various passive elements required for biasing, and the effect of those elements on the transistor noise parameters E_n and I_n. Although the limiting value for E_{ni} does not vary because of the configuration selected, E_{ni} is more importantly affected by biasing circuit noise when certain connections are employed.

Additional information regarding gain and terminal impedance properties of transistor circuits is available in texts devoted to transistor electronics.

In order to simplify the symbols necessary to represent electrical quantities in the complex circuits given in this chapter some deviations from previously used nomenclature are made. Transistor parameters are

r_x = base resistance ($r_{bb'}$)
r_π = base-emitter resistance ($r_{b'e}$ or βr_e)
r_o = output resistance ($\simeq r_{ce}$)
β = small-signal current gain (β_o)
E_n = rms value of equivalent noise voltage
I_n = rms value of equivalent noise current

Quantities R_s, E_{ni}, E_{ns}, and V_s are used as before.

Symbols for the thermal noise voltages of resistances differ from the system noted in the introductory chapter. Here we use E_A, E_B, and so forth, to represent the rms values of noise voltages in resistors R_A, R_B, and so forth. Normally a capital letter with a capital subscript stands for a dc value. However, as mentioned earlier, E represents a noise voltage. When the reader encounters E he should realize that noise is being discussed.

Where more than one transistor is used a numerical subscript indicates stage number. Thus E_{n2}^2 is the mean square value of the noise voltage generator representing stage 2. Symbol R_{I2} represents the input resistance of stage 2. All noise generators are considered to be uncorrelated.

For each of the working circuit examples given in the single stage and cascaded amplifier sections of this chapter, the bias elements and second-stage collector currents were selected for minimum total input noise parameters at 10 Hz and 10 kHz. Designing for minimum E_n and I_n provides the versatility of operating over a wider range of source resistances, but it can be more costly, for lower noise resistors and larger capacitors are required than if the amplifier were designed for a specific source resistance such as R_o.

In the performance summaries given with each practical circuit example, $\mathrm{E_{nT}}$ *and* $\mathrm{I_{nT}}$ *are used to represent the total parameters as seen from the signal source. All noise parameters are given on a per root hertz basis.*

All of the capacitors are selected to provide a minimal increase in noise voltage E_n at 10 Hz. If the lowest frequency of interest is above 10 Hz then the capacitors can all be decreased proportionally. The low-frequency cutoff f_l often is below the low-frequency point set by noise criteria because it is more difficult to bypass for noise than it is for frequency response. This point is discussed in Section 12-3.

12-2 BIASED COMMON-EMITTER STAGE

Because the CE connection provides the highest power gain, an amplifier with a CE input stage is not likely to have significant noise contributions from stages beyond the first. Input resistance is highly dependent on I_C; it is five or more times the optimum noise resistance R_o.

A practical CE stage with performance data is shown in Fig. 12-1 along with the noise equivalent circuit. *The transfer voltage gain* $\mathrm{K_t}$ *is the ratio of signal voltage at the collector to* $\mathrm{V_s}$. An expression for this gain follows from Eq. 10-5:

$$K_t \simeq \left(\frac{Z_G}{Z_s}\right) \frac{\beta R_L}{Z_G + r_x + r_\pi + \beta(Z_E + R_F)} \tag{12-1}$$

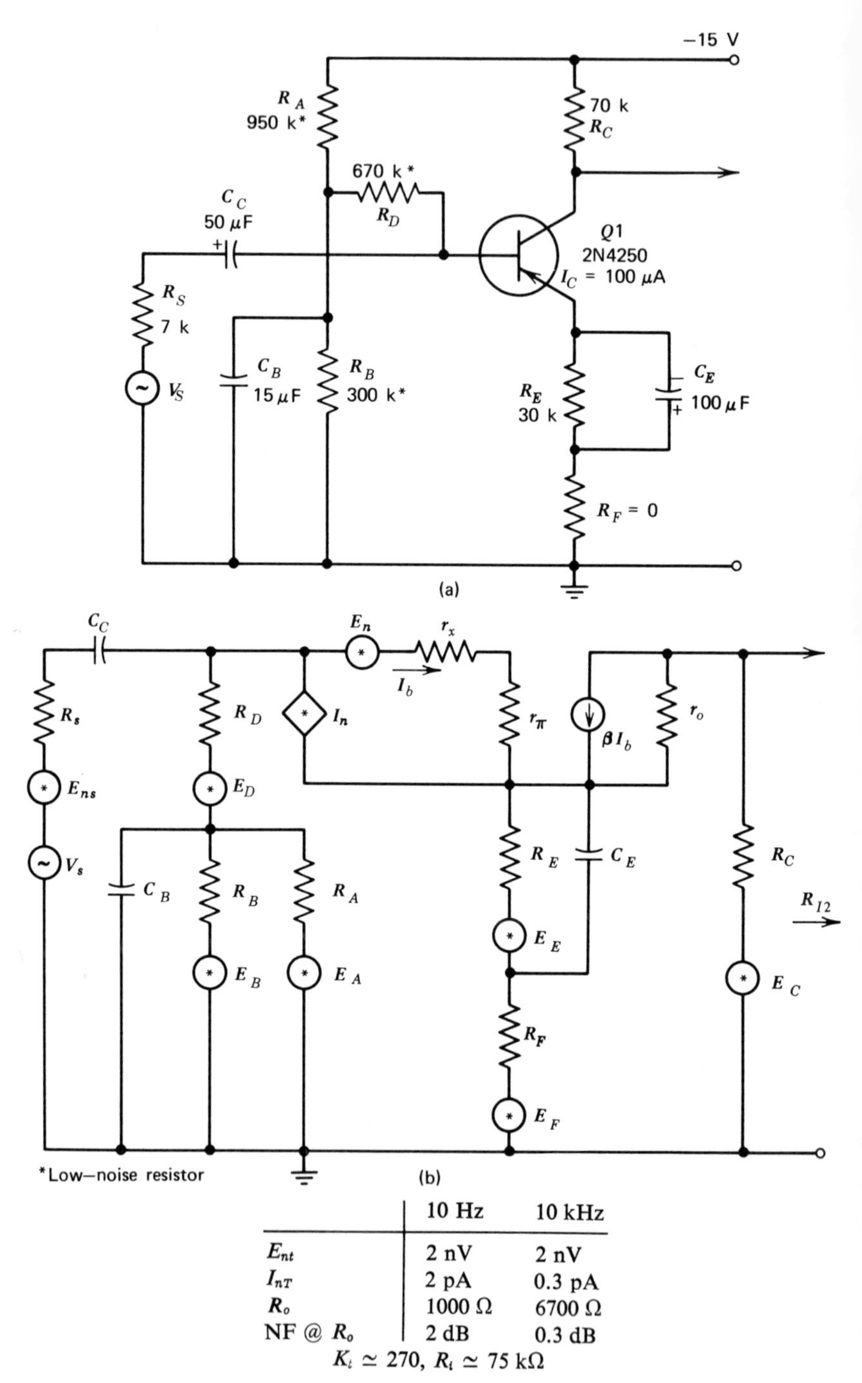

	10 Hz	10 kHz
E_{nt}	2 nV	2 nV
I_{nT}	2 pA	0.3 pA
R_o	1000 Ω	6700 Ω
NF @ R_o	2 dB	0.3 dB

$K_i \simeq 270$, $R_i \simeq 75$ kΩ

Fig. 12-1. (*a*) Biased CE stage; (*b*) noise and small-signal equivalent circuit.

The symbols used in deriving Eq. 12-1 are

$$\begin{aligned}
R_L &= R_C \| r_o \| R_{I2} \\
Z_E &= R_E \| - jX_E \\
Z_G &= Z_s \| Z_D \\
Z_s &= R_s - jX_C \\
Z_D &= R_D + R_A \| R_B \| - jX_B
\end{aligned}$$

For noise analysis purposes our interest lies with the magnitude of K_t.*

When there is negligible signal loss in biasing, coupling, and feedback elements, Eq. 12-1 simplifies to

$$K_t \simeq \frac{\beta R_L}{Z_s + r_x + r_\pi + \beta Z_E} \tag{12-2a}$$

Under the assumptions that $Z_s \to 0$ and $Z_E \to 0$, the equation can be written

$$K_t \simeq \frac{R_L}{r_e} \tag{12-2b}$$

For simplicity it is convenient to define an additional gain symbol:

$$K_t' = K_t \qquad \text{for} \qquad R_L = R_C$$

The equivalent input noise E_{ni}^2 is determined by dividing E_{no}^2/K_t^2. The result is

$$E_{ni}^2 \simeq E_{ns}^2 + E_n^2\left(\frac{R_s + R_D}{R_D}\right)^2 + I_n^2(R_s + R_F - jX_C)^2 + E_F^2$$

$$+\left[\frac{E_A^2}{(\omega R_A C_B)^2} + \frac{E_B^2}{(\omega R_B C_B)^2} + E_D^2\right]\frac{R_s^2}{R_D^2} + \frac{E_E^2}{(\omega R_E C_E)^2} + \left(\frac{E_C}{K_t'}\right)^2 \tag{12-3}$$

Several observations can be made from Eq. 12-3. It is clear that the noise from each resistor increases E_{ni}. Voltage E_n is increased by the shunting effect of R_D. Generator I_n is increased by the reactance of the coupling capacitor and by feedback element R_F if used.

A number of simplifying assumptions were made to arrive at a manageable form for Eq. 12-3. It does not hold for $\omega \to 0$ because certain of the terms will blow up. Nevertheless, the equation is useful as a guide to evaluation of noise sources in the biased CE stage.

* The phase reversal inherent in CE and common-source stages is usually designated by a minus sign preceding a gain expression such as Eq. 12-1. The gain formulas given in this chapter do not include any designation of phase inversion.

12-3 BIASED COMMON-BASE STAGE

The CB configuration offers the option of a low-input impedance with the same limiting noise as obtained from the CE connection. A CB input stage is frequently used in high-frequency amplifiers to minimize the effect of shunt capacitance.

A practical CB input stage is shown in Fig. 12-2. All biasing elements are specified, and performance data are tabulated.

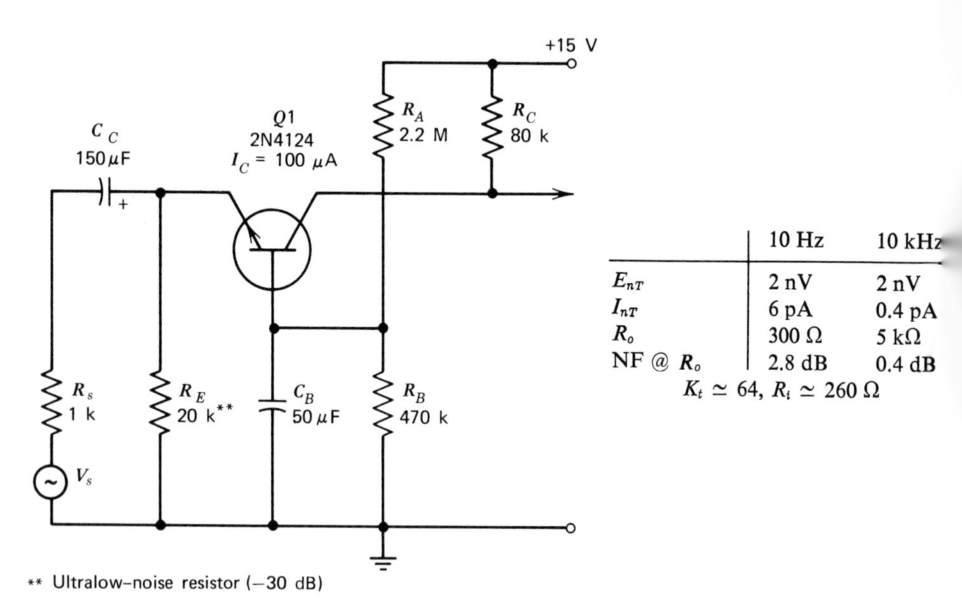

	10 Hz	10 kHz
E_{nT}	2 nV	2 nV
I_{nT}	6 pA	0.4 pA
R_o	300 Ω	5 kΩ
NF @ R_o	2.8 dB	0.4 dB

$K_t \simeq 64$, $R_i \simeq 260\ \Omega$

Fig. 12-2. Biased CB stage.

The transfer voltage gain K_t, the ratio of output signal to V_s, is given by

$$K_t \simeq \left(\frac{Z_G}{Z_s}\right) \frac{\beta R_L}{\beta Z_G + r_x + r_\pi + Z_B} \tag{12-4}$$

where

$$R_L = R_C \| r_o \| R_{I2}$$

and

$$Z_B = R_A \| R_B \| - jX_B = \frac{R_A R_B}{R_A + R_B + j\omega C_B R_A R_B}$$

$$Z_s = R_s - jX_C$$

$$Z_G = R_E \| Z_s$$

For negligible loss in biasing and coupling elements Eq. 12-4 becomes

$$K_t = \frac{\beta R_L}{R_s + r_x + r_\pi} \tag{12-5a}$$

$$K_t \simeq \frac{R_L}{r_e + R_s} \qquad (\text{for } Z_B \to 0) \tag{12-5b}$$

As before, for simplicity, we use $K_t' = K_t$ for $R_L = R_C$.

The total equivalent input noise E_{ni} is determined as before:

$$E_{ni}^2 \simeq E_{ns}^2 + E_n^2\left(\frac{R_E + R_s}{R_E}\right)^2 + I_n^2(R_s + Z_B - jX_C)^2 + \left(\frac{E_E R_s}{R_E}\right)^2 + \left(\frac{E_A}{\omega C_B R_A}\right)^2 + \left(\frac{E_B}{\omega C_B R_B}\right)^2 + \left(\frac{E_C}{K_t'}\right)^2 \tag{12-6}$$

The assumptions have been made that $X_C \ll R_E$ and $X_B/\beta \gg R_E$.

It can be seen that the noise voltage E_n is increased by the shunting of the emitter bias resistor R_E. Since R_E is part of the bias circuit it is limited in value by the available supply voltage, and, in general, R_E is smaller than R_C. If the source resistance is larger than R_E, the E_n and E_E terms are significantly increased. A low-noise resistor should be specified because of the dc flowing through R_E.

The size of the input coupling capacitor C_C is determined by both the desired low-frequency cutoff and the noise requirement. A low-frequency corner f_l is defined as the frequency at which the reactance of C_C is equal to the sum of the source resistance R_s and the amplifier input resistance R_I. From the noise standpoint, that reactance must be much less than the value of source resistance or else the noise current term of the stage, $I_n X_C$, becomes significant. When R_s is very small, $I_n X_C$ should be less than E_n. Similarly, the base capacitor C_B is also determined by the frequency response or by the noise criteria, whichever is the most stringent. From the standpoint of frequency response, X_B should be one-third of βR_s, or equal to βX_C. From the noise standpoint, capacitor C_B must bypass the noise of bias resistors R_A and R_B, as shown in the fifth and sixth terms of Eq. 12-6. Since any noise in series with the base adds directly to the input noise E_{ni}, the noise of $R_A \| R_B$ and R_E should be less than E_{ns} or even better, less than E_n.

12-4 BIASED COMMON-COLLECTOR STAGE

The CC configuration exhibits the highest input impedance of the three connections. The input impedance is limited practically by the shunting base-biasing resistor. This effect can be avoided if a dual power supply is used,

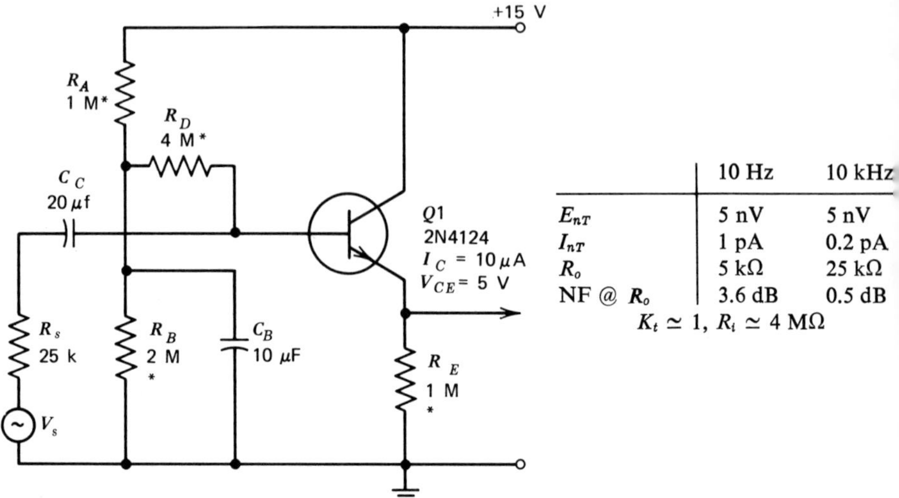

	10 Hz	10 kHz
E_{nT}	5 nV	5 nV
I_{nT}	1 pA	0.2 pA
R_o	5 kΩ	25 kΩ
NF @ R_o	3.6 dB	0.5 dB

$K_t \simeq 1$, $R_i \simeq 4\ \mathrm{M\Omega}$

* Low-noise resistor

Fig. 12-3. Biased CC stage.

and the base current supplied through the source resistance, or if low-noise biasing is used.

A practical CC stage is shown in Fig. 12-3. The transfer voltage gain is approximately unity because of the large amount of negative feedback that results from the load element R_E that is in a branch common to both input and output loops. The gain is

$$K_t \simeq \left(\frac{Z_G}{Z_s}\right) \frac{\beta R_L}{Z_G + r_x + r_\pi + \beta R_L} \tag{12-7}$$

where $R_L = R_E \| R_{I2}$

$Z_G = Z_s \| Z_D$

$Z_s = R_s - jX_C$

$Z_D = R_D + R_A \| R_B \| - jX_B$

For negligible loss in bias and coupling elements Eq. 12-7 becomes

$$K_t \simeq \frac{\beta R_L}{R_s + r_x + r_\pi + \beta R_L} \tag{12-8}$$

The total equivalent input noise E_{ni} for the CC connection is

$$E_{ni}^2 = E_{ns}^2 + E_n^2\left(\frac{R_D + R_s}{R_D}\right)^2 + I_n^2\left(R_s - jX_C + \frac{R_E}{K_t'}\right)^2 + \left[\left(\frac{E_A}{\omega C_B R_A}\right)^2 + \left(\frac{E_B}{\omega C_B R_B}\right)^2 + E_D^2\right]\left(\frac{R_s}{R_D}\right)^2 + \left(\frac{E_E}{K_t'}\right)^2 \tag{12-9}$$

where

$$K_t' = \frac{\beta R_E}{r_x + r_\pi + R_s} \tag{12-10}$$

This expression for E_{ni} is very similar to those for the other two configurations. The noise voltage E_n is again increased by the shunting effect of bias resistor R_D, and the noise current I_n is multiplied by the source resistance and the reactance of the coupling capacitor. The one significant difference is the noise contribution of the emitter load resistor R_E. The CC amplifier is the one exception to the rule that the emitter resistor should be kept as small as possible to minimize noise. The noise voltage E_E of the emitter resistor as shown in Eq. 12-9 is attenuated by a gain factor K_t', which is equivalent to the gain of a CE stage in which R_E is the load resistor. In addition, the noise current contribution $I_n R_E$ is attenuated by the same gain factor.

This apparent attenuation of the external emitter resistance noise E_E can be visualized with the aid of the figure. The contribution of E_E is clearly in series with the load resistance R_E; it also acts as an inverted voltage in series with the input. This inverted input noise is amplified by the stage gain K_t which is approximately unity and summed with the original E_E across R_E. The two voltages are out of phase and thus tend to cancel.

12-5 NOISE IN CASCADED STAGES

Amplifier stages are cascaded to obtain the desired gain, impedance characteristics, frequency response, and power level. Frequently, overall negative feedback is used to stabilize the ac and/or dc gains and to provide for the desired steady-state and transient performance. Although the noise of the first stage of a cascaded amplifier is usually dominant, subsequent stages can also contribute noise. In performing a system design the noise of each stage, including biasing elements, must be considered seriously.

A general block diagram of an amplifying system is shown in Fig. 12-4. The forward gain blocks are symbolized by K_{t1}, and so on. Noise in each stage is represented by E_n and I_n generators. A feedback connection, with transfer function β_{FB}, joins output terminal to input.

The equivalent input noise of the first stage alone is $E_{ns}^2 + E_{n1}^2 + I_{n1}^2 R_s^2$. The equivalent noise of other stages can be determined in the same manner, by summing the E_n and $I_n R_s$ terms, except that for stages beyond the first the effective source resistance is the output resistance of the preceding stage. We can calculate the total noise E_{no} at the output, including the subtractive effect of negative feedback in the calculation. For total forward gain of A_v, we can

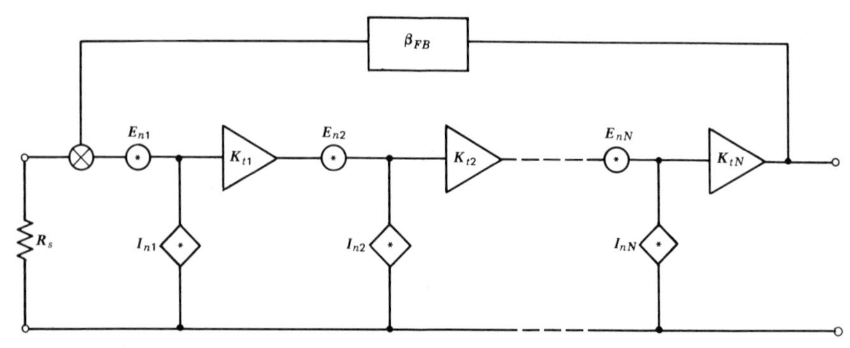

Fig. 12-4. Feedback amplifier.

determine the equivalent input noise of the system using a familiar formula from feedback theory:

$$\frac{E_{no}}{E_{ni}} = \frac{A_v}{1 - A_v\beta_{FB}}$$

With or without feedback, it can be shown that

$$E_{ni}^2 = E_{ns}^2 + E_{n1}^2 + I_{n1}^2R_s^2 + \frac{E_{n2}^2 + I_{n2}^2r_{o1}^2}{K_{t1}^2} + \frac{E_{n3}^2 + I_{n3}^2r_{o2}^2}{K_{t1}^2K_{t2}^2} + \cdots \tag{12-11}$$

Equation 12-11 can be used to lead us to more extensive knowledge of the cascade amplifier system. For simplicity, consider R_{C1} to be the dominant source resistance in the second stage. We note that for the input stage connected CE, the gain is

$$K_{t1} \simeq \frac{\beta_1 R_{C1}}{R_s + r_{x1} + r_{\pi 1}} \tag{12-12}$$

For $R_s + r_x \to 0$, $K_{t1} \simeq R_{C1}/r_{e1}$. Under these assumptions the second-stage portion of Eq. 12-11 becomes

$$E_{ni(2)}^2 = E_{n2}^2\left(\frac{r_{e1}}{R_{C1}}\right)^2 + I_{n2}^2r_{e1}^2 \tag{12-13}$$

Now we wish to design the system so that second-stage noise is insignificant compared to the noise of the input stage. This requirement, for low values of source resistance, leads to

$$E_{n1}^2 \gg E_{ni(2)}^2 \tag{12-14}$$

Let us consider the shot-noise limited region of operation. Substitute Eqs.

4-21, 4-22, and 12-13 into Eq. 12-14. The result is

$$2qI_{C1}r_{e1}^2 \gg \frac{2qI_{C2}r_{e2}^2 r_{e1}^2}{R_{C1}^2} + \frac{2qI_{C2}r_{e1}^2}{\beta_2}$$

When simplified this becomes

$$I_{C1} \gg \frac{1}{1600 I_{C2} R_{C1}^2} + \frac{I_{C2}}{\beta_2}$$

The first term on the right-hand side is quite negligible. Therefore

$$I_{C1} \gg \frac{I_{C2}}{\beta_2} \tag{12-15}$$

This can also be written $I_{C2} \simeq 10 I_{C1}$. Thus we have a guiding condition to assure that first-stage noise dominates. When one realizes that signal amplification also suggests the second-stage operating point to be at a higher level of I_C than the input stage, we find this result to be compatible with normal design techniques.

The preceding discussion applies to the shot noise region. Study of the 1/f region suggests that $I_{C2} \simeq I_{C1}$. A general rule of thumb for amplifiers operating in both 1/f and shot regions is to have collector current increase by about three times per stage.

The very nature of a system of cascaded stages implies that the problem of noise analysis is complex. We have three tools at our disposal. Given a system we can perform a noise analysis using the methods of network theory to derive performance relations; we can utilize the digital computer to determine numerical values for noise quantities; or, we can resort to an experimental analysis of the system. Usually the design engineer uses a combination of these techniques. Frequently he utilizes all three in the pursuit of knowledge about a particular system.

It may be of value at this point to suggest a "trick" useful in the analysis of complicated systems. When using a computer analysis program, the effect of a noise source at any location internal to the system on output noise can be appreciated by insertion of a noise voltage (or current) generator at that location, and by calculation of the output from that generator acting alone. The generator can have a test value of 1 V or 1 A, for example. The same technique can be used in the laboratory. The effect of noise in any component can be accentuated by inserting a signal generator in series or in parallel with the component. The generator signal must override circuit noise in order to permit easy evaluation of its effect, but it must not be so large as to overdrive the transistors and result in distortion in the output waveform. This technique is useful, for example, when one seeks the noise contribution of a Zener diode used in a power supply or as a reference.

In the studies of noise in the nine pairs treated in the next nine sections, noise originating in certain biasing elements is neglected in order to eliminate unnecessary complexity. These noise sources have been considered in Sections 12-2 through 12-4.

12-6 CE-CE PAIR

A practical CE-CE pair is shown in Fig. 12-5. This example is of a direct-coupled complementary cascade, and features a minimum component count.

The transfer voltage gain K_t of each stage is given by Eq. 12-2. For the pair, the overall cascaded gain, symbolized by K_{tc}, is simply the product of stage gains K_{t1} and K_{t2}:

$$K_{tc} \simeq \frac{\beta_1 \beta_2 R_{L1} R_{C2}}{(r_{x1} + r_{\pi 1} + R_s + \beta_1 Z_{E1})(r_{x2} + r_{\pi 2} + \beta_2 Z_{E2})} \tag{12-16}$$

where

$$R_{L1} = R_{C1} \| (r_{x2} + r_{\pi 2} + \beta_2 Z_{E2})$$

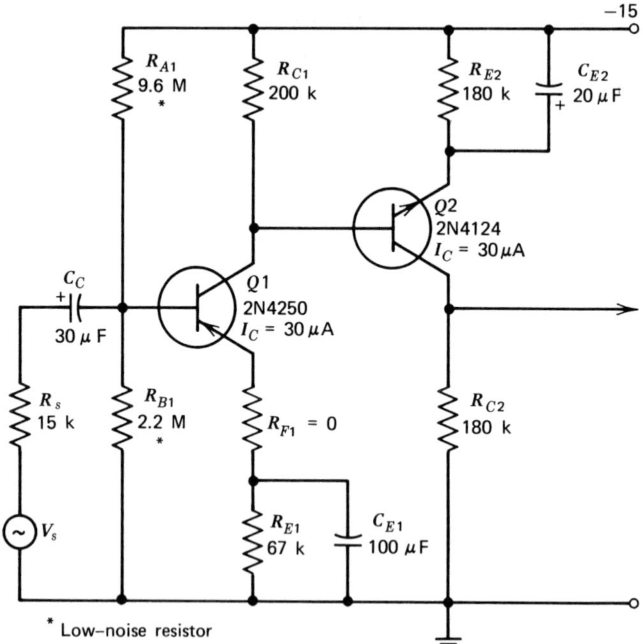

	10 Hz	10 kHz
E_{nT}	3.7 nV	3 nV
I_{nT}	0.7 pA	0.2 pA
R_o	5.3 kΩ	15 kΩ
NF @ R_o	1.4 dB	0.3 dB

$K_{tc} \simeq 30{,}000$, $R_i \simeq 250$ kΩ

Fig. 12-5. CE-CE pair.

This gain can be approximated in several ways depending on the relative sizes of collector load resistor R_{C1} and the input resistance of stage 2. For the case of R_{C1} being much smaller than $r_{\pi 2}$,

$$K_{tc} \simeq \frac{R_{C1}R_{C2}}{r_{e1}r_{e2}} \qquad (\text{for } R_{L1} = R_{C1}) \tag{12-17}$$

and, when $R_{C1} \gg r_{\pi 2}$, the expression becomes

$$K_{tc} \simeq \frac{\beta_2 R_{C2}}{r_{e1}} \qquad (\text{for } R_{L1} = r_{\pi 2}) \tag{12-18}$$

Equations 12-17 and 12-18 assume several conditions:

$$r_{\pi 1} \gg r_{x1} + R_s + \beta_1 Z_{E1} \qquad \text{and} \qquad r_{\pi 2} \gg r_{x2} + \beta_2 Z_{E2}$$

The total equivalent input noise E_{ni} is the rms sum of the first-stage noise and the second-stage noise divided by the first-stage voltage gain. In addition, noise in the second-stage load resistance is divided by K_{tc}. Therefore

$$E_{ni}^2 = E_{ns}^2 + E_{n1}^2 + I_{n1}^2 R_s^2 + \frac{1}{K_{t1}^2}(E_{C1}^2 + E_{n2}^2 + I_{n2}^2 R_{C1}^2) + \frac{E_{C2}^2}{K_{tc}^2} \tag{12-19}$$

We substitute R_{C1}/r_{e1} for K_{t1} and the relation given in Eq. 12-18 for K_{tc}. Then Eq. 12-19 simplifies to

$$E_{ni}^2 \simeq E_{ns}^2 + E_{n1}^2 + I_{n1}^2 R_s^2 + \frac{(E_{C1}^2 + E_{n2}^2) r_{e1}^2}{R_{C1}^2} + I_{n2}^2 r_{e1}^2 + \left(\frac{E_{C2} r_{e1}}{\beta_2 R_{C2}}\right)^2 \tag{12-20}$$

A large value for R_{C1} certainly reduces the effect of E_{n2}. Resistor R_{C1} drops out of the second-stage noise current expression which becomes $I_{n2} r_{e1}$, as previously predicted. To minimize the noise contribution of I_{n2}, I_{C2} should be only slightly larger than I_{C1} in this case.

In the design example shown in the figure a 30-μA collector current was selected for $Q1$ to provide low-noise operation over a wide range of source impedances centered around 10 kΩ. The collector current of $Q2$ was also set at 30 μA so the second-stage noise current term $I_{n2} r_{e1}$ did not dominate E_{n1} at 10 Hz.

The external emitter resistor should be zero if no overall negative feedback is used. At times R_{F1} has been used to stabilize the gain or raise the input impedance of an input stage. If R_{F1} is large enough to affect the gain or input resistance of $Q1$ it may also become the dominant noise source for the entire amplifier. When the gain and input resistances are adjusted with overall negative feedback, R_{F1} needs to be a very small resistance, and noise performance is not seriously degraded.

12-7 CE-CB PAIR

The CE-CB pair is widely known as the cascode circuit. Because of the low-input resistance of the second stage, the first-stage voltage gain is small, but the Miller effect resulting from $C_{b'c}$ is minimized. Thus the first-stage input capacitance is much smaller than for a regular CE stage, and the circuit has application at high frequencies.

A practical cascode is shown in Fig. 12-6. Most of the voltage gain of the pair is obtained from $Q2$. Whereas $Q1$ provides very little voltage gain, it does raise the power level of the signal.

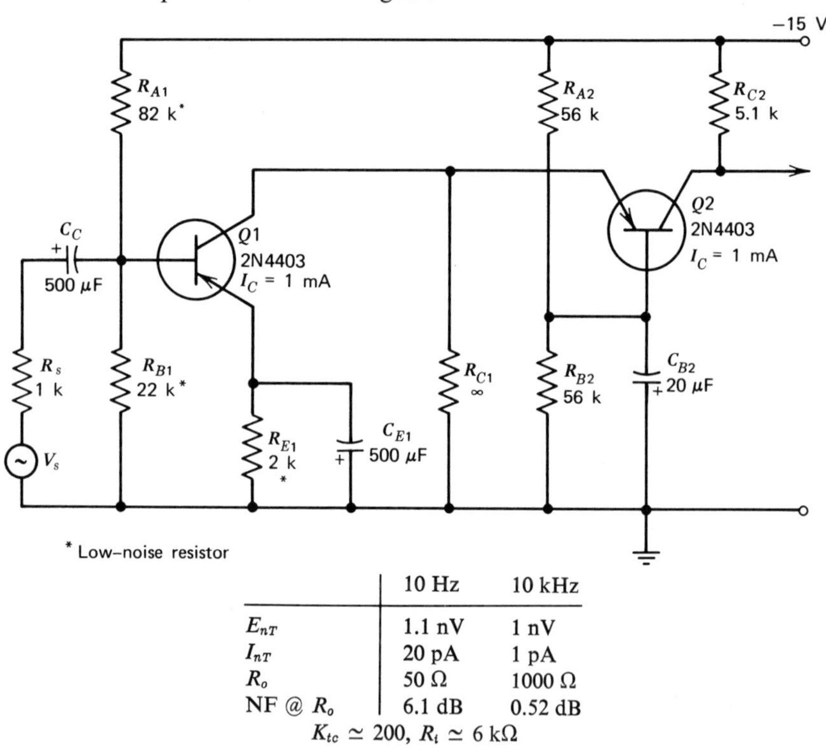

	10 Hz	10 kHz
E_{nT}	1.1 nV	1 nV
I_{nT}	20 pA	1 pA
R_o	50 Ω	1000 Ω
NF @ R_o	6.1 dB	0.52 dB

$K_{tc} \simeq 200$, $R_i \simeq 6\ \text{k}\Omega$

Fig. 12-6. CE-CB pair.

The cascaded voltage gain of the pair, K_{tc}, follows from the product of single-stage gains given in Eqs. 12-1 and 12-4:

$$K_{tc} \simeq \frac{\beta_1 \beta_2 R_{L1} R_{L2}}{(r_{x1} + r_{\pi 1} + R_s + \beta_1 Z_{E1})(r_{\pi 2} + r_{x2} + Z_{B2})} \tag{12-21}$$

where

$$R_{L1} = R_{C1} \Big\| \frac{(r_{x2} + r_{\pi 2} + Z_{B2})}{\beta_2} \simeq r_{e2}$$

For the case of $R_s \to 0$, and $R_{C1} \gg r_{\pi 2}/\beta_2$, it follows that the gain of this pair can be represented by the simple equation

$$K_{tc} \simeq \frac{R_{C2}}{r_{e1}} \tag{12-22}$$

The equivalent input noise E_{ni} is

$$E_{ni}^2 = E_{ns}^2 + E_{n1}^2 + I_{n1}^2 R_s^2 + \frac{1}{K_{t1}^2}\left[E_{C1}^2 + E_{n2}^2 + I_{n2}^2(R_{C1} + Z_{B2})^2\right] + \left(\frac{E_{A2}}{K_{t1}\omega C_{B2} R_{A2}}\right)^2 + \left(\frac{E_{B2}}{K_{t1}\omega C_{B2} R_{B2}}\right)^2 + \frac{E_{C2}^2}{K_{tc}^2} \tag{12-23}$$

Gain K_{t1} can be simply R_{L1}/r_{e1} as defined in equations 12–1 and 12–2.

In the circuit of the figure the 1-mA collector current of $Q1$ gives a low value of optimum noise resistance R_o. Because of this relatively high current the 10-Hz noise is significantly increased. The collector current of $Q2$ is also set at 1 mA to minimize E_n at 10 Hz.

In this circuit the return resistor R_{C1} is not mandatory. When it is desirable to have a smaller collector current in $Q1$ than in $Q2$, R_{C1} carries the additional current of $Q2$ to ground. For $I_{C1} > I_{C2}$, R_{C1} is connected to V_{CC}. Because of the low values selected for bias resistors R_{A1} and R_{B1} their noise contribution can be critical. This noise source can be reduced if the noiseless bias circuit of Chapter 10 is used in this application.

12-8 CE-CC PAIR

The CE-CC pair shown in Fig. 12-7 can be used when both low noise and high gain-bandwidth product are desired. The low-input capacitance of $Q2$ minimizes the loading on collector resistor R_{C1} to improve frequency response. Stage $Q2$ does not add significantly to either the noise or the phase shift.

The cascaded voltage gain K_{tc} of this pair is approximately equal to the first stage gain:

$$K_{tc} \simeq \frac{\beta_1 \beta_2 R_{L1} R_{E2}}{(R_s + r_{x1} + r_{\pi 1} + \beta_1 Z_{E1})(r_{x2} + r_{\pi 2} + \beta_2 R_{E2})} \tag{12-24}$$

where

$$R_{L1} = R_{C1} \| (r_{x2} + r_{\pi 2} + \beta_2 R_{E2})$$

An approximation for K_{tc} follows for $R_s \to 0$ and $\beta_2 R_{E2} \gg R_{C1}$:

$$K_{tc} \simeq \frac{R_{C1}}{r_{e1}} \tag{12-25}$$

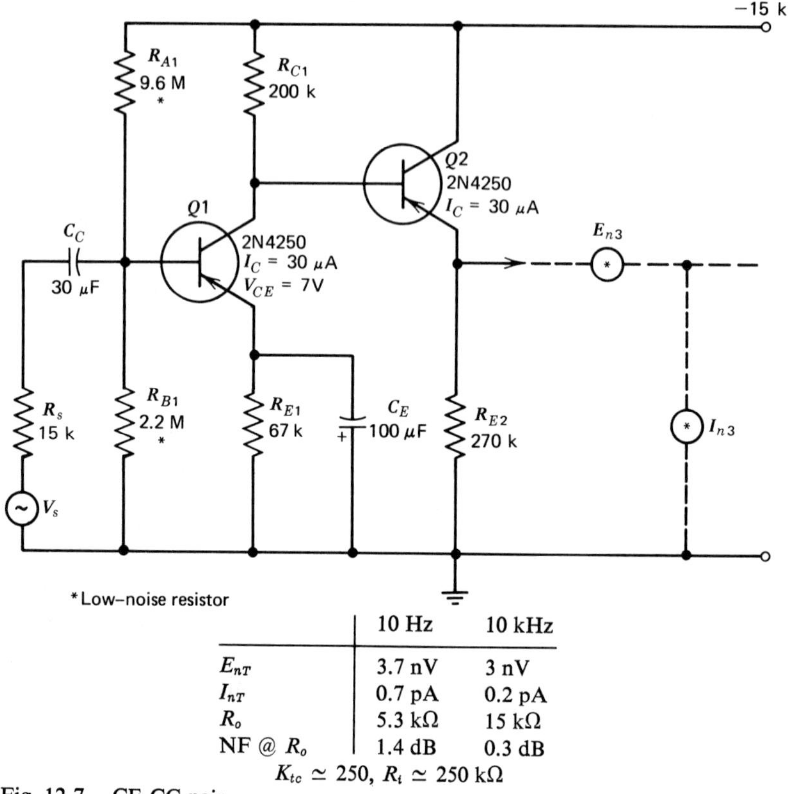

	10 Hz	10 kHz
E_{nT}	3.7 nV	3 nV
I_{nT}	0.7 pA	0.2 pA
R_o	5.3 kΩ	15 kΩ
NF @ R_o	1.4 dB	0.3 dB

$K_{tc} \simeq 250$, $R_i \simeq 250$ kΩ

Fig. 12-7. CE-CC pair.

Since $Q2$ is an emitter follower, its voltage gain is less than unity. Consequently the noise of stage 3 can become significant in certain applications of this pair. We define a noise attenuation factor in order to refer the noise in R_{E2} and the current noise of $Q3$ to the input:

$$K'_{t2} = \frac{\beta_2 R_{E2}}{r_{x2} + r_{\pi 2} + R_{C1}} \tag{12-26}$$

The total equivalent input noise E_{ni} is determined as before, but contains noise terms representing $Q3$.

$$E_{ni}^2 = E_{ns}^2 + E_{n1}^2 + I_{n1}^2 R_s^2 + \frac{1}{K_{t1}^2}(E_{C1}^2 + E_{n2}^2 + I_{n2}^2 R_{C1}^2) + \left(\frac{E_{E2}}{K_{t1}K'_{t2}}\right)^2 + \frac{E_{n3}^2}{K_{tc}^2} + \left(\frac{I_{n3}R_{E2}}{K_{t1}K'_{t2}}\right)^2 \tag{12-27}$$

K_{t1} is defined in equations 12–1 and 12–2.

For low-noise operation of the circuit in Fig. 12-7, the collector resistance R_{C1} should be large and I_{C2} not much larger than I_{C1}.

2-9 CB-CE PAIR

The CB-CE pair is useful for low-level signal conditioning because it provides low noise along with good frequency response and low-input resistance. This amplifier provides an output voltage proportional to input current when fed from a high-impedance current source such as a biased diode. A working CB-CE pair is shown in Fig. 12-8.

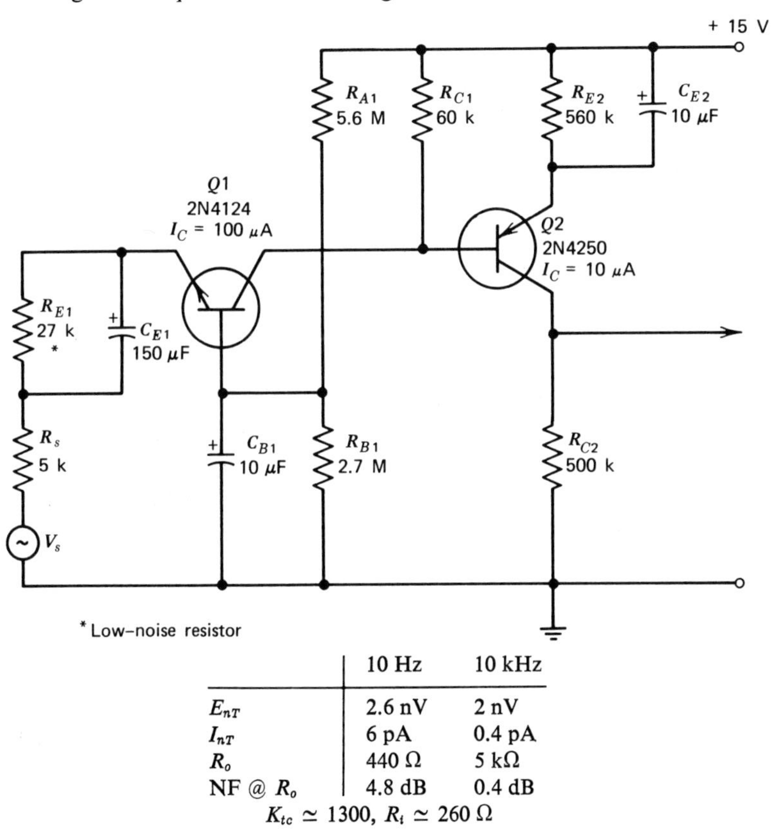

	10 Hz	10 kHz
E_{nT}	2.6 nV	2 nV
I_{nT}	6 pA	0.4 pA
R_o	440 Ω	5 kΩ
NF @ R_o	4.8 dB	0.4 dB

$K_{tc} \simeq 1300$, $R_i \simeq 260\ \Omega$

Fig. 12-8. CB-CE pair.

The gain of a CB stage was given in Eq. 12-4, and the CE gain follows from Eq. 12-1. The product of those two expressions yields the voltage gain of the pair:

$$K_{tc} \simeq \frac{\beta_1 \beta_2 R_{L1} R_{C2}}{(r_{x1} + r_{\pi 1} + Z_{B1} + \beta_1 Z_s)(r_{x2} + r_{\pi 2} + \beta_2 Z_{E2})} \tag{12-28}$$

where

$$R_{L1} = R_{C1} \| (r_{x2} + r_{\pi 2} + \beta_2 Z_{E2})$$

This expression for gain can be further simplified; its reduction depends on relative sizes of R_{C1} and $r_{\pi 2}$:

$$K_{tc} \simeq \frac{R_{C1} R_{C2}}{(r_{\pi 1} + R_s) r_{e2}} \qquad \text{for } R_{L1} = R_{C1} \tag{12-29}$$

$$K_{tc} \simeq \frac{\beta_2 R_{C2}}{r_{\pi 1} + R_s} \qquad \text{for } R_{L1} = r_{\pi 2} \tag{12-30}$$

The equivalent input noise E_{ni} is

$$E_{ni}^2 = E_{ns}^2 + E_{n1}^2 + I_{n1}^2 (R_s + Z_{B1})^2 + \frac{1}{K_{t1}^2} [E_{C1}^2 + E_{n2}^2 + I_{n2}^2 R_{C1}^2] + \frac{E_{C2}^2}{K_{tc}^2} \tag{12-31}$$

K_{t1} is defined in equation 12–4 except replace R_L with R_C.

For reasonable values of R_s, K_{t1} can be approximated by R_{C1}/R_s. When K_{t1} is substituted into Eq. 12-31, the I_{n2} term becomes $I_{n2}^2 R_s^2$ and therefore can be quite significant. Since I_{C2} is often greater than I_{C1}, $I_{n2}^2 R_s^2$ is the dominant input noise current contribution. This problem can be somewhat alleviated if it is practical to make $I_{C2} < I_{C1}$. In the circuit design of Fig. 12-8 the collector current of $Q1$ is set at 100 μA and I_{C2} designed for 10 μA to insure that I_n is dominated by I_{n1}. The output resistance of the pair, 500 kΩ, is excessively high for certain applications; consequently this pair may have to be coupled to the load with an emitter-follower stage.

12-10 CB-CB PAIR

The CB-CB pair offers low-input impedance and good-frequency response, but relatively low voltage gain. It is useful as a wideband current amplifier.

In the practical circuit design of Fig. 12-9, R_{E1} is bypassed by a large capacitance and R_{E2} is unnecessary. We modify Eq. 12-4 for the conditions of this pair. The overall voltage gain becomes

$$K_{tc} \simeq \frac{\beta_1 \beta_2 R_{L1} R_{C2}}{(r_{x1} + r_{\pi 1} + \beta_1 R_s + Z_{B1})(r_{x2} + r_{\pi 2} + Z_{B2})} \tag{12-32}$$

where

$$R_{L1} = \frac{R_{C1} \| (r_{\pi 2} + r_{x2} + Z_{B2})}{\beta_2}$$

Simplification of Eq. 12-32 is possible for the case of $\beta_1 R_s \gg r_{x1} + Z_{B1}$, $r_{\pi 2} \gg r_{x2} + Z_{B2}$, and $R_{C1} \gg r_{\pi 2}/\beta_2$. Then

$$K_{tc} = \frac{R_{C2}}{R_s + r_{e1}}$$

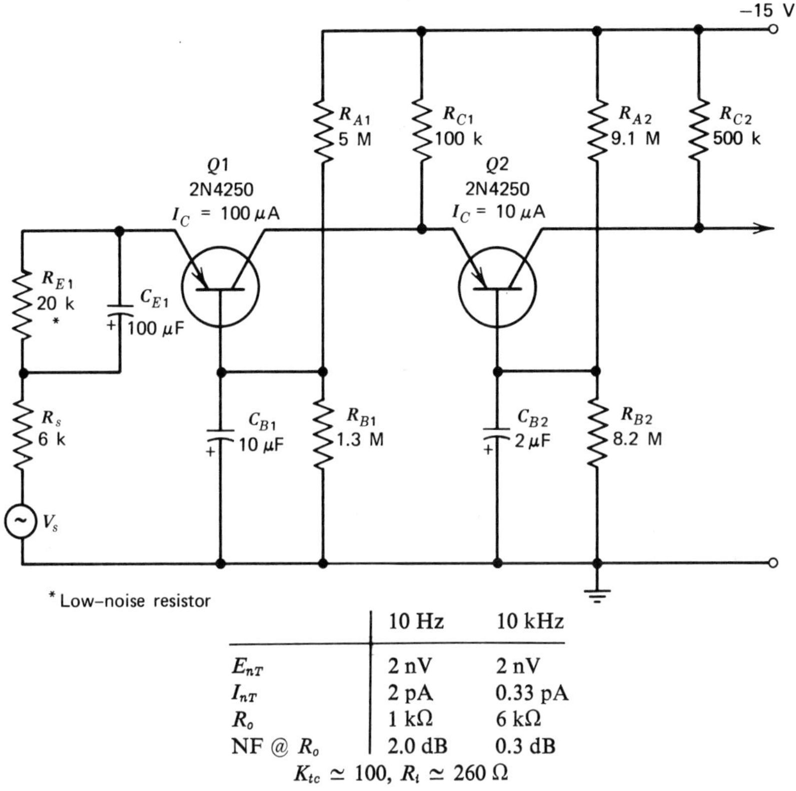

	10 Hz	10 kHz
E_{nT}	2 nV	2 nV
I_{nT}	2 pA	0.33 pA
R_o	1 kΩ	6 kΩ
NF @ R_o	2.0 dB	0.3 dB

$K_{tc} \simeq 100$, $R_t \simeq 260\ \Omega$

Fig. 12-9. CB-CB pair.

The equivalent input noise E_{ni} is

$$E_{ni}^2 = E_{ns}^2 + E_{n1}^2 + I_{n1}^2(Z_s + Z_{B1})^2 + \frac{1}{K_{t1}^2}\left[E_{n2}^2 + I_{n2}^2(R_{C1} + Z_{B2})^2 + E_{C1}^2\right] + \left(\frac{E_{B2}}{K_{t1}\omega C_{B2}R_{B2}}\right)^2 + \left(\frac{E_{A2}}{K_{t1}\omega C_{B2}R_{A2}}\right)^2 + \frac{E_{C2}^2}{K_{tc}^2} \tag{12-33}$$

Equation 12-33 does not include the noise contributions from first-stage biasing elements; they were studied in Section 12-3. To minimize the E_{n2} contributions R_{C1} must be large. As noted in the preceding section, because of low first-stage gain, noise current I_{n2} directly adds to I_{n1}. First stage gain K_{t1} is approximately $R_C/(r_e + R_s)$ as defined in equation 12–5 with $R_L = R_C$.

In the circuit of Fig. 12-9 I_{C1} is set at 100 μA for operation with a source resistance of a few thousand Ω. Current I_{C2} is 10 μA to keep I_n^2 low. The additional collector current of $Q1$ is carried by resistor R_{C1}.

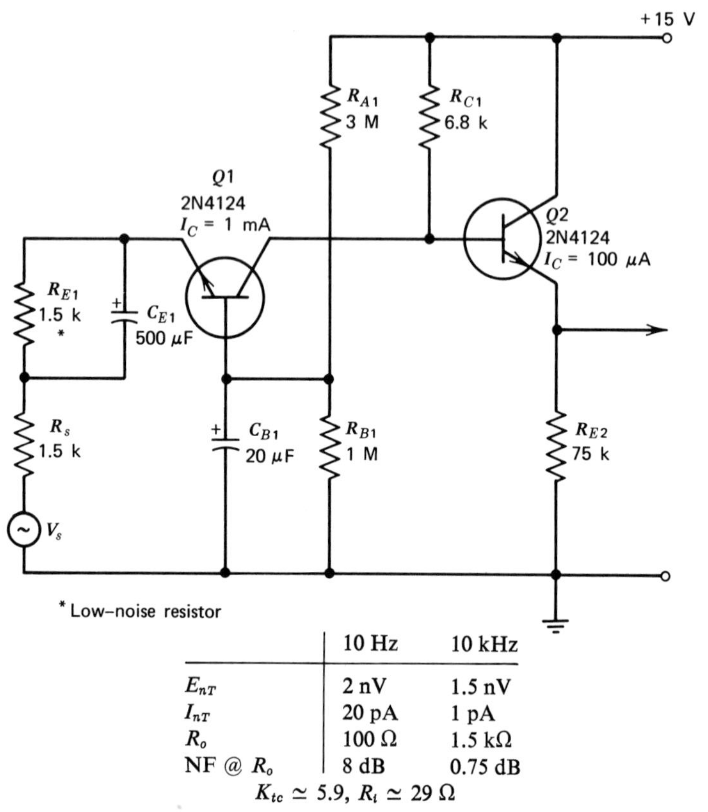

	10 Hz	10 kHz
E_{nT}	2 nV	1.5 nV
I_{nT}	20 pA	1 pA
R_o	100 Ω	1.5 kΩ
NF @ R_o	8 dB	0.75 dB

$K_{tc} \simeq 5.9$, $R_i \simeq 29\ \Omega$

Fig. 12-10. CB-CC pair.

12-11 CB-CC PAIR

The CB-CC pair offers low input impedance with low voltage gain and no phase inversion. A practical direct coupled example is given in Fig. 12-10 along with performance data. The transfer voltage gain of the pair is

$$K_{tc} \simeq \frac{\beta_1 \beta_2 R_{L1} R_{E2}}{(r_{x1} + r_{\pi 1} + Z_{B1} + \beta_1 Z_s)(r_{x2} + r_{\pi 2} + \beta_2 R_{E2})} \tag{12-34}$$

where

$$R_{L1} = R_{C1} \| (r_{x2} + r_{\pi 2} + \beta_2 R_{E2})$$

Equation 12-34 can be simplified for the case where $R_{C1} \ll \beta_2 R_{E2}$. The result is

$$K_{tc} \simeq \frac{R_{C1}}{r_{e1} + Z_s} \tag{12-35}$$

The equivalent input noise is

$$E_{ni}^2 \simeq E_{ns}^2 + E_{n1}^2 + I_{n1}^2(Z_s + Z_{B1})^2 + \frac{1}{K_{t1}^2}(E_{C1}^2 + E_{n2}^2 + I_{n2}^2 R_{C1}^2)$$
$$+ \left(\frac{E_{E2}}{K_{t1}K'_{t2}}\right)^2 + \left(\frac{E_{n3}}{K_{tc}}\right)^2 + \left(\frac{I_{n3}R_{E2}}{K_{t1}K'_{t2}}\right)^2 \quad (12\text{-}36)$$

where

$$K'_{t2} = \frac{\beta_2 R_{E2}}{R_{C1} + r_{x2} + r_{\pi 2}} \simeq \frac{R_{E2}}{r_{e2}} \quad (12\text{-}37)$$

Again, because of low gain we include noise parameters of a third stage.

The gain of the CB stage is low, approximately R_{C1}/R_s. Therefore, second-stage noise is important. In this design $I_{C1} = 1$ mA and $I_{C2} = 100\ \mu$A in order to assure that the total effective I_n is low.

12-12 CC-CE PAIR

The CC-CE pair offers low-input noise with E_n only slightly larger than for a CE input stage, along with significantly higher input resistance and lower input capacitance. For general-purpose instrumentation amplifier applications, this pair, or the comparable pair, a FET source follower in cascade with a CE bipolar transistor, is an excellent compromise to achieve both low noise and high input impedance. When the collector of $Q1$ is joined to the collector of $Q2$, this pair is referred to as a compound connection.

A sample design is shown in Fig. 12-11. Noiseless biasing is used for $Q1$. Stage $Q1$ provides voltage gain of no greater than unity. The transfer voltage gain of the pair is

$$K_{tc} \simeq \frac{\beta_1 \beta_2 R_{L1} R_{C2}}{(R_s + r_{x1} + r_{\pi 1} + \beta_1 R_{L1})(r_{x2} + r_{\pi 2} + \beta_2 Z_{E2})} \quad (12\text{-}38)$$

where

$$R_{L1} = R_{E1} \| (r_{x2} + r_{\pi 2} + \beta_2 Z_{E2})$$

Equation 12-38 can be simplified for the practical case of $\beta_1 R_{L1} \gg (R_s + r_{x1} + r_{\pi 1})$ and $r_{\pi 2} \gg r_{x2} + \beta_2 Z_{E2}$. Then we have

$$K_{tc} \simeq \frac{R_{C2}}{r_{e2}} \quad (12\text{-}39)$$

The equivalent input noise is

$$E_{ni}^2 \simeq E_{ns}^2 + E_{n1}^2 + I_{n1}^2 R_s^2 + \frac{E_{n2}^2}{K_{t1}^2} + \left(\frac{I_{n2}R_{E1}}{K'_{t1}}\right)^2 + \left(\frac{E_{E1}}{K'_{t1}}\right)^2 + \frac{E_{C2}^2}{K_{tc}^2} \quad (12\text{-}40)$$

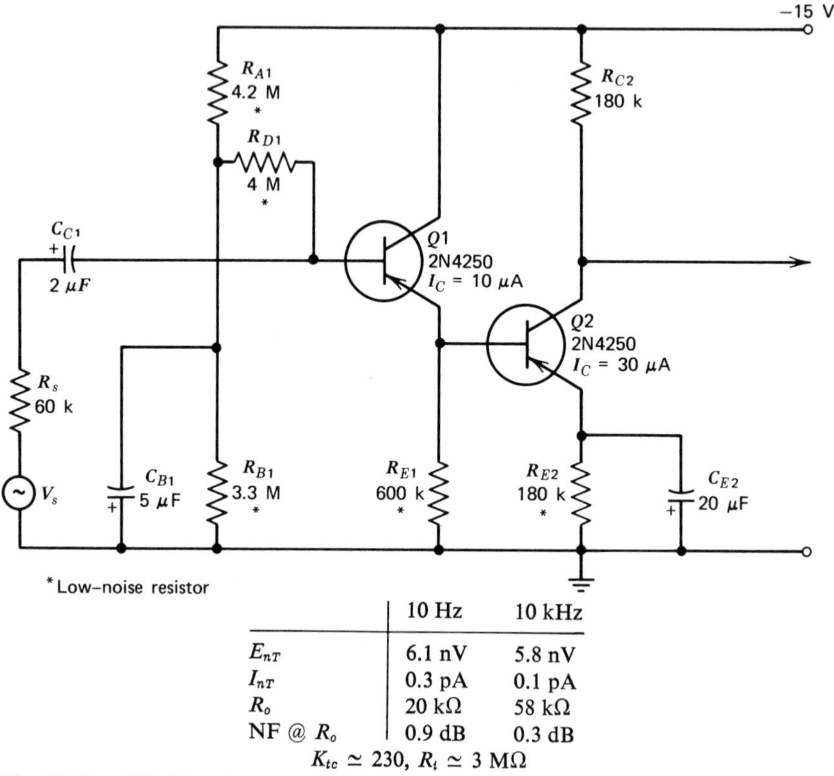

	10 Hz	10 kHz
E_{nT}	6.1 nV	5.8 nV
I_{nT}	0.3 pA	0.1 pA
R_o	20 kΩ	58 kΩ
NF @ R_o	0.9 dB	0.3 dB

$K_{tc} \simeq 230$, $R_i \simeq 3$ MΩ

Fig. 12-11. CC-CE pair.

where K'_{t1}, as given before, is the gain of a CE stage with R_{E1} as its load resistance:

$$K'_{t1} = \frac{\beta_1 R_{E1}}{R_s + r_{x1} + r_{\pi 1}} \simeq \frac{R_{E1}}{r_{e1} + R_s/\beta_1} \tag{12-41}$$

The gain of $Q1$, K_{t1}, is about unity. Consequently we see from Eq. 12-40 that E_{n2} is as important to input noise as E_{n1}. On the other hand, the coefficient of I_{n2}, $r_{e1} + R_s/\beta_1$, is much smaller than the I_{n1} term.

In this sample design $I_{C1} = 10\ \mu$A to present $R_o = 50$ kΩ. A choice of 30 μA for I_{C2} assures that E_{n2} and I_{n2} are not significant.

12-13 CC-CB PAIR

The CC-CB pair is often referred to as an emitter-coupled amplifier. It is widely used as a linear IC. A sample design is presented in Fig. 12-12.

The pair is a differential amplifier with a zero collector load resistance in the first stage; hence, the effective C_{in} is reduced. This differential type am-

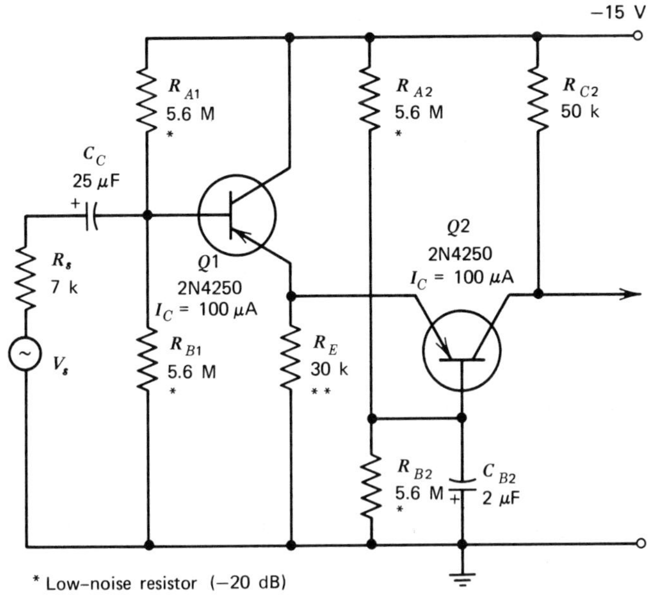

	10 Hz	10 kHz
E_{nT}	2 nV	2 nV
I_{nT}	1 pA	0.3 pA
R_o	2 kΩ	6.7 kΩ
NF @ R_o	1 dB	0.33 dB

$K_{tc} \simeq 200$, $R_i \simeq 50$ kΩ

Fig. 12-12. CC-CB pair.

plifier gives full temperature compensation so that the circuit is useful for amplifying small dc signals. Ideally, the external base resistance and the transistors should be matched for low offset and dc stability. If not integrated, a dual transistor is desirable, for the two transistors should be maintained at the same temperature.

The transfer voltage gain of this pair is

$$K_{tc} \simeq \frac{\beta_1 R_{L1} \beta_2 R_{C2}}{(R_s + r_{x1} + r_{\pi 1} + \beta_1 R_{L1})(r_{\pi 2} + r_{x2} + Z_{B2})} \tag{12-42}$$

where

$$R_{L1} = R_{E1} \left\| \frac{(r_{\pi 2} + r_{x2} + Z_{B2})}{\beta_2} \right. \simeq r_{e2}$$

For the case of $\beta_1 R_{L1} \gg R_s + r_{x1} + r_{\pi 1}$ and $r_{\pi 2} \gg r_{x2} + Z_{B2}$, it follows that an approximate form for K_{tc} is

$$K_{tc} \simeq \frac{R_{C2}}{r_{e2}} \tag{12-43}$$

The equivalent input noise is

$$E_{ni}^2 = E_{ns}^2 + E_{n1}^2 + I_{n1}^2 R_s^2 + \frac{E_{n2}^2}{K_{t1}^2} + \frac{E_{E1}^2}{K_{t1}'^2} + I_{n2}^2\left(\frac{R_{E1}}{K_{t1}'} + \frac{Z_{B2}}{K_{t1}}\right)^2 + \frac{E_{C2}^2}{K_{tc}^2} + \left(\frac{E_{A2}}{K_{t1}\omega R_{A2}C_{B2}}\right)^2 + \left(\frac{E_{B2}}{K_{t1}\omega R_{B2}C_{B2}}\right)^2 \quad (12\text{-}44)$$

where K_{t1}' is as given previously in Eq. 12-41.

It can be seen that E_{n2} again is important to input noise, for $K_{t1} \simeq 1$. Similarly, resistor noise in stage 2 should be examined carefully in the design process.

The sample design has $I_{C1} = I_{C2} = 100\ \mu\text{A}$. The relatively large direct voltage drop across R_{E1} can result in a large amount of excess noise; it is likely that a selected low-noise resistor may be necessary for that element.

12-14 CC-CC PAIR

The CC-CC pair can be referred to as the compound connection or Darlington pair. It can be used when high-input impedance is desired; the active devices are available as an integrated pair. It is difficult to make this pair as quiet as the other configurations because of the low gain. Since the CC-CC amplifier is used with high source resistances, a low value for I_{C1} increases the input resistance and helps to optimize the noise behavior of the system.

A sample design is shown in Fig. 12-13. The transfer voltage gain K_{tc} is

$$K_{tc} \simeq \frac{\beta_1 \beta_2 R_{L1} R_{E2}}{(R_s + r_{x1} + r_{\pi 1} + \beta_1 R_{L1})(r_{x2} + r_{\pi 2} + \beta_2 R_{E2})} \quad (12\text{-}45)$$

where

$$R_{L1} = R_{E1} \| (r_{x2} + r_{\pi 2} + \beta_2 R_{E2})$$

For K_{tc} to be approximately unity it is necessary for $\beta_1 R_{L1} \gg R_s + r_{x1} + r_{\pi 1}$ and $\beta_2 R_{E2} \gg r_{x2} + r_{\pi 2}$.

The equivalent input noise E_{ni} is

$$E_{ni}^2 = E_{ns}^2 + E_{n1}^2 + I_{n1}^2 R_s^2 + \left(\frac{E_{E1}}{K_{t1}'}\right)^2 + \left(\frac{E_{n2}}{K_{t1}}\right)^2 + \left(\frac{I_{n2} R_{E1}}{K_{t1}'}\right)^2 + \left(\frac{E_{E2}}{K_{t2}' K_{t1}}\right)^2 + \left(\frac{E_{n3}}{K_{tc}}\right)^2 + \left(\frac{I_{n3} R_{E2}}{K_{t1} K_{t2}'}\right)^2 \quad (12\text{-}46)$$

where

$$K_{t1}' = \frac{\beta_1 R_{E1}}{R_s + r_{x1} + r_{\pi 1}} \simeq \frac{R_{E1}}{r_{e1}} \quad (12\text{-}47)$$

and

$$K_{t2}' = \frac{\beta_2 R_{E2}}{r_{x2} + r_{\pi 2}} \simeq \frac{R_{E2}}{r_{e2}} \quad (12\text{-}48)$$

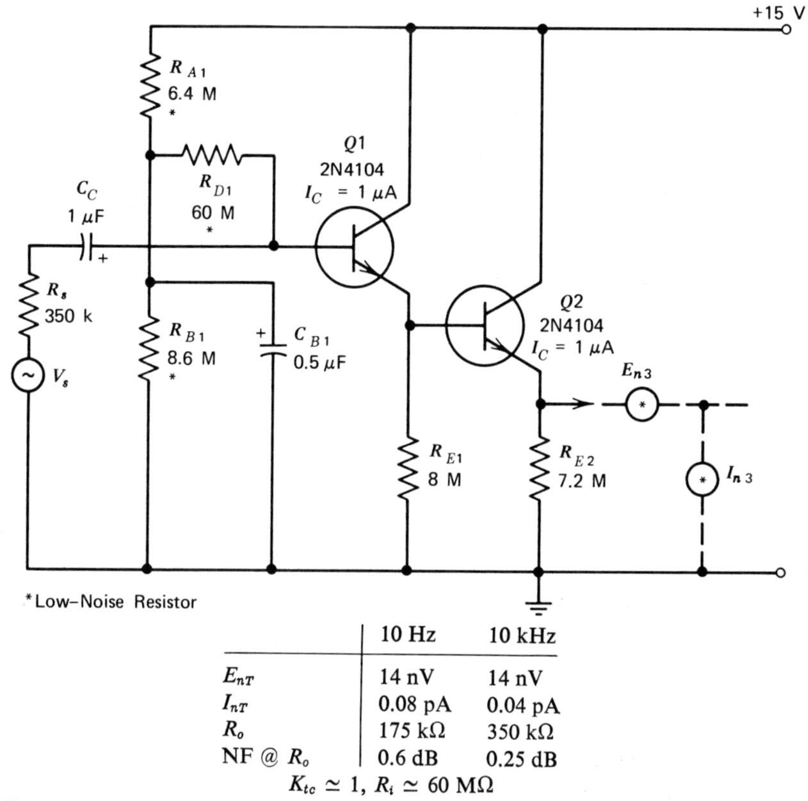

	10 Hz	10 kHz
E_{nT}	14 nV	14 nV
I_{nT}	0.08 pA	0.04 pA
R_o	175 kΩ	350 kΩ
NF @ R_o	0.6 dB	0.25 dB

$K_{tc} \simeq 1,\ R_i \simeq 60\ \text{M}\Omega$

Fig. 12-13. CC-CC pair.

The noise contribution of the third stage is also included as E_{n3} and I_{n3} in E_{ni}. The noise currents I_{n2} and I_{n3} are multiplied by emitter resistances r_{e1} and r_{e2} of the preceding stages. These terms are probably negligible compared to $I_{n1}R_s$.

In the circuit of the figure, $I_{C1} = 1\ \mu\text{A}$ to provide a high-input resistance and optimize the noise for a source resistance of 150 kΩ. Current I_{C2} is also 1 μA so that $I_{n2}r_{e1}$ is less than E_{n1}. The noise voltage and current of the third stage must be similarly adjusted.

12-15 FET BIPOLAR PAIRS

Noise in FET devices was introduced in Chapter 6. For certain applications the FET is an attractive input device. Our present interest is in the performance of pairs. From the nine usable FET-FET pairs and the nine practical

FET bipolar pairs we have selected the three most widely used for discussion in this section.

The examples given show JFETs as the input active device. A suitable MOSFET could just as well apply to each example. The bipolar stage is connected CE in all cases. Details regarding the design of the CE stage are left to the reader. For convenience, capacitance coupling is shown in the diagrams. Noise contributions of biasing elements are considered earlier in the chapter and are not repeated here.

In all three designs presented in this section, the static drain current of $Q1$ is at a relatively high level, 0.5 or 1.0 mA. The reason for operating at high levels of I_D was given in Chapter 6, to minimize the E_n parameter. Unfortunately, this selection for I_D results in relatively low values for the drain load resistor, R_D. Since voltage gain in the common-source and common-gate connections is proportional to R_D, we are not able to develop high gain in the input stage, as was accomplished in the preceding discussions on bipolar transistors.

CS-CE Pair

In the common-source (CS) configuration, the FET stage provides its highest voltage gain, along with almost infinite input impedance. Our example of this connection is the CS-CE pair shown in Fig. 12-14*a*. The gate-to-channel diode of this *n*-channel FET is reverse biased at -1 V. This bias results from the $I_D R_{S1}$ drop of 5 V and the R_{A1}-R_{B1} voltage divider that provides 4 V at the junction with R_{G1}. Because of the exceptionally high input resistance of the FET, virtually no dc flows through R_{G1}.

Figure 12-14*b* shows the small-signal ac and noise equivalent circuit for this pair. Almost no attenuation of the signal occurs in the gate biasing elements. The voltage gain provided by the CS stage is

$$K_{t1} \simeq \frac{g_m R_{L1}}{1 + g_m Z_{S1}} \tag{12-49}$$

where

$$R_{L1} = R_{D1} \| r_o \| R_{I2}$$

$$Z_{S1} = R_{S1} \| -jX_{S1}$$

Again, phase reversal is ignored in the gain equation.

For the pair, the overall gain from $Q2$ collector to V_s is

$$K_{tc} \simeq \frac{g_m R_{L1} \beta R_{C2}}{(1 + g_m Z_{S1})(r_{x2} + r_{\pi 2})} \tag{12-50}$$

For $r_{\pi 2} \gg R_{D1}$ a simple form for K_{tc} is

$$K_{tc} \simeq \frac{g_m R_{D1} R_{C2}}{r_{e2}} \tag{12-51}$$

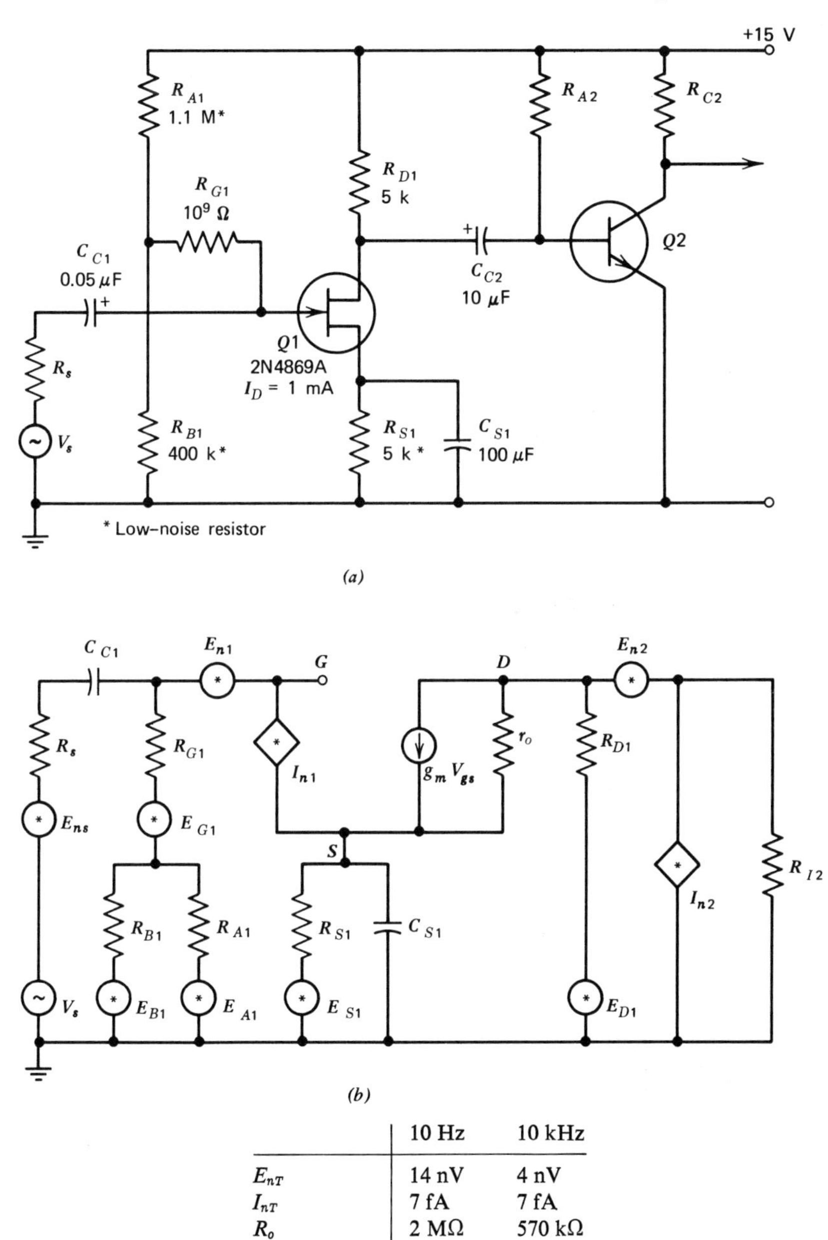

	10 Hz	10 kHz
E_{nT}	14 nV	4 nV
I_{nT}	7 fA	7 fA
R_o	2 MΩ	570 kΩ
NF @ R_o	0.05 dB	0.02 dB

$K_{t1} \simeq 10$ (FET only), $R_i \simeq 10^9\ \Omega$

Fig. 12-14. Practical CS-CE pair, performance summary, and equivalent electrical circuit.

Using typical numbers of $g_m = 0.002$ mho, $R_{D1} = 5\ \text{k}\Omega$, $R_{C2} = 50\ \text{k}\Omega$, and $r_{e2} = 250\ \Omega$ yield $K_{t1} \simeq 10$, $K_{t2} \simeq 200$, and $K_{tc} \simeq 2000$.

The equivalent input noise becomes

$$E_{ni}^2 = E_{ns}^2 + E_{n1}^2\left(\frac{R_{G1} + R_s}{R_{G1}}\right)^2 + I_{n1}^2(R_s - jX_{C1})^2$$
$$+ \left[E_{A1}^2\left(\frac{R_{B1}}{R_{A1}}\right)^2 + E_{B1}^2\left(\frac{R_{A1}}{R_{B1}}\right)^2 + E_{G1}^2\right]\frac{R_s^2}{R_{G1}^2} + \left(\frac{E_{S1}}{\omega R_{S1} C_{S1}}\right)^2$$
$$+ \frac{1}{K_{t1}^2}(E_{D1}^2 + E_{n2}^2 + I_{n2}^2 R_{D1}^2) + \frac{E_{C2}^2}{K_{tc}^2} \qquad (12\text{-}52)$$

Should R_{G1} be made very large, no particular noise problems result from the E_{n1} or the bias element terms. Rather large voltages appear across R_{A1} and R_{B1} and this may require the use of low-noise resistors in those locations.

Excellent noise performance is evident from the performance summary given in the figure.

CD-CE Pair

The common-drain (CD) FET configuration is analogous to the emitter-follower bipolar connection. The CD stage provides high input impedance, low output impedance, and voltage gain of less than unity. Input capacitance does not suffer from the Miller effect, and therefore this orientation can be used in critical applications such as the PIN photodiode example of Section 3-7.

The $Q1$ stage in Fig. 12-15 is self-biased by the $I_D R_{S1}$ drop of 1 V. Element

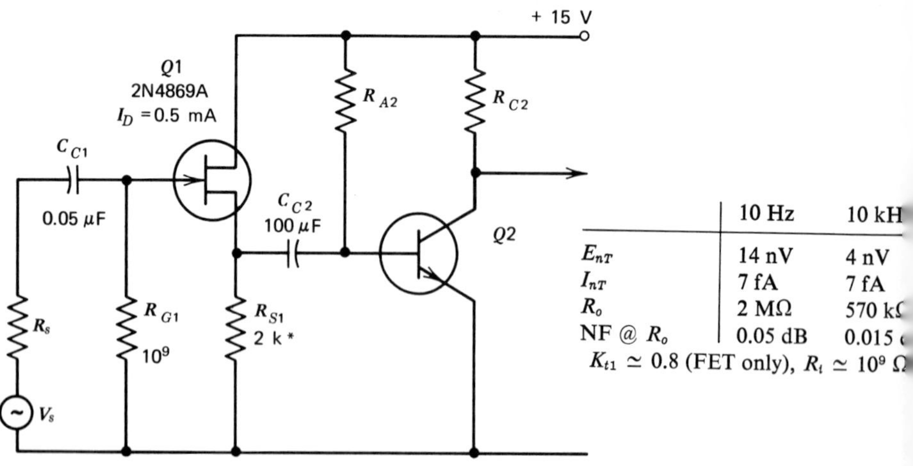

	10 Hz	10 kH
E_{nT}	14 nV	4 nV
I_{nT}	7 fA	7 fA
R_o	2 MΩ	570 k
NF @ R_o	0.05 dB	0.015

$K_{t1} \simeq 0.8$ (FET only), $R_i \simeq 10^9\ \Omega$

Fig. 12-15. CD-CE pair and performance summary.

R_{G1} is necessary to complete the gate-drain path, although no dc flows through that resistor.

The voltage gain of the CD stage is

$$K_{t1} \simeq \frac{g_m r_o R_{L1}}{r_o + R_{L1}(1 + g_m r_o)} \tag{12-53}$$

where

$$R_{L1} = R_{S1} \| R_{I2}$$

This value of this gain is always less than unity.

The amplification of the CD-CE pair is

$$K_{tc} \simeq \frac{g_m r_o R_{L1} \beta_2 R_{C2}}{[r_o + R_{L1}(1 + g_m r_o)](r_{x2} + r_{\pi 2})} \tag{12-54}$$

An approximate expression for K_{tc} is R_{C2}/r_{e2}. We also define a gain K'_{t1}:

$$K'_{t1} = g_m R_{S1} \tag{12-55}$$

The corresponding equation for E_{ni} is

$$E_{ni}{}^2 = E_{ns}{}^2 + E_{n1}{}^2\left(\frac{R_s + R_{G1}}{R_{G1}}\right)^2 + I_{n1}{}^2(R_s - jX_C)^2 + E_{G1}{}^2\left(\frac{R_s}{R_{G1}}\right)^2 + \frac{E_{n2}{}^2}{K_{t1}{}^2}$$

$$+ I_{n2}{}^2\left(\frac{R_{S1}}{K'_{t1}}\right)^2 + \frac{E_{S1}{}^2}{(K'_{t1})^2} + \frac{E_{C2}{}^2}{K_{tc}{}^2} \tag{12-56}$$

As is expected, E_{n2} and I_{n2} can be important factors in E_{ni}. The performance summary given in the figure shows noise parameters identical with the CS-CE pair.

CG-CE Pair

In the common-gate (CG) configuration, the FET exhibits low-input resistance. This connection is analogous to the CB bipolar case. Biasing for the stage shown in Fig. 12-16 is identical to the CS stage discussed earlier.

The low-frequency voltage gain of a CG stage, including signal source impedance, is given by

$$K_{t1} \simeq \left(\frac{Z_G}{Z_s}\right) \frac{R_{L1}(1 + g_m r_o)}{R_{L1} + r_o + Z_G(1 + g_m r_o)} \tag{12-57}$$

where

$$R_{L1} = R_{D1} \| R_{I2}$$

$$Z_s = R_s - jX_{C1}$$

$$Z_G = R_{S1} \| Z_s$$

A simplified form of Eq. 12-57, applicable when biasing and coupling losses are negligible, has the multiplier $Z_G/Z_s = 1$.

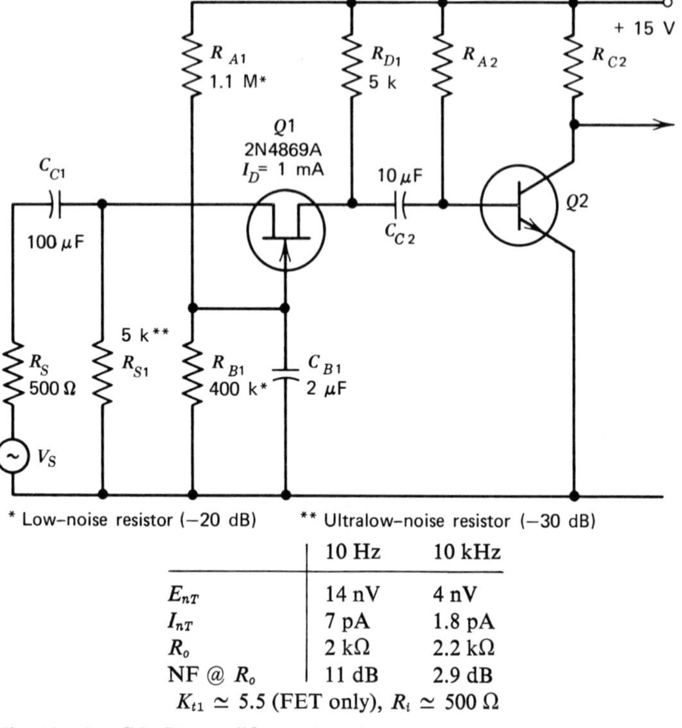

	10 Hz	10 kHz
E_{nT}	14 nV	4 nV
I_{nT}	7 pA	1.8 pA
R_o	2 kΩ	2.2 kΩ
NF @ R_o	11 dB	2.9 dB

$K_{t1} \simeq 5.5$ (FET only), $R_i \simeq 500\ \Omega$

Fig. 12-16. CG-CE amplifier and performance summary.

Voltage amplification of the CG-CE pair is

$$K_{tc} \simeq \frac{R_{L1}(1 + g_m r_o)\beta_2 R_{C2}}{[R_{L1} + r_{o1} + Z_G(1 + g_m r_o)](r_{x2} + r_{\pi 2})} \tag{12-58}$$

The expression for equivalent input noise is

$$E_{ni}^2 = E_{ns}^2 + E_{n1}^2\left(\frac{R_s + R_{S1}}{R_{S1}}\right)^2 + I_{n1}^2(Z_G + Z_B)^2 + E_{S1}^2\left(\frac{R_s}{R_{S1}}\right)^2 + \frac{1}{K_{t1}^2}[E_{D1}^2 + E_{n2}^2 + I_{n2}^2 R_{D1}^2] + \frac{E_{C2}^2}{K_{tc}^2} \tag{12-59}$$

In Eq. 12-59 $Z_B = R_{A1} \| R_{B1} \| - jX_B$. The performance summary given in the figure indicates that the NF at R_o is poorer than for the other two FET bipolar pairs. The input resistance of this pair is only 500 Ω.

12-16 PARALLEL AMPLIFIER NOISE

When the sensor resistance is very low valued, less than 100 Ω, an input coupling transformer is often utilized to match the source resistance to the R_o of the amplifier. Another method to accomplish matching is to reduce

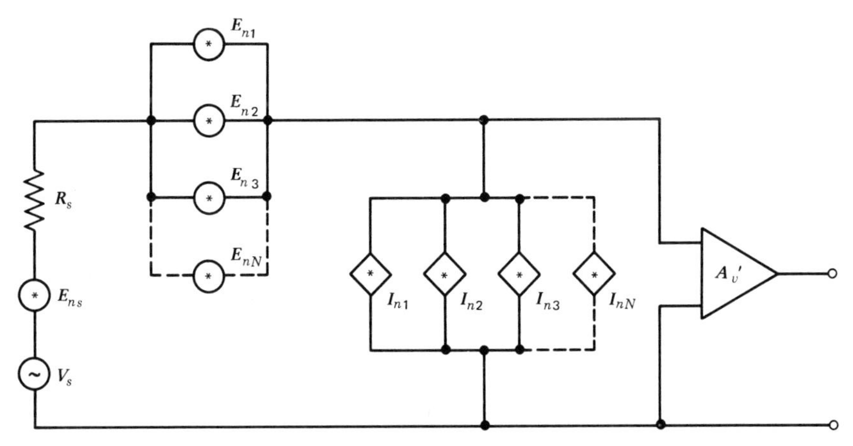

Fig. 12-17. Noise model of parallel-stage amplifier.

the amplifier's optimum noise resistance by paralleling several amplifying devices. This technique can provide matching for source impedances as small as 10 Ω.

Paralleling amplifier stages is equivalent to paralleling their E_n and I_n generators as illustrated in Fig. 12-17. The equivalent noise voltage E_n' and the equivalent noise current I_n' of a parallel system consisting of N identical stages are given by

$$E_n' = \frac{E_n}{\sqrt{N}} \tag{12-60}$$

and

$$I_n' = \sqrt{N} I_n \tag{12-61}$$

A new optimum noise resistance, E_n'/I_n', can now be defined. Its relation to R_o is

$$R_o' = \frac{E_n'}{I_n'} = \frac{R_o}{N} \tag{12-62}$$

Thus R_o can be lowered in proportion to the number of parallel stages. A practical limit, determined by cost, size, and input capacitance, is about ten transistors in parallel.

The minimum NF, F_{opt}, is proportional to the product of E_n and I_n. Since this product is unchanged by paralleling,

$$F_{\text{opt}}' = F_{\text{opt}}$$

The voltage gain A_v' of the parallel amplifier is increased in approximate proportion to the number of stages:

$$A_v' = NA_v \tag{12-63}$$

The limiting noise voltage E_n of a bipolar transistor amplifier is determined by the base resistance r_x and emitter resistance r_e (or emitter current I_E). From Chapter 4,

$$E_n^2 = 4kT\left(r_x + \frac{r_e}{2}\right) \tag{12-64}$$

Since the emitter resistance r_e is inversely proportional to I_E, it can be reduced by increasing emitter current. For $I_E > 1$ mA, r_e is usually less than r_x, and the thermal noise of the base resistance becomes the dominant noise voltage contribution. Base resistance can be controlled in transistor design, but it is difficult to reduce this parameter much below 50 Ω. Paralleling transistors, in effect, parallels the base resistances, and the resultant r_x is

$$r_x' = \frac{r_x}{N} \tag{12-65}$$

The stage input capacitance C_π' and the Miller effect increase in proportion to N:

$$C_\pi' = NC_\pi \tag{12-66}$$

The output resistance r_0' is decreased by the number of parallel stages:

$$r_o' = \frac{r_o}{N} \tag{12-67}$$

Although we have been discussing bipolar transistors a similar analysis applies to FETs and ICs.

Consider the practical circuit shown in Fig. 12-18. The gain of the parallel stages, K_{t1}, is approximately N times the gain of a single CE stage as given in Eq. 12-1. For no loading by $Q5$, we have

$$K_{t1} \simeq \frac{N\beta_1 R_{C1}'}{R_s + r_{x1} + r_{\pi 1} + \beta_1 Z_{E1}} \tag{12-68}$$

where

$$R_{C1}' = R_{C1} \left\| \frac{r_{o1}}{N} \right.$$

The second stage of the figure is connected CB. The cascaded gain is given by the product of Eqs. 12-4 and 12-68.

$$K_{tc} \simeq \frac{N\beta_1\beta_5 R_{L1} R_{C5}}{(r_{x1} + r_{\pi 1} + R_s + \beta_1 Z_{E1})(r_{\pi 5} + r_{x5} + Z_{B5})} \tag{12-69}$$

where

$$R_{L1} = R_{C1}' \left\| \frac{(r_{x5} + r_{\pi 5} + Z_{B5})}{\beta_5} \right.$$

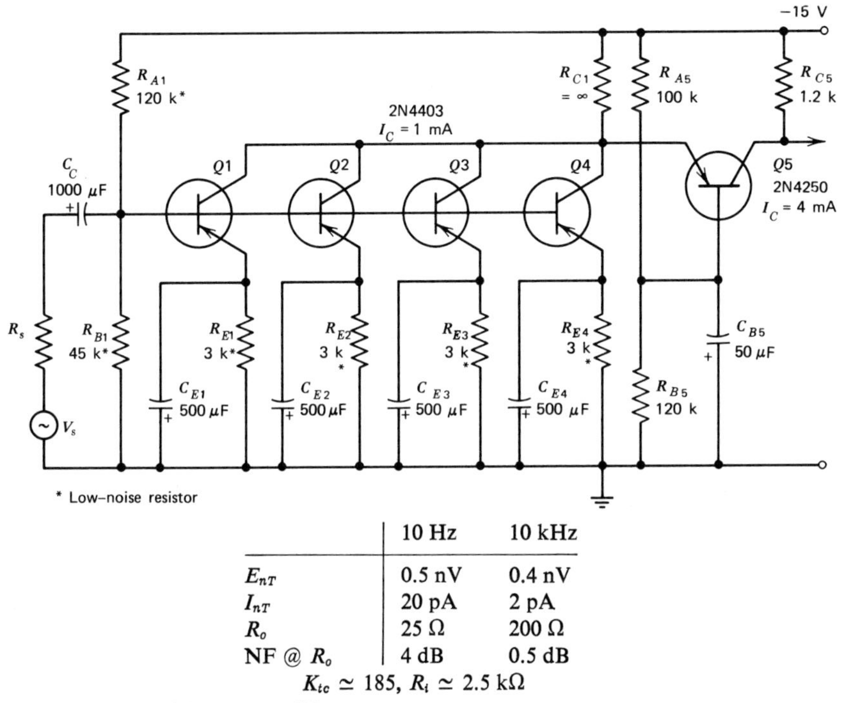

	10 Hz	10 kHz
E_{nT}	0.5 nV	0.4 nV
I_{nT}	20 pA	2 pA
R_o	25 Ω	200 Ω
NF @ R_o	4 dB	0.5 dB

$K_{tc} \simeq 185,\ R_i \simeq 2.5\ \text{k}\Omega$

Fig. 12-18. Parallel-stage amplifier.

An approximate form of Eq. 12-69 follows from the assumptions that $r_{\pi 1}$ and $r_{\pi 5}$ dominate, and $R_{C1} \gg r_{\pi 5}/\beta_5$:

$$K_{tc} \simeq \frac{NR_{C5}}{r_{e1}} \tag{12-70}$$

The equivalent input noise is

$$E_{ni}^2 = E_{ns}^2 + E_n'^2 + (I_n'R_s)^2 + \frac{E_{EN}^2}{N(\omega R_{E1}C_{E1})^2} + \frac{E_{C5}^2}{K_{tc}^2} + \frac{1}{K_{t1}^2}\left[\left(\frac{E_{C1}r_{o1}}{R_{C1}}\right)^2 + E_{n5}^2 + I_{n5}^2(R_{C1} + Z_{B5})^2\right] \tag{12-71}$$

where E_{EN} is the noise voltage of any one of the emitter resistors R_1 to R_N. Note that the rms noise of the emitter resistors is reduced in proportion to the square root of the number of parallel stages.

In the circuit of Fig. 12-18 the collector current of each of the parallel stages is set at 1 mA to give a low value of E_n and maintain minimum excess noise at 10 Hz. The cascade configuration is used in this example to minimize the Miller capacitance which is increased by N. If transistors $Q1$ to $Q4$ are matched for gain and temperature coefficient, it is possible to use a single resistor in place of R_{E1} to R_{E4}.

Overall negative feedback can be added to the circuit. An inverting stage can be connected in the forward path or in the feedback loop for the correct phasing. The feedback signal can be brought to a resistor connected between the common side of the four biasing resistors R_E and ground.

The achieved midband noise voltage E_{nT} of 0.4 nV is equivalent to the thermal noise of a 10-Ω resistor.

12-17 DIFFERENTIAL AMPLIFIER NOISE

A widely used two transistor circuit is the differential amplifier or diff amp. The diff amp was discussed in Chapter 6 in conjunction with noise in ICs. The treatment here concentrates on the noise behavior of the pair.

In the system diagram shown in Fig. 12-19*a*, the diff amp is represented by noise generators E_{n1}, E_{n2}, I_{n1}, and I_{n2}. One side of the signal source V_s is grounded, and it has noisy series resistance R_{s1}. The inverting (negative) terminal of the diff amp is returned to ground through R_{s2}. Resistances R_{s1} and R_{s2} need not be equal.

The equivalent input noise is the sum of the equivalents of the two halves of the pair:

$$E_{ni}^2 = E_{ns1}^2 + E_{ns2}^2 + E_{n1}^2 + E_{n2}^2 + I_{n1}^2 R_{s1}^2 + I_{n2}^2 R_{s2}^2 \quad (12\text{-}72)$$

Should $R_{s2} = 0$, the final term is eliminated, but E_{n2} is still present in E_{ni}.

Let us now suppose that the source is ungrounded (a floating source).The noise model of Fig. 12-19*b* applies. The noise voltages E_{n1} and E_{n2} are summed to give the total amplifier noise voltage E_{nT}:

$$E_{nT}^2 = E_{n1}^2 + E_{n2}^2 \simeq 2E_{n1}^2 \quad (12\text{-}73)$$

The noise current contributions of I_{n1} and I_{n2} are halved since each effectively sees half of the source resistance, and the total noise current I_{nT} becomes

$$I_{nT}^2 = \left(\frac{I_{n1}}{2}\right)^2 + \left(\frac{I_{n2}}{2}\right)^2 \simeq \frac{I_{n1}^2}{2} \quad (12\text{-}74)$$

The equivalent input noise E_{ni} follows from Eqs. 12-73 and 12-74:

$$E_{ni}^2 = E_{ns}^2 + E_{n1}^2 + E_{n2}^2 + \left(\frac{I_{n1}R_s}{2}\right)^2 + \left(\frac{I_{n2}R_s}{2}\right)^2 \quad (12\text{-}75)$$

If the amplifier noise mechanisms are identical, E_{ni} for the ungrounded source case simplifies to

$$E_{ni}^2 \simeq E_{ns}^2 + 2E_{n1}^2 + \frac{I_{n1}^2 R_s^2}{2} \quad (12\text{-}76)$$

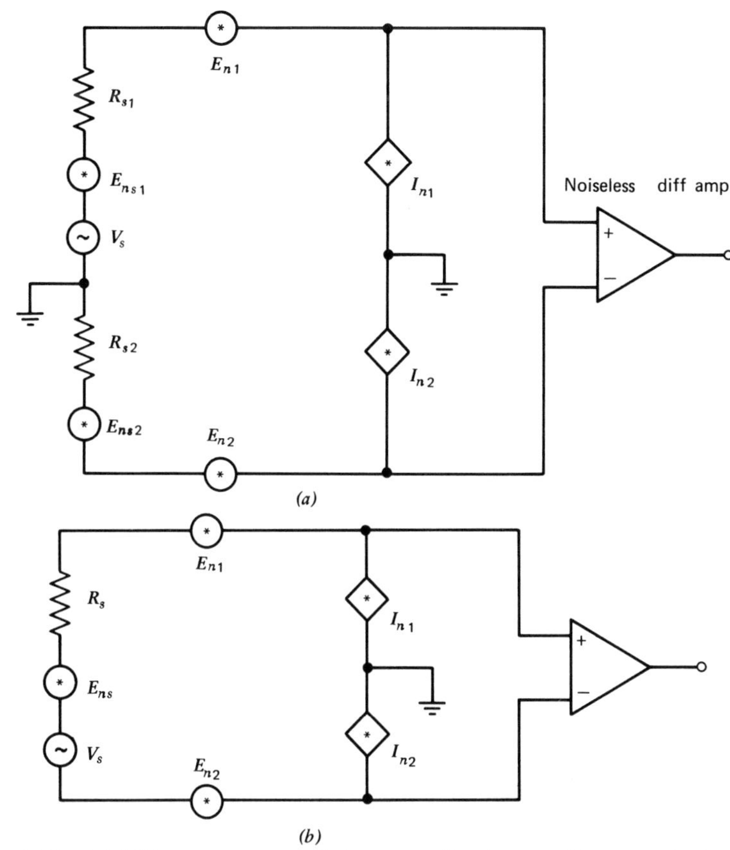

Fig. 12-19. (*a*) Diff amp with grounded source; (*b*) diff amp with source floating.

For the ungrounded source, the noise current I_{nT} is 0.7 of the noise current for the grounded input case, and the noise voltage E_{nT} is 1.4 times the single-stage noise voltage E_{n1}.

Let us now concentrate on the noise sources within the diff amp. We wish to develop the equivalent input noise model for design purposes for the circuits shown in Figs. 12-20 and 12-21.

It is necessary initially to define several gain expressions. This approach is based on the work of Meindl [1]. *The differential gain* K_{dd} *is the ratio of the differential output signal* $V_{c2} - V_{c1}$ *to the differential input signal* $V_{s2} - V_{s1}$. The output is taken between collectors. From Fig. 12-20 we obtain

$$K_{dd} = \frac{V_{c2} - V_{c1}}{V_{s2} - V_{s1}} \simeq \frac{R_C R_L/(R_C + R_L)}{1/g_m + R_E + R_s/\beta} \tag{12-77}$$

for $R_s < R_B$. The transistors are assumed to have identical parameters, and

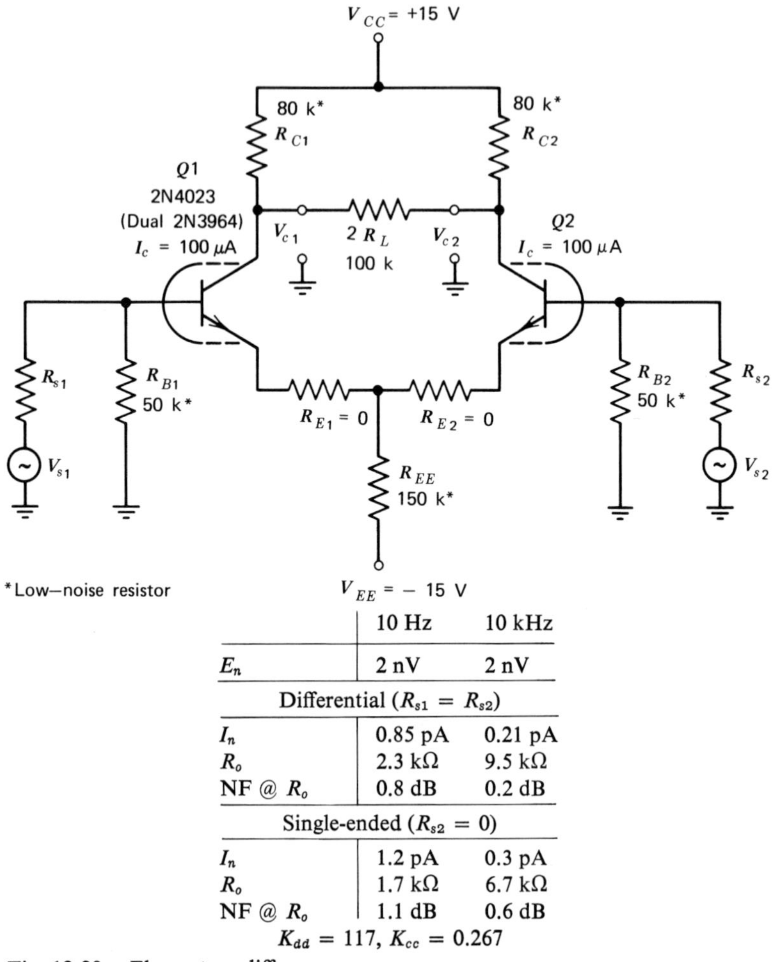

	10 Hz	10 kHz
E_n	2 nV	2 nV
Differential ($R_{s1} = R_{s2}$)		
I_n	0.85 pA	0.21 pA
R_o	2.3 kΩ	9.5 kΩ
NF @ R_o	0.8 dB	0.2 dB
Single-ended ($R_{s2} = 0$)		
I_n	1.2 pA	0.3 pA
R_o	1.7 kΩ	6.7 kΩ
NF @ R_o	1.1 dB	0.6 dB

$K_{dd} = 117$, $K_{cc} = 0.267$

Fig. 12-20. Elementary diff amp.

circuit elements for each channel are equal: $R_{C1} = R_{C2} = R_C$; $R_{E1} = R_{E2} = R_E$; and $R_{s1} = R_{s2} = R_s$.

A noise gain, independent of R_L is

$$K'_{dd} \simeq \frac{R_C}{1/g_m + R_E + R_s/\beta} \tag{12-78}$$

The common-mode voltage gain K_{cc} *is defined as the common-mode output signal for a common-mode input.* The two signal sources V_{s1} and V_{s2} are equal and in phase. Then

$$K_{cc} = \frac{V_{c2} + V_{c1}}{V_{s2} + V_{s1}} \simeq \frac{R_C}{1/g_m + R_E + R_s/\beta + 2R_{EE}} \simeq \frac{R_C}{2R_{EE}} \tag{12-79}$$

Again, $R_L \to \infty$. The simple form results when R_{EE} is very large.

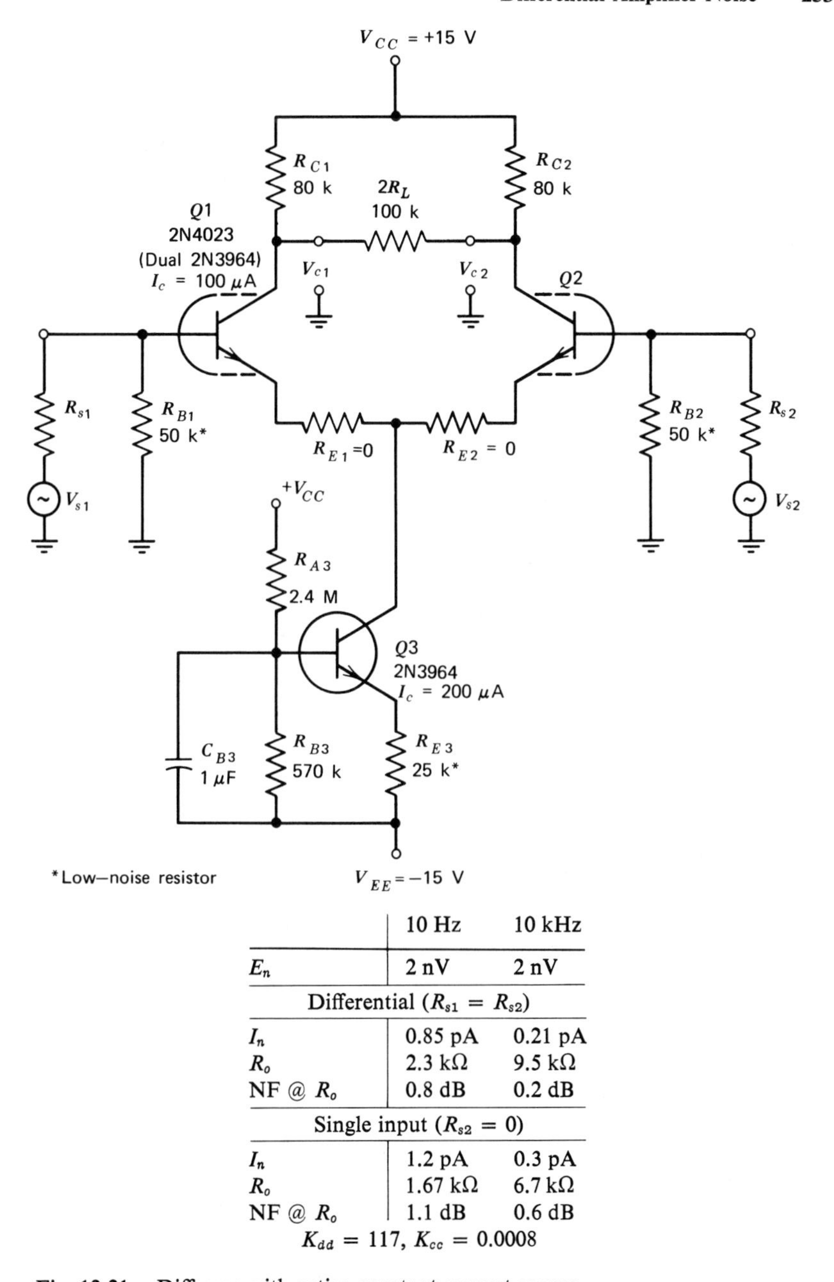

	10 Hz	10 kHz
E_n	2 nV	2 nV
Differential ($R_{s1} = R_{s2}$)		
I_n	0.85 pA	0.21 pA
R_o	2.3 kΩ	9.5 kΩ
NF @ R_o	0.8 dB	0.2 dB
Single input ($R_{s2} = 0$)		
I_n	1.2 pA	0.3 pA
R_o	1.67 kΩ	6.7 kΩ
NF @ R_o	1.1 dB	0.6 dB

$K_{dd} = 117$, $K_{cc} = 0.0008$

Fig. 12-21. Diff amp with active constant-current source.

More important is the gain K_{dc} *that produces a differential output signal for a common-mode input signal.* This gain is highly dependent on circuit balance.

$$K_{dc} = \frac{V_{c2} - V_{c1}}{2V_s} \simeq \frac{R_{p1}\left(\dfrac{R_{s2}}{\beta_2} + \dfrac{1}{g_{m2}} + R_{E2}\right) - R_{p2}\left(\dfrac{R_{s1}}{\beta_1} + \dfrac{1}{g_{m1}} + R_{E1}\right)}{2R_{EE}\left[\dfrac{R_{s2}}{\beta_2} + \dfrac{1}{g_{m2}} + R_{E2} + \dfrac{R_{s1}}{\beta_1} + \dfrac{1}{g_{m1}} + R_{E1}\right]} \tag{12-80}$$

where

$$V_{s1} = V_{s2} = V_s$$

$$R_{p1} = R_{C1} \| R_L$$

$$R_{p2} = R_{C2} \| R_L$$

If the parameters of each channel are equal, $K_{dc} \to 0$. We note from Eq. 12-80 that K_{dc} is highly dependent on R_{EE}, and on the equality of R_{C1} and R_{C2}.

The expressions presented in this section indicate the gain for signals applied at the input terminals. They also apply to common-mode voltages in series with the common-emitter resistance R_{EE}, and partially to noise from the supply voltage V_{CC} of Fig. 12-20. It is difficult to predict the exact noise contribution from the power supply V_{CC} and V_{EE}. K_{dc} predicts the maximum value, but it is possible to reduce the V_{CC} noise by coupling a portion of the noise to the inverting input or even to the base of the transistor acting as a constant-current source ($Q3$).

The preceding gain equations apply to input signals that are correlated; that is, either dc signals or ac signals from the same source. The uncorrelated noise voltage E_n in series with each input is not attenuated by the gain K_{dc}. It is amplified by the differential gain K_{dd}. Each input noise source produces an output independent of the others.

Equations 12-78 and 12-80 do not apply to common-mode noise developed in the constant-current source $Q3$ of Fig. 12-21 and in its biasing elements. *An expression for* K_{cE}, *the common-mode output for emitter input to* Q3, *can be derived from the discussion of the CB-CB pair considered earlier*. In essence, stages $Q3$ and $Q1$ form a CB-CB cascade to a signal inserted in series with R_{E3}. From Eq. 12-32 we obtain

$$K_{cE} \simeq \frac{\beta_3 R_{p2}}{r_{x3} + r_{\pi 3} + \beta_3 R_{E3} + Z_{B3}} \tag{12-81}$$

where

$$R_{p2} = R_{C2} \| R_L$$

$$Z_{B3} = R_{A3} \| R_{D3} \| - jX_{CB3}$$

A more important expression is *the gain* K_{dE} *that produces a differential mode output voltage for input at the emitter of* $Q3$:

$$K_{dE} \simeq \frac{R_{p2} - R_{p1}}{2R_{E3}} \tag{12-82}$$

for the case of $\beta_3 R_{E3} \gg r_{x3} + r_{\pi 3} + Z_{B3}$. K_{dE} is similar to the expression for the common-mode input K_{dc} with the substitution of emitter resistor R_{E3} for the common-mode resistor R_{EE} of Eq. 12-80.

The expression for equivalent input noise of the circuits of Figs. 12-20 and 12-21 can be written in terms of the preceding gain expressions

$$\begin{aligned}
E_{ni}^2 = {} & E_{s1}^2 + E_{s2}^2 + \left(\frac{E_{n1}R_{B1}}{R_{s1} + R_{B1}}\right)^2 + \left(\frac{E_{n2}R_{B2}}{R_{s2} + R_{B2}}\right)^2 \\
& + I_{n1}^2 R_{s1}^2 + I_{n2}^2 R_{s2}^2 \\
& + \left(\frac{E_{B1}R_{s1}}{R_{B1}}\right)^2 + \left(\frac{E_{B2}R_{s2}}{R_{B2}}\right)^2 + E_{E1}^2 + E_{E2}^2 \\
& + \frac{E_{C1}^2 + E_{C2}^2}{(K'_{dd})^2} + \left(\frac{E_{EE}^2 + E_{VEE}^2 + E_{VCC}^2}{K_{dd}^2}\right) K_{dc}^2 \\
& + \left[\left(\frac{E_{A3}}{\omega R_{A3} C_{B3}}\right)^2 + \left(\frac{E_{B3}}{\omega R_{B3} C_{B3}}\right)^2 + E_{n3}^2 + (I_{n3} Z_{B3})^2 + E_{E3}^2\right] \frac{K_{dE}^2}{K_{dd}^2}
\end{aligned} \tag{12-83}$$

Where E_{VEE} and E_{VCC} are the noise variations on power supply lines, as felt at the R_{EE} location. If a transistor with output resistance r_{o3} replaces R_{EE}, the following substitution is made in the equation for K_{dc}:

$$R_{EE} = r'_{o3} \simeq r_{\mu 3} \| (r_{o3} + 2\beta_3 R_{E3})$$

For low-noise operation a diff amp should be followed by a second differential stage. If the second stage is single ended (output taken from one side of R_L) the differential gain expressions are replaced by their common-mode expressions in Eq. 12-83, and there may be little or no rejection of the common-mode noise voltages E_{EE}, E_{E3}, E_{VEE}, and E_{VCC}. Thus K_{cc} is substituted for K_{dc} and K_{cE} is substituted for K_{dE}. This reduces the common-mode rejection by 10 to 100.

The noise of the second diff amp stage is less critical because of the first-stage gain. The second-stage noise can be analyzed as in Eq. 12-83 and divided by the first-stage differential gain K_{dd}.

The circuits in Figs. 12-20 and 12-21 use a dual 2N4023 transistor consisting of two matched low-noise 2N3964 chips on a single substrate. This results in both low noise and low dc offset. Although both circuits have

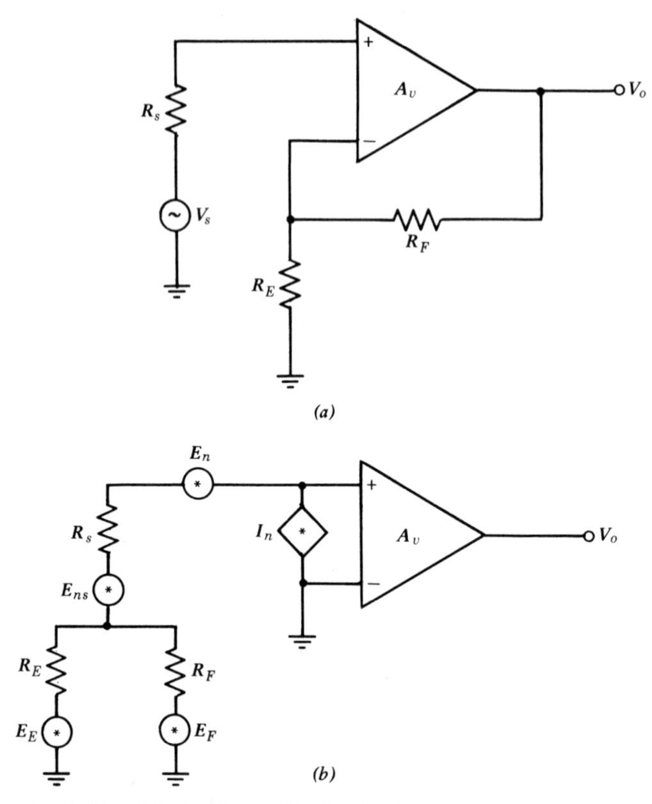

Fig. 12-22. "Series-" negative feedback.

approximately the same noise, the circuit of Fig. 12-21 with its active constant-current source gives higher common-mode rejection.

Resistors R_{A3} and R_{B3} are bypassed with capacitor C_{B3} to decrease their noise contribution and to increase the output resistance r'_{o3}. Bypassing the noise of R_{E3} does not help, since it increases the common-mode gain in proportion to the decrease in noise. It is important for both R_{E3} and R_{EE} to be low noise.

12-18 NOISE IN FEEDBACK AMPLIFIERS

One of the most powerful tools in the circuits designer's hands is negative feedback. Negative feedback can be used to raise or lower terminal impedance levels, decrease gain, increase bandwidth, and stabilize operating performance.

Negative feedback does not change the equivalent input noise of an amplifier at frequencies for which the internal feedback capacitance of the active device provides isolation [2,3,4]. This does not mean that the output noise of an amplifier is unaffected by feedback; we know that output noise is linearly related to gain. Only the signal-to-noise ratio remains unchanged by negative feedback. Although feedback does not add noise, the resistors of the feedback network can contribute to E_{ni} because of their thermal and excess noise mechanisms.

Two examples of negative voltage feedback are illustrated in Figs. 12-22 and 12-23. In the first figure, feedback is in series with the input port; Fig. 12-23 shows the feedback signal in parallel with the input port. In Fig. 12-22 the gain block is depicted to be a differential amplifier or an operational amplifier; this is equivalent to feedback to the emitter of a CE input stage.

Series voltage feedback (Fig. 12-22) increases the input impedance of the amplifier, and is useful when operating from a wide range of source resistances or when measuring the signal from a voltage source. The overall gain K_f is defined by the classic expression for the gain of a feedback amplifier noted earlier:

$$K_f = \frac{A_v}{1 - A_v\beta_{FB}} \tag{12-84}$$

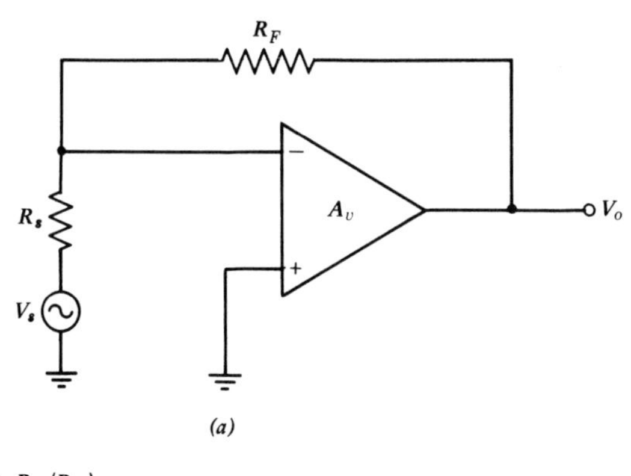

(a)

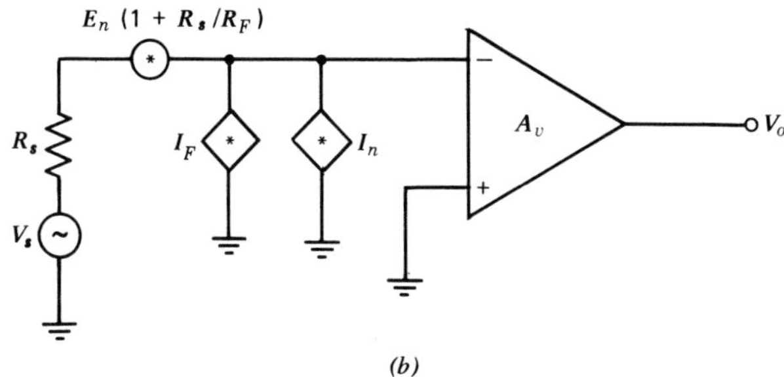

(b)

Fig. 12-23. "Parallel-" negative feedback.

The gain of the forward path is A_v. For this example,

$$\beta_{FB} = -\frac{R_E}{R_F + R_E} \tag{12-85}$$

when r_o of the amplifier is low valued.

An expression for E_{ni} is now presented. We obtain additional terms due to the product of I_n and $R_E \| R_F$, and the noise voltages of R_E and R_F.

$$E_{ni}^2 \simeq E_{ns}^2 + E_n^2 + I_n^2\left(R_s + \frac{R_E R_F}{R_E + R_F}\right)^2 + E_E^2\left(\frac{R_F}{R_E + R_F}\right)^2 + E_F^2\left(\frac{R_E}{R_E + R_F}\right)^2 \tag{12-86}$$

Equation 12-86 assumes $r_o \to 0$ and $A_v > 1/\beta_{FB}$. The noise equivalent circuit of this amplifier, including feedback elements, is shown in Fig. 12-22*b*.

An amplifier with negative feedback paralleling its input port is shown in Fig. 12-23. This type of feedback reduces the input resistance of the amplifier, and minimizes the effect of shunt capacitance. When operated from a high-impedance source, this is a current amplifier with output voltage proportional to the input signal current. We can again invoke Eq. 12-84 to predict performance. However, for this example, $\beta_{FB} = -R_s/R_F$. The closed-loop gain equals $-R_F/R_s$ for large A_v. For best performance it is desirable to have amplifier R_I greater than R_s.

The equivalent input noise for this system is

$$E_{ni}^2 \simeq E_{ns}^2 + E_n^2\left(1 + \frac{R_s}{R_F}\right)^2 + I_n^2 R_s^2 + \left(\frac{E_F R_s}{R_F}\right)^2 \tag{12-87}$$

A noise equivalent circuit is shown in Fig. 12-23*b*.

Examples of several amplifiers with overall negative feedback are given in Chapter 13.

SUMMARY

a. Selection of the configuration for a bipolar input stage can be based on nonnoise characteristics.

b. E_{ni}, E_n, and I_n are basically identical for all bipolar configurations. E_{ni} is affected by noise from biasing components and second-stage contributions.

c. Overall negative feedback does not affect E_{ni}, except for the added thermal and excess noise in feedback elements. Hence there is no change in R_o.

d. Selection of the appropriate bipolar pair depends on requirements for gain, input impedance, and frequency behavior. A low-noise figure is achievable for any pair if the proper precautions are included during design.

e. The FET bipolar pair can provide low-noise parameters, and is especially useful with sources of high-internal resistance.

f. The parallel-stage amplifier is useful when a low R_o must be present.

g. Noise in the diff amp is the result of many causes, including differential output for common-mode input. A second diff amp pair is recommended to minimize this type of noise.

PROBLEMS

1. Show that the gain expression relating E_c^2 to E_{ni}^2 in the CE circuit of Fig. 12-1 is K_t', where $K_t' = K_t$ when $R_L = R_C$.

2. Assume E_E to be the only noise source in Fig. 12-3. Determine E_{no} caused by this noise, and show that $E_{ni} = E_E/K_t'$, where K_t' is given by Eq. 12-9.

3. Design an emitter-coupled pair (CC-CB) to meet the following requirements:

$$V_{CC} = +10\text{ V} \qquad Q1 = Q2 = \text{2N4250}$$

$$R_s = 2\text{ k}\Omega \qquad V_s = 1\ \mu\text{V (max)}$$

$$K_t = 100\text{ (min)} \qquad f_L = 1\text{ kHz}$$

Determine values for all bias elements after selecting appropriate operating points. Calculate E_{nT}, I_{nT}, E_{ni}, and K_{tc} for your design.

4. The three pairs with CC input stages are found to have 3 MΩ, 50 kΩ, and 60 MΩ input resistances. Explain why the input resistance of the CC-CB pair is so much lower than the other two.

5. Equation 12-63 states that the voltage gain of the parallel-stage amplifier is proportional to the product of the number of stages (N) and the gain of each stage (A_v).
   (a) Show a simple model for each transistor in order for this to be true.
   (b) What approximations are being made in your model?

REFERENCES

1. Meindl, J. D., *Micropower Circuits*, Wiley, New York, 1969, p. 145.
2. Nielsen, E. G., "Behavior of Noise Figure and Junction Transistors," *Proc. IRE*, **45**, 1 (July 1957), 957–963.
3. Middlebrook, R. D., "Optimum Noise Performance of Transistor Input Circuits," *Semiconductor Products* (July-August, 1958), 14–20.
4. Bogner, R. E., "Feedback and Noise Performance," *Electron. Eng.*, **37**, 444 (February 1965), 115–117.

Chapter 13

LOW-NOISE AMPLIFIER DESIGNS

The design of low-noise amplifiers follows the techniques discussed in Chapters 7 through 10. The analyses given in Chapter 12 provide guides for calculation of the equivalent input noise parameters of cascaded stages, and supply us with knowledge about the important noise generators in the electronic system.

By presenting examples of complete amplifiers, this chapter intends to suggest to the reader a set of successful low-noise designs. The CE-CE cascade is discussed as an open-loop and feedback amplifier. Designs using the cascode connection with and without feedback are also presented. A FET-input amplifier design is shown. The final example is that of a low-noise differential-input-stage amplifier.

For specific applications where the sensor characteristics are known, the designs shown in this chapter can be modified in order to marry the amplifier to the sensor. Any of the given designs can be constructed as presented for use where a general-purpose low-noise amplifier is required.

13-1 CASCADE AMPLIFIER

As the first example consider the complementary cascade shown in Fig. 13-1. The voltage gain of this amplifier is dependent on the CE-CE pair noted as $Q1$ and $Q2$. This pair was analyzed in Section 12-6. The output stage is an emitter follower to provide a low output impedance.

Direct coupling is employed between stages. Capacitor C_{C1} is shown at the

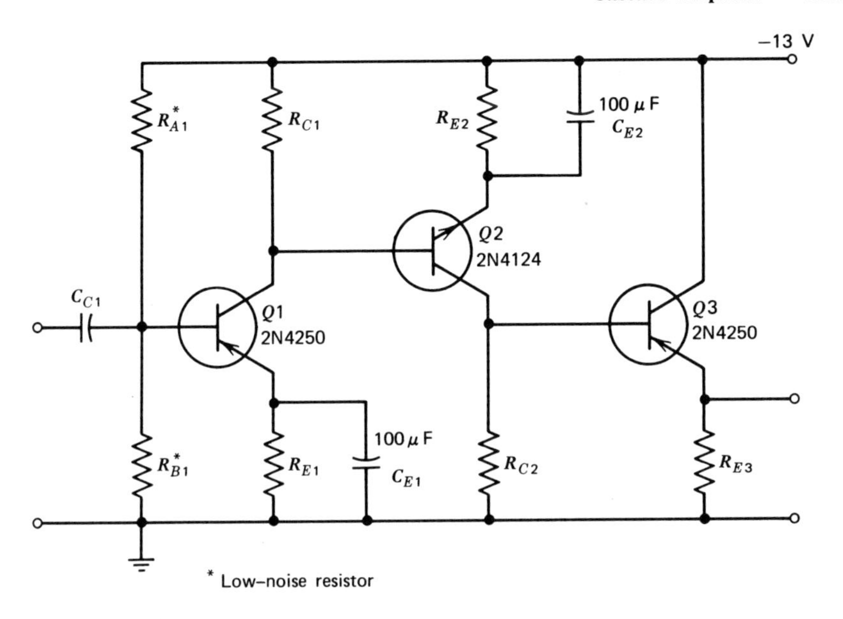

Fig. 13-1. Complementary cascade example.

input port in order to isolate the amplifier direct voltage level from the signal source. No value is given for that element, since selection of its size is dependent on the product of the source resistance and the noise current I_n of the first stage at the lowest frequency of interest. To assure that I_n results in negligible system noise, it is desirable that

$$I_n|R_s - jX_c| \ll E_n$$

at that frequency.

In order to show the potential of this amplifier, performance data are tabulated in Table 13-1. The four columns represent four values for quiescent I_C ranging from high I_C (1 and 3 mA) to low (1 and 10 μA). To achieve these variations values of biasing elements must change. For a reduction of 10:1 in I_C, it is desirable that R_C and R_E increase by a factor of ten times.

Certain trends are clear from the data. The voltage-gain upper-cutoff frequency f_2 declines with decreased I_C, as does the input capacitance C_i and the voltage gain. Input resistance, as expected, increases.

From the noise parameter listing we find that I_{nT} is reduced by a factor of about ten, and E_{nT} is increased by that same amount when I_C is lowered. The optimum source resistance R_o increases by more than 100:1.

Table 13-1. Performance Data on Complementary Cascade of Fig. 13-1. (Noise values are given on a per root hertz basis.)

I_{C1}	1 mA	100 μA	10 μA	1 μA
I_{C2}	3 mA	300 μA	30 μA	10 μA
R_{A1}	470 kΩ	4.7 MΩ	30 MΩ	30 MΩ
R_{B1}	220 kΩ	2.2 MΩ	22 MΩ	22 MΩ
R_{E1}	3 kΩ	30 kΩ	300 kΩ	3 MΩ
R_{C1}	3.6 kΩ	36 kΩ	360 kΩ	3 MΩ
R_{E2}	820	8.2 kΩ	82 kΩ	240 kΩ
R_{C2}	1.8 kΩ	18 kΩ	180 kΩ	500 kΩ
R_{E3}	1 kΩ	1 kΩ	100 kΩ	100 kΩ
A_v	6700	5250	2900	600
f_2	240 kHz	36.5 kHz	7.5 kHz	4.0 kHz
R_i (1 kHz)	11 kΩ	90 kΩ	825 kΩ	3 MΩ
C_i (1 kHz)	310 pF	280 pF	160 pF	62 pF
E_{nT} (10 Hz)	2.0 nV	2.5 nV	5.4 nV	14 nV
E_{nT} (10 kHz)	1.5 nV	2 nV	5 nV	14 nV
I_{nT} (10 Hz)	6.0 pA	1.2 pA	0.4 pA	0.1 pA
I_{nT} (10 kHz)	0.9 pA	0.3 pA	0.1 pA	0.06 pA
R_o (10 Hz)	330 Ω	2.4 kΩ	13.5 kΩ	140 kΩ
R_o (10 kHz)	1.7 kΩ	6.7 kΩ	50 kΩ	230 kΩ

Table 13-2. Performance Data on Cascode Amplifier of Fig. 13-2

I_{C1}	1 mA	100 μA	10 μA	1 μA
I_{C2}	1 mA	100 μA	10 μA	1 μA
R_{A1}	300 kΩ	3 MΩ	20 MΩ	20 MΩ
R_{B1}	150 kΩ	1.5 MΩ	10 MΩ	10 MΩ
R_{E1}	3 kΩ	30 kΩ	300 kΩ	3 MΩ
R_{C2}	3 kΩ	30 kΩ	300 kΩ	3 MΩ
R_{E2}	10 kΩ	100 kΩ	1 MΩ	1 MΩ
A_v	97	115	120	145
f_2	5 MHz	500 kHz	47 kHz	5.3 kHz
R_i (1 kHz)	11.1 kΩ	75 kΩ	730 kΩ	4.0 MΩ
C_i (1 kHz)	130 pF	34 pF	35 pF	27 pF
E_{nT} (10 Hz)	2 nV	2.5 nV	5.4 nV	14 nV
E_{nT} (10 kHz)	1.5 nV	2 nV	5 nV	14 nV
I_{nT} (10 Hz)	6 pA	1.2 pA	0.4 pA	0.1 pA
I_{nT} (10 kHz)	0.9 pA	0.3 pA	0.1 pA	0.06 pA
R_o (10 Hz)	330 Ω	2.4 kΩ	13.5 kΩ	140 kΩ
R_o (10 kHz)	1.7 kΩ	6.7 kΩ	50 kΩ	230 kΩ

13-2 CASCODE AMPLIFIER

The cascode amplifier shown in Fig. 13-2 is presented for comparison with the cascade amplifier of Section 13-1. Transistors *Q*1 and *Q*2 provide the gain and *Q*3 is an emitter follower useful for impedance matching.

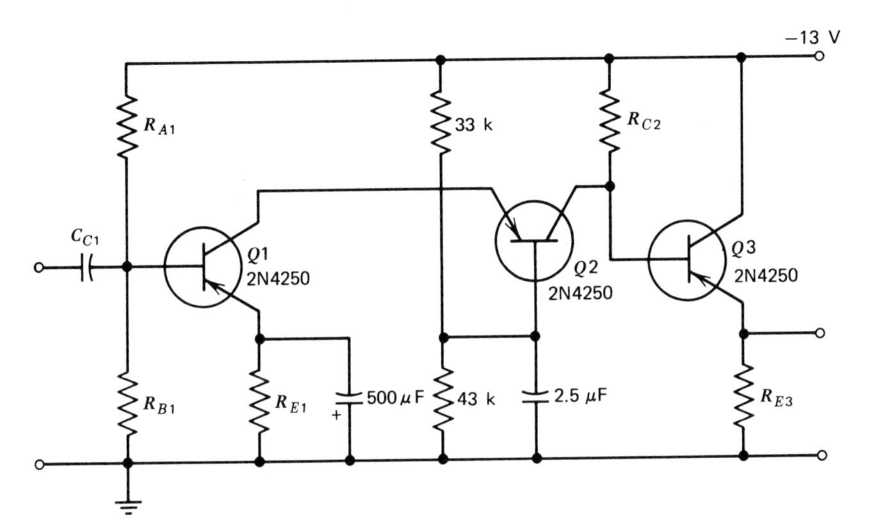

Fig. 13-2. Cascode amplifier example.

A performance summary is given in Table 13-2. These data can be compared with Table 13-1. The total voltage gain A_v of the cascode circuit is lower than the CE-CE configuration. Noise parameters are nearly equal. A major reduction is achieved in C_i and the voltage-gain upper-cutoff frequency is clearly extended.

13-3 HIGH-IMPEDANCE AMPLIFIER

When an extremely high input resistance and low input capacitance are required, the common-drain FET configuration can be considered for the input device. It is possible to obtain $R_i \simeq 10^9\ \Omega$ and $C_i \simeq 1.5$ pF with the circuit shown in Fig. 13-3.

Transistors *Q*2 and *Q*3 are a cascode of the type discussed in the preceding section. The emitter-follower *Q*4 has a 100-Ω resistance in its collector branch to protect the collector-base diode if the output is shorted.

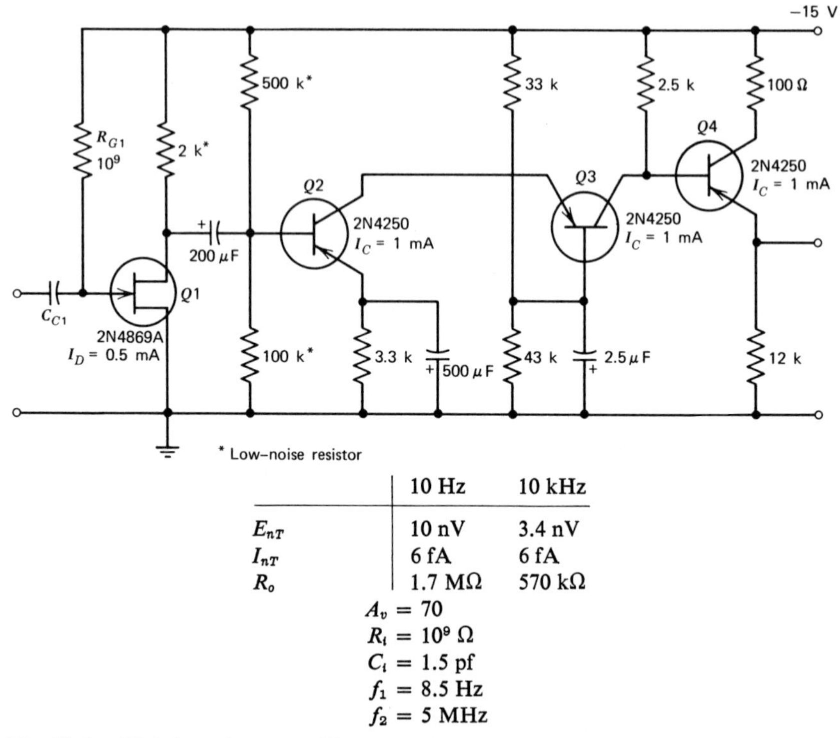

	10 Hz	10 kHz
E_{nT}	10 nV	3.4 nV
I_{nT}	6 fA	6 fA
R_o	1.7 MΩ	570 kΩ

$A_v = 70$
$R_i = 10^9\ \Omega$
$C_i = 1.5$ pf
$f_1 = 8.5$ Hz
$f_2 = 5$ MHz

Fig. 13-3. High-impedance amplifier example.

A performance summary is shown in the figure. The circuit has a 3-dB bandwidth from 8.5 Hz to 5 MHz. The noise parameters are low; E_{nT} can be reduced if I_D is selected closer to the value of I_{DSS}.

13-4 CASCADE FEEDBACK AMPLIFIER

A widely used low-noise amplifier design is the complementary cascade circuit with overall feedback shown in Fig. 13-4. The feedback path connects the collector of $Q4$ to the emitter circuit of $Q2$. $R_{C4} \| C_4$ forms the feedback impedance Z_F, and R_{F2} is the impedance Z_I. Voltage gain of the stages within the feedback loop is approximately Z_F/Z_I. In this example that gain is 100.

Transistors $Q1$, $Q3$, and $Q5$ are connected as emitter followers. The overall gain is therefore about 100. Input resistance is determined by the base-biasing elements of $Q1$. Three silicon diodes are shown to protect the first three transistors from being avalanched by input and turn-on transients.

To minimize the possibility of oscillation I_{C2} and I_{C4} are selected as shown

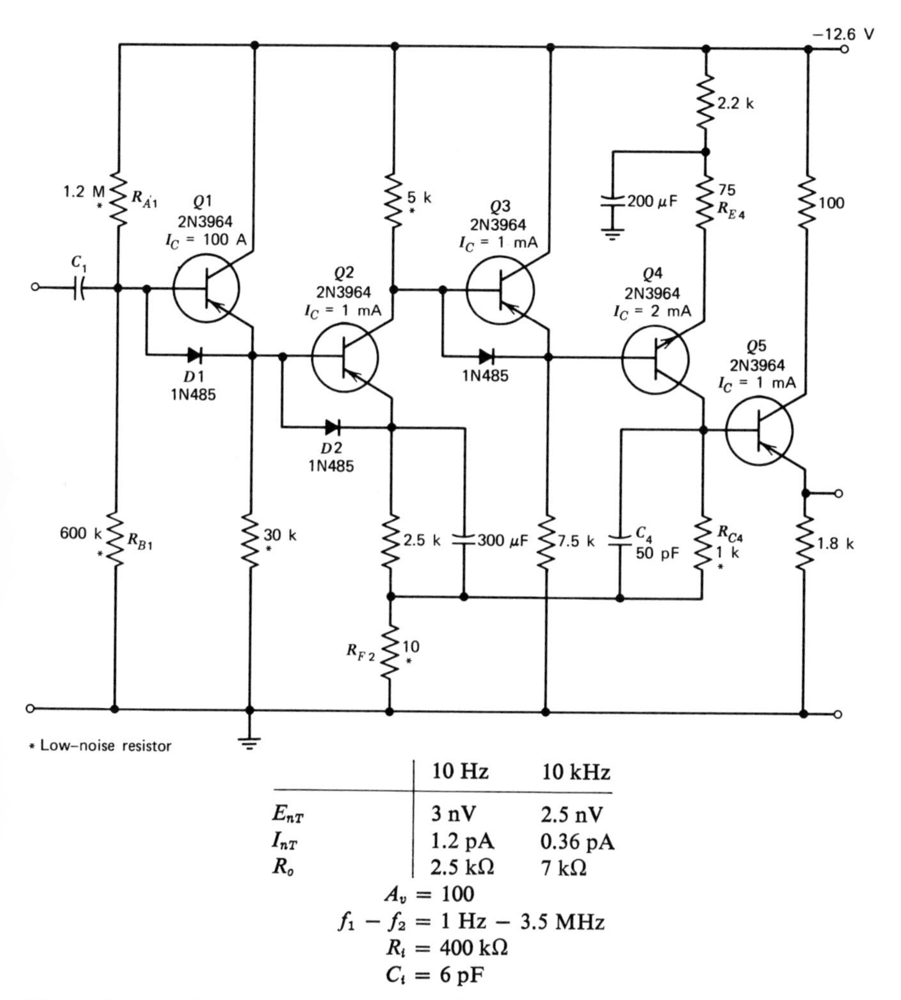

	10 Hz	10 kHz
E_{nT}	3 nV	2.5 nV
I_{nT}	1.2 pA	0.36 pA
R_o	2.5 kΩ	7 kΩ

$A_v = 100$

$f_1 - f_2 = 1\text{ Hz} - 3.5\text{ MHz}$

$R_i = 400\text{ k}\Omega$

$C_i = 6\text{ pF}$

Fig. 13-4. Complementary cascade with feedback.

in the figure, and C_4 increases the feedback at high frequencies where the total phase shift around the forward-feedback loop is large.

We now present a sample calculation of the input noise parameters for the circuit of Fig. 13-4. The first two stages are a CC-CE pair. The expression for E_{ni} is available as Eq. 12-40:

$$E_{ni}^2 \simeq E_{ns}^2 + E_{n1}^2 + I_{n1}^2 R_s^2 + \frac{E_{n2}^2}{K_{t1}^2} + \left(\frac{I_{n2} R_{E1}}{K'_{t1}}\right)^2 + \left(\frac{E_{E1}}{K'_{t1}}\right)^2 + \frac{E_{C2}^2}{K_{tc}^2} \tag{12-40}$$

We consider $K_{t1} \simeq 1$, $K'_{t1} \simeq R_{E1}/r_{e1}$, and $K_{tc} \simeq R_{C2}/r_{e2}$. Equation 12-40 can be separated into two parts, E_{nT} and I_{nT}. Source resistance is not specified; we use 30 μF for C_1. A term E_{F2}^2 is added to include noise in feedback resistor R_{F2}. Then, from Eq. 12-40, for $R_s \to 0$,

$$E_{nT}^2 \simeq E_{n1}^2 + I_{n1}^2 X_{C1}^2 + E_{n2}^2 + I_{n2}^2 r_{e1}^2 + \left(\frac{E_{E1} r_{e1}}{R_{E1}}\right)^2 + \left(\frac{E_{C2} r_{e2}}{R_{C2}}\right)^2 + E_{F2}^2 \tag{13-1}$$

and, for $R_s \to \infty$,

$$I_{nT}^2 = I_{n1}^2 + \frac{E_{A1}^2}{R_{A1}^2} + \frac{E_{B1}^2}{R_{B1}^2} \tag{13-2}$$

where E_{A1} and E_{B1} are the noise voltages in first-stage bias elements. An additional term, $(I_{n3} r_{e2})^2$, is sometimes included in Eq. 13-1.

Let us evaluate Eqs. 13-1 and 13-2 for the present example. At 10 Hz, the data for the 2N3964 (same as 2N4250) shown in Appendix I predict $E_n/\sqrt{\sim} =$ 2 nV and $I_n/\sqrt{\sim} = 1.2$ pA at $I_C = 100\ \mu$A, the operating point of $Q1$. For $Q2$ the corresponding values are 1.5 nV and 6 pA. The value of $X_{C1} = 530\ \Omega$ at that frequency. From the values for I_{C1} and I_{C2}, we obtain $r_{e1} = 250\ \Omega$ and $r_{e2} = 25\ \Omega$.

The thermal noise levels in R_{E1}, R_{A1}, R_{B1}, R_{F2}, and R_{C2} are 22 nV, 150 nV, 100 nV, 0.4 nV, and 9 nV, respectively. Assume that each of these resistors has a NI of -20 dB. This corresponds to 0.02 μV/V at 10 Hz from Fig. 9-3. The dc drops across these elements are 3, 9.0, 3.6, 0.02, and 5 V, respectively. Hence

$$E_{E1} = [(22 \times 10^{-9})^2 + (0.02 \times 10^{-6} \times 3)^2]^{1/2} = 64 \text{ nV}$$

Similarly, at 10 Hz, $E_{F2} = 0.56$ nV, $E_{A1} = 235$ nV, $E_{B1} = 133$ nV, and $E_{C2} = 100$ nV.

The terms in Eq. 13-1 are

$$E_{n1} = 2 \text{ nV} \qquad I_{n2} r_{e1} = 1.5 \text{ nV} \qquad \frac{E_{C2} r_{e2}}{R_{C2}} = 0.5 \text{ nV}$$

$$I_{n1} X_{C1} = 0.64 \text{ nV} \qquad \frac{E_{E1} r_{e1}}{R_{E1}} = 0.55 \text{ nV}$$

$$E_{n2} = 1.5 \text{ nV} \qquad E_{F2} = 0.56 \text{ nV}$$

The sum of the squares of these terms is 9.7×10^{-18}. *The square root of this number is* 3.1 *nV, which is* $\mathrm{E_{nT}}$.

The terms in Eq. 13-2 are

$$I_{n1} = 1.2 \text{ pA}$$

$$\frac{E_{A1}}{R_{A1}} = 0.2 \text{ pA}$$

$$\frac{E_{B1}}{R_{B1}} = 0.2 \text{ pA}$$

Squaring these terms, summing, and then taking the square root gives $I_{nT} =$ 1.23 pA.

Similar calculations are possible for other frequencies provided that the necessary data are available. At 10 kHz we find $E_{n1} = 2$ nV, $E_{n2} = 1.5$ nV, $I_{n1} = 0.3$ pA, and $I_{n2} = 0.9$ pA. The $I_{n1}{}^2 X_{C1}{}^2$ term in Eq. 13-1 has been eliminated from importance, and transistor noise is less than before. Excess noise in the resistors is 1/30 of the 10-Hz values. New calculations yield

$$E_{nT} = (2^2 + 1.5^2 + 0.23^2 + 0.18^2 + 0.5^2 + 0.4^2)^{1/2} = 2.6 \text{ nV}$$

and

$$I_{nT} = (0.3^2 + 0.13^2 + 0.17^2)^{1/2} = 0.4 \text{ pA}$$

These calculations are in agreement with the measured values shown in the figure.

13-5 CASCODE FEEDBACK AMPLIFIER

A circuit that provides a wider bandwidth and the same gain as the preceding example (100) is the cascode feedback amplifier shown in Fig. 13-5. Stages $Q2$ and $Q3$ are a cascode of the type discussed in Section 13-2. Transistors $Q1$, $Q4$, and $Q6$ are emitter followers for impedance matching. Transistor $Q5$ is a CE amplifying stage.

Overall feedback connects the collector of $Q5$ with the emitter circuit of $Q2$. The gain is determined primarily by R_{C5} and R_{F2}. Currents I_{C2} and I_{C3} are chosen to be rather large in order to increase f_T of those stages. Element C_{B5} lowers the gain of $Q5$ at high frequencies, and, along with C_{F5}, helps stabilize the amplifier.

Calculations for E_{nT} and I_{nT} are based on the following:

$$E_{n1} = 1.5 \text{ nV} \qquad E_{n2} = 1 \text{ nV}$$

$$I_{n1} = 6 \text{ pA} \qquad I_{n2} = 50 \text{ pA}$$

These values follow from the information given in Appendix I for 10 Hz. Again, the $I_{n3} r_{e2}$ term may be considered since the gain of $Q2$ is low.

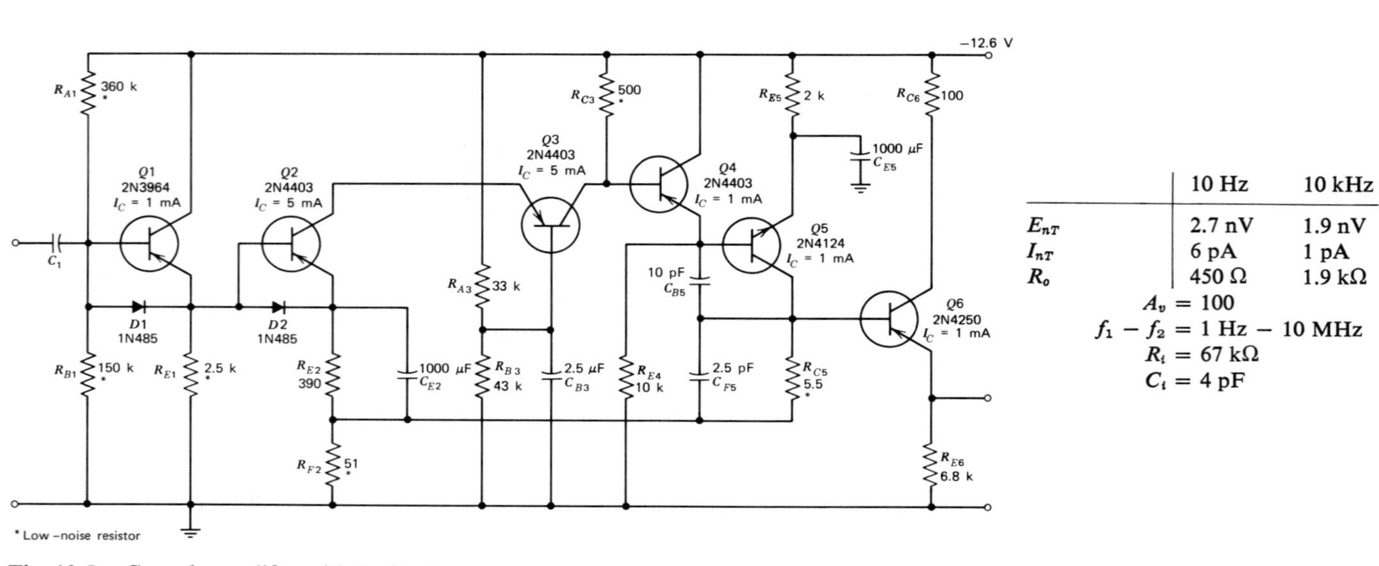

	10 Hz	10 kHz
E_{nT}	2.7 nV	1.9 nV
I_{nT}	6 pA	1 pA
R_o	450 Ω	1.9 kΩ

$A_v = 100$

$f_1 - f_2 = 1 \text{ Hz} - 10 \text{ MHz}$

$R_i = 67 \text{ k}\Omega$

$C_i = 4 \text{ pF}$

Fig. 13-5. Cascode amplifier with feedback.

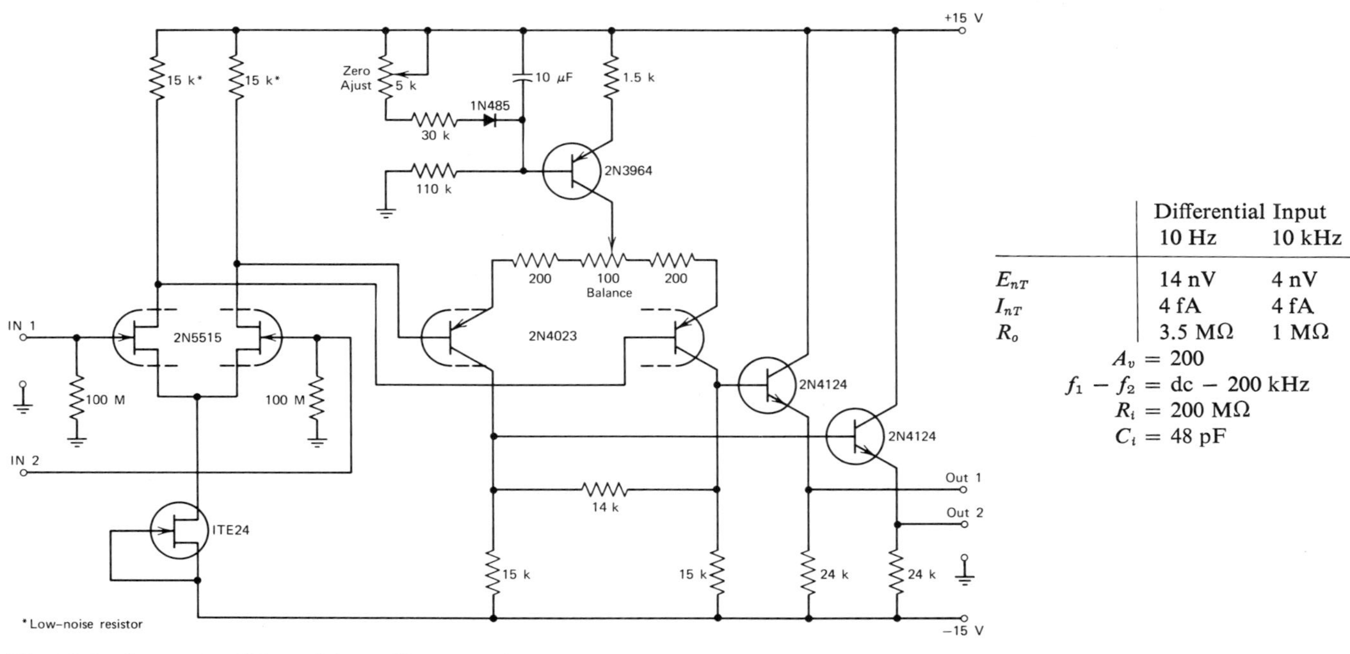

	Differential Input 10 Hz	10 kHz
E_{nT}	14 nV	4 nV
I_{nT}	4 fA	4 fA
R_o	3.5 MΩ	1 MΩ

$$A_v = 200$$
$$f_1 - f_2 = \text{dc} - 200 \text{ kHz}$$
$$R_i = 200 \text{ M}\Omega$$
$$C_i = 48 \text{ pF}$$

Fig. 13-6. Low-noise differential amplifier example.

Calculations for E_{nT} and I_{nT} at 10 Hz and 10 kHz are in agreement with the values noted in Fig. 13-5.

13-6 DIFFERENTIAL AMPLIFIER

The sample design shown in Fig. 13-6 is entirely direct coupled and can be used over the frequency range from dc to 200 kHz. The input stage is a JFET differential amplifier. A dual JFET is used to obtain low drift, low noise, and high-input impedance. The second differential amplifier stage uses dual bipolars, type 2N4023. The double-sided output is taken from emitter followers.

Transistor types ITE24 and 2N3964 perform the constant-current source functions. Balance and zero-adjust controls are available in the second differential pair. A performance summary is given in the figure.

PROBLEMS

1. Calculate E_{nT} and I_{nT} for the circuit of Fig. 13-1 for $I_{C1} = 1$ mA and $I_{C2} = 3$ mA at $f = 10$ Hz and 10 kHz.

2. Use the information given in Table 13-2 to calculate E_{nT} for the cascode circuit of Fig. 13-2 for $I_{C1} = I_{C2} = 10\ \mu$A at $f = 10$ kHz. Give a reason for the difference between your answer and the measured value of E_{nT} in Table 13-2.

3. Confirm that the voltage gain of the cascode of Fig. 13-2 at $I_C = 100\ \mu$A is about 115 by simply dividing R_{C2} by r_{e2}.

4. Equation 13-1 can legitimately contain a term $(I_{n3}r_{e2})^2$ because the first two stages do not provide a great deal of amplification. For the example of Section 13-4, evaluate this term and show that it can be omitted from this example.

5. Verify the values listed for the cascode feedback amplifier in Fig. 13-5 for E_{nT} and I_{nT} at 10 Hz.

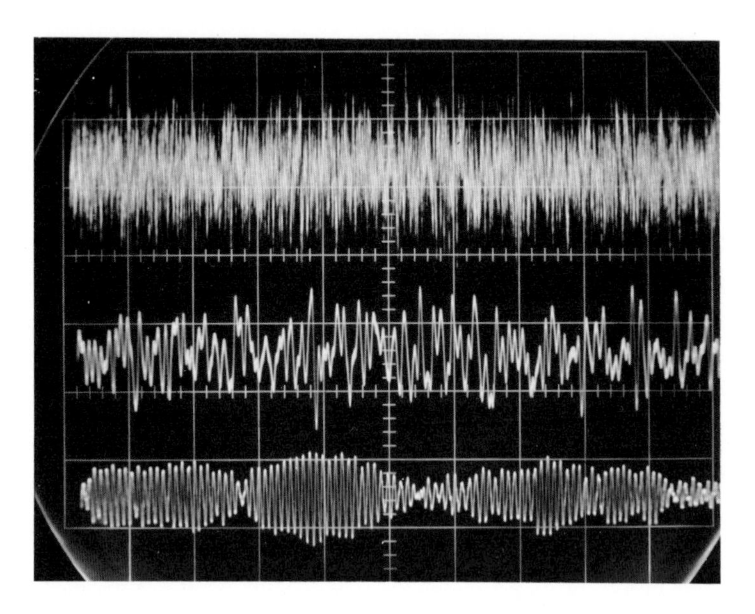

Broadband white noise is shown in the top trace. When passed through a 2-kHz crystal filter with $\Delta f = 50$ Hz we obtain the middle trace. In the bottom trace, the same noise is passed through a 2-kHz RLC high-Q filter with $\Delta f = 50$ Hz. Note the "ringing" effect in the bottom trace. Horizontal sensitivity is 5 msec/cm.

Chapter 14

NOISE MEASUREMENT

When designing low-noise electronics, we must be able to measure the noise to evaluate performance or to compare alternate designs. Noise is measured in much the same manner as other electrical quantities; the most significant difference is the voltage level. Several specific noise parameters have been derived such as noise voltage E_n, noise current I_n, and noise figure NF. These are useful unambiguous quantities capable of measurement.

Since noise voltages are frequently in the nV region, it is virtually impossible to measure noise directly at its source. We cannot put a sensitive voltmeter at the input of the amplifier and say, "Here is the noise." Usually, noise generation is not physically located at the input, but is distributed throughout the system. The total noise is the sum of contributions from all noise generators. In any case, the signal-to-noise ratio at the output is the main concern, for that is where the relay, meter, display, or other output device is located. Noise is measured at the output port where the level is highest.

Two general techniques for noise measurement are the *sine wave method* and the *noise generator method*. In the sine wave method we measure the rms noise at the output of the amplifier; we measure the transfer voltage gain with a sine wave signal, and finally, divide the output noise by the gain to obtain equivalent input noise. In this way, both the noise and the gain can be measured at high levels. The noise generator method uses a calibrated broadband noise generator. A measurement is made of the total noise power at the output of the amplifier; then a calibrated noise voltage large enough to double the output noise power is inserted at the input. This makes the noise generator voltage equal to the equivalent input noise of the amplifier.

Sine wave and noise generator methods each have certain areas of application, as well as specific limitations. The choice between methods depends on the frequency range and the equipment available. The sine wave method requires more measurements, but it uses common instruments and is more applicable at low frequencies; the noise generator method is usually simpler and more applicable at high frequencies. These methods are contrasted in Section 14-6.

14-1 NOISE MEASUREMENT: SINE WAVE METHOD

When considering the noise of a component such as a resistor we are often concerned with a single noise mechanism. With a sensor-amplifier system we seek not only the noise of an amplifier composed of many noise sources, but the signal-to-noise ratio of the entire system. Since the signal is located at the input of the amplifier, it is logical to sum all of the amplifier and input network noise into an equivalent input noise parameter.

We have defined the equivalent input noise E_{ni} as a Thevenin equivalent noise voltage generator located in series with the sensor impedance and equal to the sum of the sensor and amplifier noise as in Fig. 14-1. This refers all of

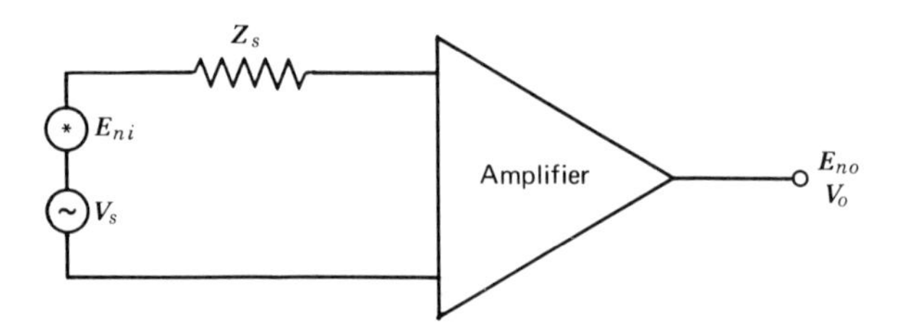

Fig. 14-1. Measurement of equivalent input noise—sine wave method.

the amplifier and input noise to the signal point. Since the signal and noise generators are located at the same point and the same transfer function applies, the equivalent input noise is inversely proportional to the signal-to-noise ratio.

The measurement of equivalent input noise is basic to both the determination of NF and the characterization of the amplifier noise voltage and noise current parameters. Either the sine wave or the noise generator method can be used to measure E_{ni}.

The sine wave method requires measurement of both output noise E_{no} and transfer voltage gain K_t. The procedure for measuring equivalent input noise is

1. Measure the transfer voltage gain K_t.
2. Measure the total output noise E_{no}.
3. Calculate the equivalent input noise E_{ni} by dividing the output noise by the transfer voltage gain.

Referring to Fig. 14-1, the transfer voltage gain K_t can be defined as

$$K_t = \frac{V_o}{V_s} \tag{14-1}$$

where V_s is the input sine wave signal and V_o is the output sine wave signal. The equivalent input noise E_{ni} is then the total output noise divided by the gain;

$$E_{ni} = \frac{E_{no}}{K_t} \tag{14-2}$$

where E_{no} is the output noise of the amplifier.

To measure the transfer voltage gain K_t we insert a voltage generator V_s in series with the source impedance Z_s and measure the signal V_o. Transfer voltage gain K_t is the ratio of V_o to V_s. *Since this gain must be measured at a signal level higher than the noise level, insure that the amplifier is not saturating* by doubling and halving the input signal; the output signal should double and halve proportionately. Note that K_t is dependent on source impedance and amplifier input impedance.

The transfer voltage gain must be measured using a generator impedance equal to the impedance of the signal source. Do not use the voltage gain of the amplifier. If the equivalent input noise is to be measured at various frequencies or source impedances, both the transfer voltage gain K_t and the output noise E_{no} must be remeasured each time with each source impedance at each frequency.

The next step in calculating E_{ni} is measurement of the total output noise E_{no}. Remove the signal generator and replace it with a shorting plug. Do not remove the source impedance Z_s. The output noise E_{no} is now measured with an rms voltmeter as described in Section 14-9. Equivalent input noise E_{ni} is the ratio E_{no}/K_t.

The signal generator must be removed from the noise test circuit before measuring the noise. This is necessary whether the signal generator is ac or battery operated. In either case its capacitance to ground can result in noise pickup. Also, if it is plugged into the line or grounded, there is a possibility of ground-loop pickup.

It has been noted that if the output noise is measured on an average responding rather than an rms meter, multiply by 1.13 to get rms. This is discussed in more detail in Section 14-9. To obtain the noise spectral density divide the equivalent input noise by the square root of the noise bandwidth Δf. When measuring the noise of a tuned amplifier, the bandwidth of the analyzer must be narrower than the amplifier. For testing a broadband untuned amplifier, a bandwidth of one-third the frequency or a bandwidth equal to the frequency provides sufficient accuracy (see Section 14-8).

Measurement of E_n and I_n

The amplifier noise voltage E_n and noise current I_n parameters are calculated from the equivalent input noise for two source resistance values. As defined previously, the equivalent input noise is

$$E_{ni}^2 = E_t^2 + E_n^2 + I_n^2 R_s^2 + 2CE_n I_n R_s \quad (14\text{-}3)$$

Measurement gives the total equivalent input noise E_{ni}. To determine each of the three quantities, E_n, I_n, and E_t, make one term dominant or subtract the effects of the other two. In general, the correlation coefficient C is zero and can be neglected. Measurement of C is discussed later in this section.

To measure the noise voltage E_n measure equivalent input noise with a small value of source resistance. When the source resistance is zero, the thermal noise of the source E_t is zero and the noise current term $I_n R_s$ is also zero; therefore, *the total equivalent input noise is the noise voltage* E_n. How small should R_s be for E_n measurement? The thermal noise E_t of R_s should be much less than E_n. Usually 5 Ω is adequate, 50 Ω can add some noise, and 500 Ω is too large.

To measure the noise current I_n remeasure E_{ni} with a large source resistance. Measure the output noise and transfer voltage gain to calculate the equivalent input noise E_{ni}. Assuming the $I_n R_s$ term to be dominant, I_n *is simply the equivalent input noise* E_{ni} *divided by the source resistance* R_s. If R_s is large, the $I_n R_s$ term dominates the E_n term, and it also dominates the thermal noise since the thermal noise voltage E_t increases as the square root of the resistance, whereas the $I_n R_s$ term increases linearly with resistance. When the $I_n R_s$ term cannot be made dominant, the thermal noise voltage (obtained from Fig. 1-2) can be subtracted from the equivalent input noise. Since this is an rms subtraction, a thermal noise of one-third the noise current term only adds 10% to the equivalent input noise.

The source resistor should be shielded to prevent pickup of stray signals. The resistor used for R_s need not be low noise as long as there is no direct voltage drop across it. When the source resistance is also used for biasing, a low-noise resistor is used. How large a source resistance should be used for the I_n measurement? Try doubling and halving the resistance. Remeasure the output noise and transfer voltage gain with these new values. The equivalent input noise current I_n should be the same in each case.

The correlation coefficient C can be measured after the other three quantities, E_n, I_n, and E_t, have been determined. Select a value of source resistance R_o such that $I_n R_o$ is equal to E_n. For this resistance the correlation term $2CE_n I_n R_s$ has maximum effect. To determine C measure the equivalent input noise a third time with the source resistance R_o. The correlation coefficient C can be calculated from Eq. 14-3 by subtracting the contributions of E_n,

I_n, and E_t. This is a difficult measurement of E_{ni} since the correlation term at most can increase E_{ni} by 40% when $R_s = R_o$ and $C = 1$.

The value of the source resistance necessary to measure the I_n of a bipolar transistor can be calculated. Refer to Eq. 4-16, which mathematically expresses E_{ni} in terms of transistor parameters and operating conditions. For the midband frequency region we select those terms that best approximate the noise current and noise voltage generators. To determine the noise current of the transistor the I_nR_s term must be much larger than the noise voltage. This criteria is expressed in the following relation (the left side is the noise current term and the right side is the noise voltage):

$$2qI_B(r_{bb'} + R_s)^2 \gg \frac{2qI_C}{\beta_o^2}(r_{bb'} + R_s + r_{b'e})^2 \qquad (14\text{-}4)$$

Assume that the source resistance and the transistor input resistance are much larger than the base resistance $r_{bb'}$. The common $2q$ term can be cancelled. Then we have

$$\frac{I_C}{\beta_o} R_s^2 \gg \frac{I_C(R_s + r_{b'e})^2}{\beta_o^2}$$

Table 14-1. Table for Recording Measured Noise Values

	Frequency			
	$f_1 =$	$f_2 =$	$f_3 =$	$f_4 =$
I_C (dc collector current)				
V_g (ac generator terminal voltage)				
V_s (V_g/attenuation)				
V_o (output signal)				
$K_t = (V_o/V_s)$				
R_s (source resistance)				
E_{no} (total output noise)				
E_{nA} (post amp and wave analyzer noise at measurement frequency)				
$E'_{no} = (E_{no}^2 - E_{nA}^2)^{1/2}$				
$E_{ni} = (E'_{no}/K_t)$				
Δf (noise bandwidth)				
M (meter correction factor)				
$E_{ni}/\sqrt{\sim} = ME_{ni}/\sqrt{\Delta f} = E_n$, if R_s is 0				
E_t (source resistance thermal noise)				
$I_n^2R_s^2 = (E_{ni}/\sqrt{\sim})^2 - (E_t/\sqrt{\sim})^2 - E_n^2$				
$I_n = I_nR_s/R_s$				

After canceling the common terms the criterion for I_n measurement is

$$R_s \gg \frac{R_s + r_{b'e}}{\sqrt{\beta_o}} \tag{14-5}$$

Therefore, to measure the I_n of a bipolar transistor, the source resistance R_s should be much greater than $(R_s + r_{b'e})/\sqrt{\beta_o}$. This criterion is easily met if R_s is about equal to $r_{b'e}$, and is larger than $r_{bb'}$. For I_n measurements at low and high frequencies, R_s must be much larger than the base resistance $r_{bb'}$. If there is any question about the size of the source resistance, remeasure with a larger source resistance; the same I_n should result.

A table for systematically recording measured noise data and calculating E_n and I_n is shown in Table 14-1. This table serves as a reminder to include the various correction factors.

FET Measurements

When measuring the noise current of a FET, it is difficult to use a source resistor R_s large enough to meet the criterion that $I_n R_s$ dominate the thermal noise E_t of the resistor. For example, a device with $I_n = 2 \times 10^{-15}$ A/$\sqrt{\text{Hz}}$ and R_s of 10^8 Ω would produce a noise voltage $I_n R_s = 2 \times 10^{-7}$ V/$\sqrt{\text{Hz}}$. The thermal noise of a 10^8-Ω resistor is 1.26×10^{-6} V/$\sqrt{\text{Hz}}$, six times greater than the $I_n R_s$ value. For accurate I_n measurements, $I_n R_s$ should be three times larger than E_t.

To circumvent the problem noted in the preceding paragraph we proceed to eliminate the thermal noise by using a reactive source. Typically, a lossless 100-pF mica capacitor is adequate. Now the $I_n X_c$ term is very large, and, since the reactive impedance has no thermal noise, the equivalent input noise is

$$E_{ni}^2 = I_n^2 X_c^2 + E_n^2 \tag{14-6}$$

where X_c is the reactance of the source impedance at the measurement frequency. Parameter I_n is calculated from

$$I_n^2 X_c^2 = E_{ni}^2 - E_n^2 \tag{14-7}$$

The only unknown in Eq. 14-7 is I_n. This method is useful only at frequencies below 100 Hz.

In FET measurements of I_n the source impedance should be large enough to allow detection of the shot noise current of the gate leakage current. The shot noise must dominate the thermal noise of the source E_t and the amplifier E_n. MOSFETs exhibit a high value of E_n that dominates the total noise and makes I_n difficult to measure. In this case it is possible to obtain approximate values for I_n by measuring the gate leakage current with an electrometer and assuming that the I_n is shot noise of the gate current I_G

$$I_n^2 = 2qI_G \Delta f \tag{14-8}$$

Noise Figure Measurement

An amplifier's NF must be measured with a specific source resistance. The determination of NF is straightforward once you have the equivalent input noise E_{ni} for that value of source resistance. Noise figure is the ratio of the total amplifier and sensor noise to the thermal noise of the sensor alone:

$$\text{NF} = 10 \log \frac{E_{ni}^2}{E_t^2} \tag{14-9}$$

Another definition of NF is the degradation in signal-to-noise power ratio when the signal is passed through a network. This definition in equation form is

$$\text{NF} = 10 \log \frac{S_i/N_i}{S_o/N_o} = 10 \log \frac{S_i}{N_i} - 10 \log \frac{S_o}{N_o} \tag{14-10}$$

where S_i and N_i are the signal and noise powers at the input, and S_o and N_o are the signal and noise at the output. In terms of voltages Eq. 14-10 is

$$\text{NF} = 20 \log \frac{V_s}{E_t} - 20 \log \frac{V_{so}}{E_{no}} \tag{14-11}$$

The NF of a transistor can be *measured* using the definition in Eq. 14-11. Make the input signal level equal 100 times the thermal noise. Then Eq. 14-11 becomes

$$\text{NF} = 40\text{ dB} - 20 \log V_{so} + 20 \log E_{no} \tag{14-12}$$

Adjust the gain of the amplifier until the output meter reads 40 dB with the input signal connected. Substituting 40 dB for the V_{so} term in Eq. 14-12 gives

$$\text{NF} = 20 \log E_{no}$$

The meter reading, after removing the input signal generator, is equal to the NF in dB.

If the amplifier considered in the preceding discussion were perfectly noiseless then the input and output signal-to-noise ratios would be 40 dB. On removal of the input signal, the output meter would decrease by 40 to 0 dB. In a practical case, the amplifier is not noiseless and the output does not decrease by 40 dB, but by some lesser value such as 35 dB. The NF is, therefore, 40 minus 35 or 5 dB. Thus the output meter reading is easily converted to NF for the specific source resistance and noise bandwidth of the system under study.

14-2 NOISE MEASUREMENT EQUIPMENT: SINE WAVE GENERATOR METHOD

The sine wave generator method of noise measurement requires measurement of total output noise and transfer voltage gain. The instruments needed are sine wave oscillator, attenuator, amplifier, wave analyzer or filter, and rms voltmeter or power meter. With the exception of the wave analyzer, there is no special equipment required. Most engineering labs have this equipment available for general measurements.

A photo of the noise measurement equipment is shown in Fig. 14-2. A diagram of the noise measurement instrumentation is shown in Fig. 14-3. We discuss the system shown in the diagram stage by stage.

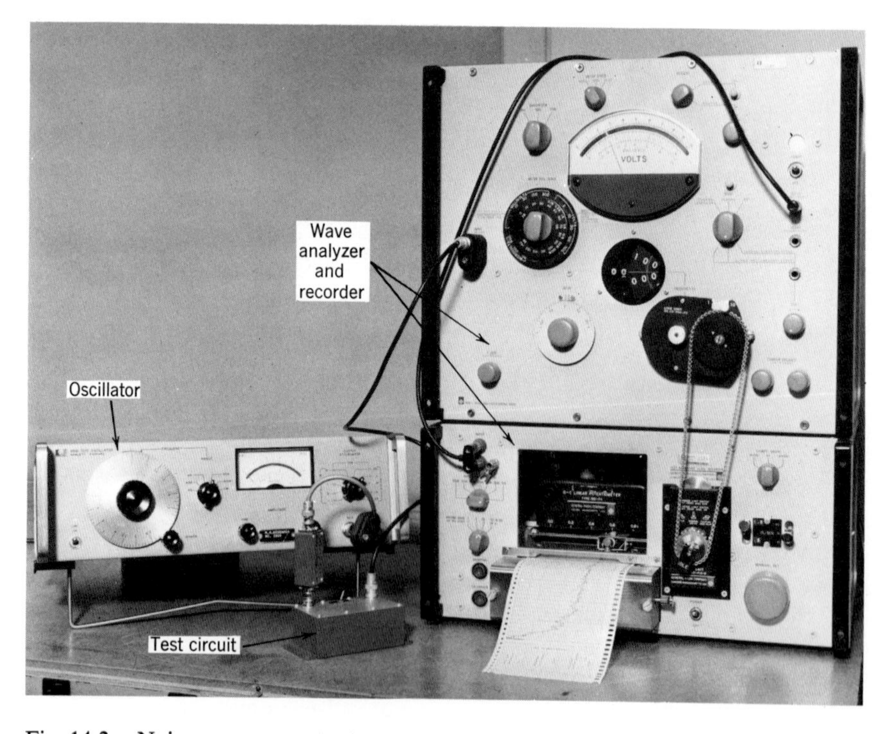

Fig. 14-2. Noise measurement setup.

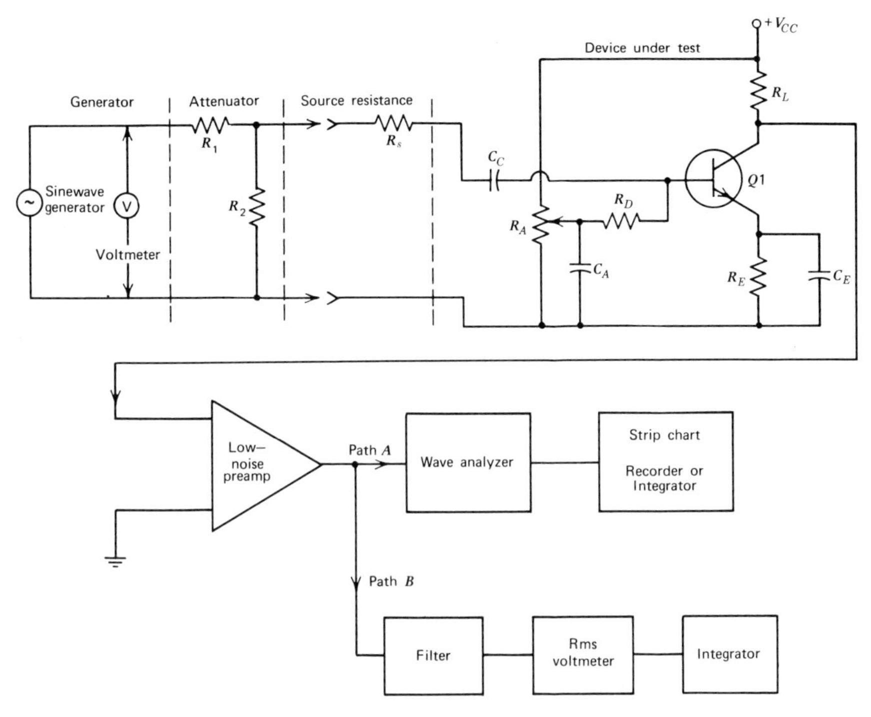

Fig. 14-3. Noise measurement instrumentation.

Signal Generator

The input signal generator is a sine wave oscillator. An oscillator with a built-in voltmeter, such as the Hewlett-Packard Model 651B, is convenient. If the meter is not built-in, measure the signal level before attenuation with a broadband ac voltmeter or an oscilloscope.

The oscillator can be line or battery operated since it is disconnected before making any noise measurements. One precaution: some of the older tube-type signal generators have excessive radiated signals that can yield poor results because of pickup.

Attenuator

Since low-noise amplifiers often have high gain, it is necessary to reduce the oscillator signal below its minimum setting to avoid overdriving the amplifier. This is accomplished by an attenuator following the signal generator. The attenuator serves two functions: it reduces the oscillator signal by a known

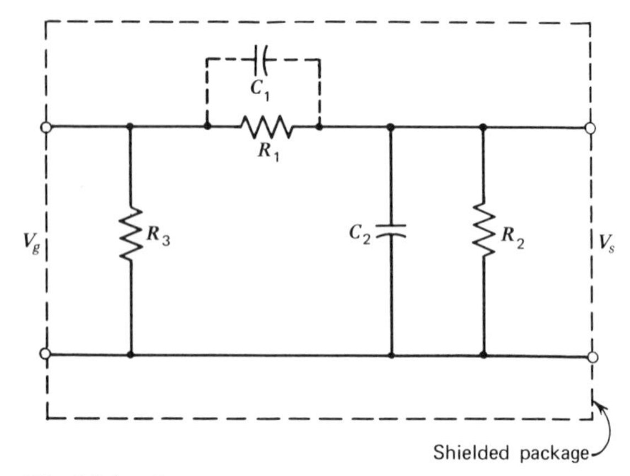

Fig. 14-4. Attenuator.

factor and it provides a low-impedance voltage source for measuring the amplifier gain.

A calibrated attenuator can be built as shown in Fig. 14-4. The attenuation is $R_2/(R_1 + R_2)$ at low frequencies. If there is no dc in the divider low-noise resistors are not required, although metal-film resistors with their lower capacitance and higher stability are recommended. Resistor R_3 provides the matching load for the signal generator, if needed.

For operation above a few kHz, the divider must be frequency compensated. All resistors such as R_1 have a shunt capacitance C_1 of 0.1 to 1 pF. At high frequencies, the attenuation ratio approaches $C_1/(C_1 + C_2)$. To compensate for a flat frequency response connect a capacitor C_2 across R_2 so that $C_1/(C_1 + C_2) = R_2/(R_1 + R_2)$. Now the attenuation is equal for all frequencies. To test for frequency compensation insert a square wave at V_g and look for peaking at V_s. Compensate as with a scope probe. Attenuations of more than 1000 to 1 are difficult to attain in a single package because stray capacitances cause feedthrough at high frequencies.

Although the attenuator is usually removed before measuring noise, it should be shielded because we are concerned with low-level signals and high-gain amplifiers. Construct the attenuator in a small shielded package that can be plugged into the amplifier input terminals.

Source Impedance

To measure an amplifier noise current or NF a resistor R_s or impedance Z_s equivalent to the source impedance is required. Following the attenuator in

Fig. 14-3 is a low-capacitance resistor R_s that simulates the source resistance. It is not necessary for R_s to be a low-noise resistor if there is no dc drop across it.

Any capacitance shunting R_s causes two kinds of measurement errors. An apparent increase in gain K_t at high frequencies results, and a reduction in thermal noise is apparent. Such capacitance increases the transfer voltage gain by causing a decrease in the series source impedance. In addition, shunt capacitance across R_s shorts-out part of the thermal noise voltage and therefore reduces the noise. Both of these effects cause the equivalent input noise E_{ni} to appear too low. (This may be good if you have trouble meeting a spec, but it is a headache when trying to predict performance.)

For noise measurement the attenuator can be replaced by a shielded resistor of value equal to attenuator output resistance. If R_s is built into the test circuit, the attenuator can be replaced by a shorting plug; thus either the oscillator signal or a short circuit is connected.

Transistor Test Circuit

The fourth stage in the diagram of Fig. 14-3 represents the device or circuit being tested. The sine wave measurement method applies equally well for measuring the noise of a single transistor, an IC, or a complete amplifier. A complete amplifier or IC can be substituted in place of $Q1$ and its bias circuitry.

Although the noise of a transistor is about the same for all three configurations, it is normally measured in the common-emitter configuration. The biasing network must not contribute additional noise. When measuring noise over a wide range of collector currents and frequencies, biasing becomes a problem. The low-noise biasing method used in Fig. 14-3 is discussed in Chapter 10. Element R_A sets the bias current which can best be measured by the voltage drop across R_L. R_L should be large enough for adequate gain and small enough for acceptable frequency response. A variable V_{CC} supply may be required.

When testing an integrated op-amp circuit we can use overall negative feedback to set the gain of the op-amp system to some nominal value such as 100. As discussed in Chapter 13 this does not change the basic noise voltage and current measurements as long as the feedback resistors do not add additional noise.

Careful packaging of the test circuit is important. If the test circuit is constructed as a breadboard and operated on a bench, significant low-frequency pickup is very likely. This is particularly true when the transistor gain is low at μA collector currents. As a rule of thumb, construct the test circuitry to be as compact as possible and place it in a small shielded box. There is a positive correlation between the size of the box and pickup. In fact, the pickup seems to increase exponentially with increasing package size. We have found that 2 in. × 3 in. × 5-in. aluminum boxes such as those made by the Pomona Electronic Company do a very good job of shielding. It does require extra effort to construct the test circuitry carefully and package it compactly, but this pays off later in ease of measurement and freedom from pickup.

Preamplifier

Frequently, the output of the test stage in Fig. 14-3 is as low as 20 nV. Since wave analyzers and voltmeters require several μV, the addition of a low-noise decade amplifier with a gain of 10, 100, or 1000 after the test stage is called for. These amplifiers are available from several companies or you can build your own as described in Chapter 13.

Ideally, the noise of a decade amplifier does not contribute to the noise of the device under test. To test for added noise terminate the amplifier input in R_L (the output impedance of the test stage) and turn the test transistor "off." The remaining output noise comes from the amplifier and wave analyzer. If the amplifier is contributing noise, it can be subtracted as the difference of the squares of the voltages as noted in Table 14-1. Similarly, any noise contribution of the wave analyzer can also be removed. This noise contribution should be checked at each test frequency.

Wave Analyzer

For spectral or spot noise measurement, bandwidth limiting is required. The rms noise is measured in a specified noise bandwidth Δf. Two methods of measuring noise versus frequency are illustrated in Fig. 14-3 as paths *A* and *B*. Path *A* uses a wave analyzer or tuned voltmeter and path *B* uses a filter. The wave analyzer is, in effect, a narrowband voltmeter with bandwidth about equal to 1 Hz.

A general purpose wave analyzer is the General Radio Type 1900-A. It has bandwidths of 3, 10, and 50 Hz, and works over a frequency range of 10 to 54,000 Hz. A lower frequency wave analyzer is the Quan-Tech Model 304 which covers the frequency range from 1 Hz to 5 kHz. For frequencies from 1 kHz to 1.5 MHz, the Hewlett Packard Model 310A works well. This unit has noise bandwidths of 200, 1000, and 3000 Hz. At these higher frequencies a wider bandwidth can be used for faster response and reduced jitter on the meter. These three instruments are pictured in Fig. 14-5. Since all of these wave analyzers use average responding meters, multiply the noise reading by 1.13 as described in Section 14-9.

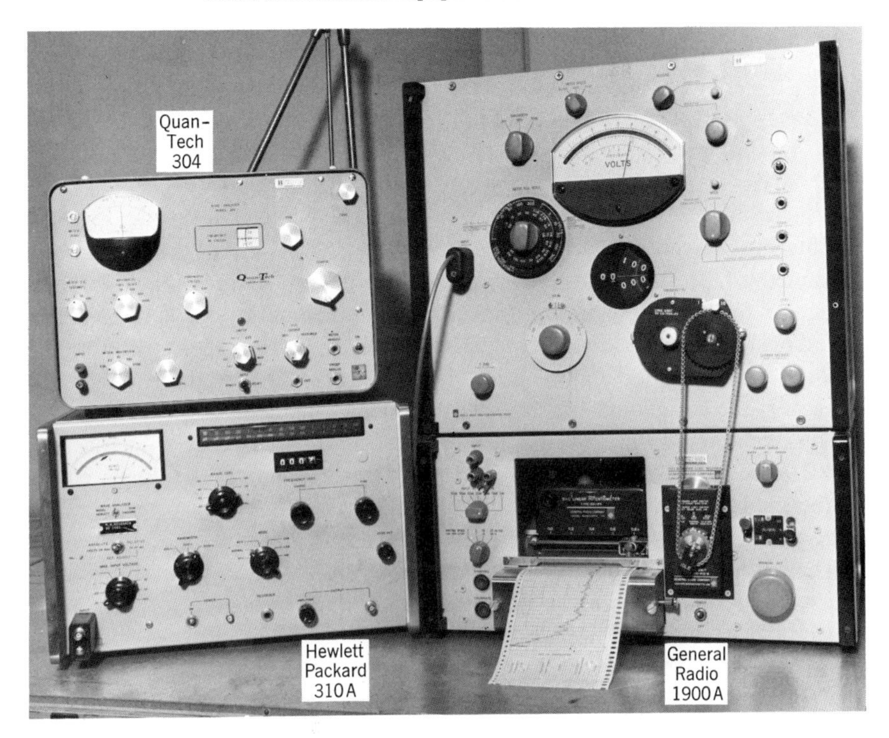

Fig. 14-5. Wave analyzer.

Tuned voltmeters such as the Princeton Applied Research Model 121 also can be used as a variable frequency filter. This unit has a variable Q to change the noise bandwidth. Since the bandwidth is variable, accurate relative readings can be taken, but it is necessary to determine the noise bandwidth and noise spectral density. Techniques for measuring noise bandwidths are described in Section 14-10.

Path B of Fig. 14-3 illustrates an alternate system. A bandpass filter, such as an RLC filter, determines the noise bandwidth. If a commercial filter is not available, the filter of the system being designed can often be used. The signal is read on an rms meter with a bandwidth wider than the filter.

Averaging Device

Although the long-term rms value of noise is a constant, the instantaneous amplitude is totally random, and therefore the meter jitters. For an accurate noise measurement we can smooth the meter fluctuations by averaging over a long period of time. Three principal methods of smoothing are

1. Use a long *R-C* time constant.
2. Integrate the signal over a period of time.
3. Record the signal and average its value.

Filtering or averaging with a long *R-C* time constant is the most commonly used method. The time constant is increased by placing a large capacitor across the meter terminals. Accuracy is inversely proportional to the square root of time constant and bandwidth as defined in Section 14-7. In addition to the theoretical averaging time, wait several time constants for the capacitor to charge. If the bandwidth is greater than 100 Hz, the meter response provides adequate damping. Usually it is only necessary to add an external capacitor. The charge and discharge time constants of the capacitor should be equal. If the charge time constant is shorter than discharge, a peak responding instrument results and the ratio of average to rms no longer holds.

There are two methods for integration. One technique uses an analog integration circuit such as an operational amplifier with capacitor feedback, or simply a capacitor at the input of a high-impedance amplifier. In operation, noise current charges the capacitor for a period of time such as 1 min. Since the voltage across a capacitor is $\int i\,dt/C$, the noise current is given by $I_{no} = CV_C/t$, where C is the capacity in F, V_C is the end voltage in V, and t is the time in sec. After discharging the capacitor is ready for the next reading.

A second true-integration technique digitizes the noise signal with a voltage-to-frequency converter and sums the total number of cycles on a digital counter. The total number of cycles divided by the number of secs gives the average frequency and the average noise voltage. These integration methods are fast and give unambiguous readings, but may have the disadvantages of poor stray noise rejection and higher cost.

For a simple laboratory method of averaging we can record the rectifier noise output on a dc strip-chart recorder and read the average value. The speed and time constant of the recorder are adjusted to see the fluctuations in signal amplitude. If the pen moves too fast, the time constant is too short; there will result a solid band of ink and the variations cannot be seen. This trace is the short-term noise at the output power meter. To read the average value of the noise a straight edge (such as a wire) is laid on the recorded trace. To obtain a correct reading the wire is moved back and forth until there is approximately equal area or equal color on each side. The recording may have high narrow positive excursions and low broad negative ones so that half of the peak-to-peak value is not rms. A typical recording of noise is shown in Fig. 14-6. To verify the reading use a calibrated noise signal.

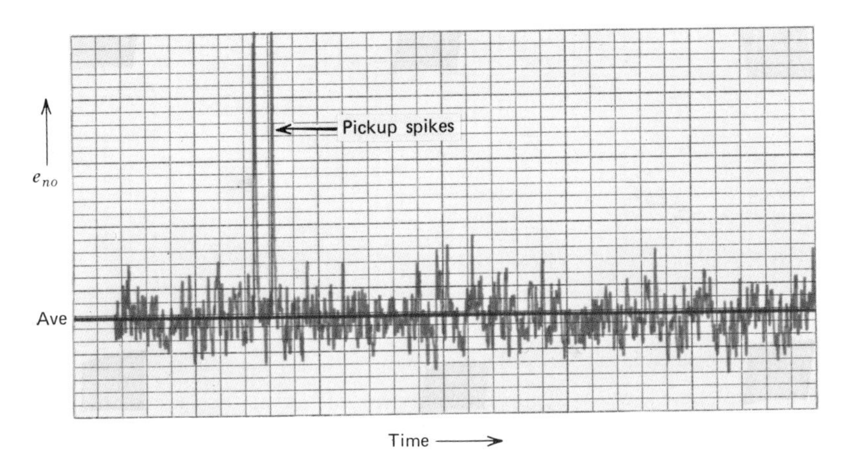

Fig. 14-6. Reading of noise on a recorder.

For a wave analyzer with a 5-Hz noise bandwidth, a 2-min recording time is used to obtain 5% repeatability. This is higher accuracy than necessary for most engineering noise work. It may be needed when measuring the noise contribution of an amplifier that is dominated by a large thermal noise component. As a rule of thumb, 10% is good accuracy for noise measurements, considering that the excess noise of a transistor type varies by 300 to 1000%.

In theory the *R-C* method produces the same accuracy as the recorder, but in practice the recorder has an advantage. In a lab environment there are sporadic noise spikes. An integrating or averaging circuit sums these with the circuit noise. With the recorder method, noise spikes or a sudden change in noise level show on the record. These spikes can often be traced to an interference source, or a record for a longer period of time can be made until the signal becomes more typical. Essentially, this is using the operator's brain as an additional selective filter. The recorder and *R-C* methods are laboratory techniques since they use operator judgment. For production testing the true integration method is both easier and faster.

14-3 NOISE MEASUREMENT: NOISE GENERATOR METHOD

Amplifier noise can be measured by comparison using a calibrated noise source E_{ns} and rms output noise meter located at E_{no} in Fig. 14-7. The unknown noise level of the amplifier is compared with the known amplitude of the noise generator. Accuracy is determined primarily by calibration of the noise generator. It must have a uniform noise spectral density over the bandwidth

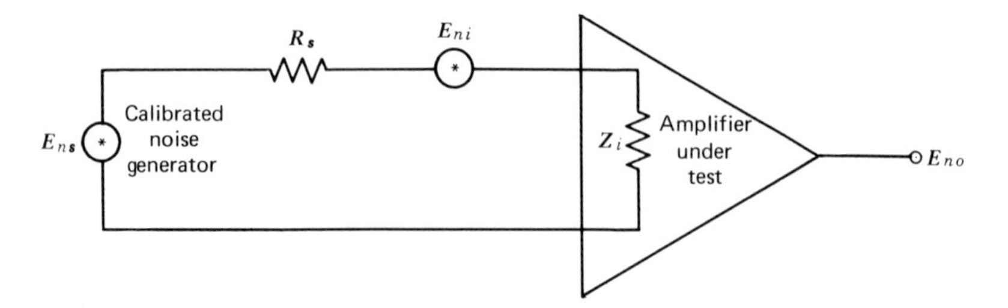

Fig. 14-7. Noise measurement—noise generator method.

of measurement (white noise). Although this criteria may be difficult to meet at low frequencies where 1/f noise dominates, at higher frequencies the noise generator method is the easiest form of noise measurement.

The calibrated noise source is shown in Fig. 14-7 as a noise voltage generator in series with the sensor resistance R_s. The system equivalent input noise is summed at E_{ni}. The amplifier and generator noise are measured as E_{no}. Alternately, a high-impedance noise current generator can be connected in parallel with the source impedance to measure the equivalent input noise current I_{ni}.

The purpose of the noise measurement is to determine the value of equivalent input noise E_{ni} or I_{ni}. The procedure is to measure the noise at the output twice as E_{no1} and E_{no2}. Output noise with the noise generator E_{ng} connected is E_{no1}, and output noise with the noise generator disconnected is E_{no2}. These measurements are described by two equations:

$$E_{no1}^2 = K_t^2(E_{ni}^2 + E_{ng}^2) \tag{14-13a}$$

and

$$E_{no2}^2 = K_t^2 E_{ni}^2 \tag{14-13b}$$

where K_t is the transfer voltage gain. For good accuracy E_{no1} should be much larger than E_{no2}. From these two noise measurements and the value of the noise generator E_{ng}, the transfer voltage gain K_t can be calculated:

$$K_t^2 = \frac{E_{no1}^2 - E_{no2}^2}{E_{ng}^2} \tag{14-14}$$

The equivalent input noise E_{ni} can be calculated for any known noise generator level as follows:

$$E_{ni}^2 = \frac{E_{no2}^2}{K_t^2} = \frac{E_{no2}^2 E_{ng}^2}{E_{no1}^2 - E_{no2}^2} \tag{14-15}$$

The technique most commonly employed is to increase E_{ng} to double the output noise power. So

$$E_{no1}^2 = 2E_{no2}^2$$

By substituting this relation into the general expression for equivalent input noise, Eq. 14-15, we find

$$E_{ni}^{\ 2} = \frac{E_{no2}^2 E_{ng}^{\ 2}}{2E_{no2}^2 - E_{no2}^2} = E_{ng}^{\ 2} \tag{14-16}$$

Therefore, *the noise voltage of the noise generator necessary to double output noise power is equal to the equivalent input noise of the amplifier.*

In summary, the measurement procedure is

1. Measure the total output noise.
2. Insert a calibrated noise signal at the input to increase the output noise voltage by 3 dB.
3. The noise generator signal is now equal to the amplifier equivalent input noise.

An uncalibrated power meter can be used to measure the noise. First the noise of the amplifier is measured, then the amplifier gain attenuated by 3 dB, and, finally, the noise generator increased until the output power meter returns to its original level. The added noise is equal to the amplifier equivalent input noise.

14-4 NOISE MEASUREMENT EQUIPMENT: NOISE GENERATOR METHOD

There are several types of dispersed signal generators. Although most are noise generators, some produce a nonrandom swept-frequency sine wave. Either type serves as a broadband noise source. There is one basic criterion: for noise measurement the generator must have a flat noise spectral density.

Random noise signals are generated by temperature-limited vacuum diodes, zener diodes, heated resistors, and gas discharge tubes. One of the most common is the temperature-limited vacuum diode commonly called a noise diode. As a noise diode, it is operated with an anode voltage large enough to collect all the electrons emitted by the cathode, hence the name temperature-limited diode. There is no space-charge region around the filament to smooth out the electron emission, so the anode current shows full shot noise I_{sh}, as given in Chapter 1:

$$I_{sh}^{\ 2} = 2qI_{\mathrm{DC}}\,\Delta f \tag{14-17}$$

where q is electronic charge, I_{DC} is direct anode current, and Δf is noise bandwidth. The noise is Gaussian and independent of frequency from a few kHz to several hundred MHz. At low frequencies, l/f noise dominates. In the high-frequency region, the electron transit time can become significant and the output noise decreases. The anode current is controlled by varying the temperature of the filament. Since the emission current is an exponential function of filament temperature, it is important to control the filament current very accurately. A regulated supply is desirable.

A circuit diagram of the vacuum noise diode used as a noise current generator is shown in Fig. 14-8. Filament temperature and, therefore, anode current

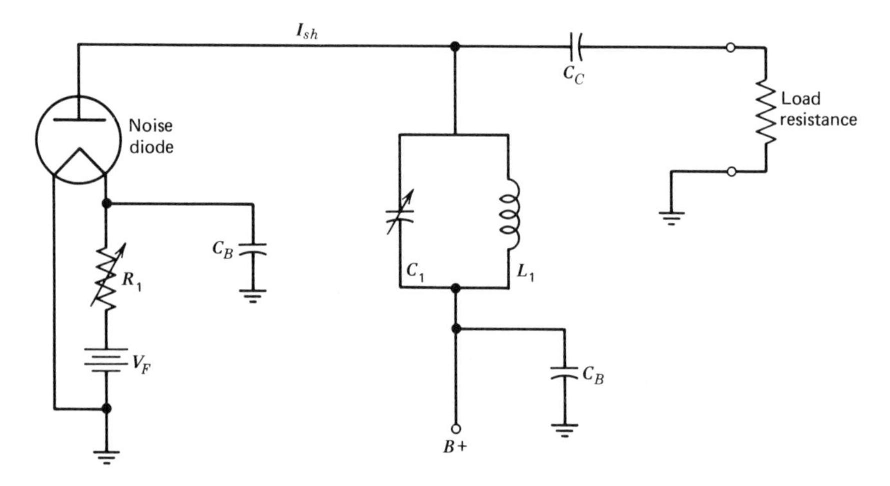

Fig. 14-8. Circuit diagram of vacuum noise diode.

is controlled by variable resistor R_1. Ac load impedance is set by the tuned circuit C_1L_1. Using a resonant load minimizes the effects of anode and wiring capacitance. The noise current I_{sh} is coupled through capacitor C_C. Capacitors C_B are bypass capacitors.

Refer to the noise equivalent circuit in Fig. 14-9. The noise generator is a current source I_{sh} shunted by the plate resistance r_p. The plate resistance of a noise diode is typically from 25,000 Ω to 1 MΩ. The tuned circuit is represented by C_1, L_1, and R_p in parallel, where R_p is the effective parallel resistance of the tuned circuit. R_p has a parallel thermal noise current generator I_t. Total noise current is the sum of the shot noise and thermal noise components:

$$I_{tot}^2 = I_{sh}{}^2 + I_t{}^2 \tag{14-18}$$

For a calibrated noise source the shot noise term must dominate the thermal noise. This requires a large R_p or high circuit Q.

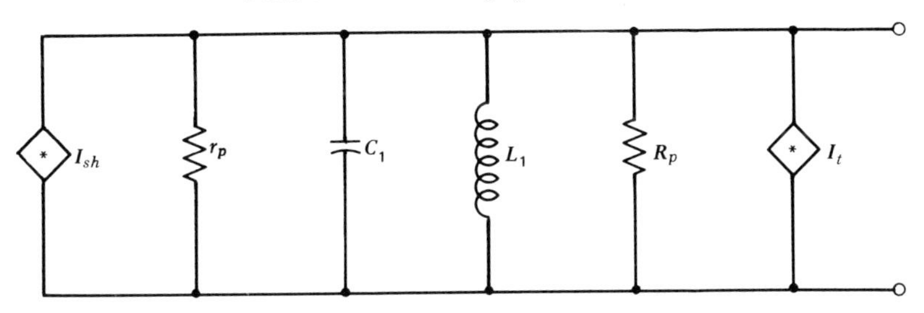

Fig. 14-9. Noise equivalent circuit of noise diode circuit.

If a noise voltage source with a specific output resistance is desired, a resistor is placed in parallel with the noise diode output. This resistor must be much smaller than the plate resistance of the diode. The output noise voltage is the shot noise current I_{sh} times the shunt resistance R_L plus the thermal noise voltage of the resistor. When using a noise generator the output cable capacitance must not seriously degrade the noise signal.

When a resistive load R_L is used in place of L_1-C_1, the output noise has a flat frequency distribution. Although required for broadband measurements, there are two disadvantages of a resistive load: the thermal noise of R_L is likely to be significant, and the shunt capacitance of the tube and wiring causes a roll-off in noise at high frequencies. In general, a reactive load is useful to keep the losses low and the impedance high.

Another common calibrated noise generator is the low-voltage Zener diode. The Zener mechanism is shot noise limited. Selected Zener diodes can have a 1/f noise corner as low as a few Hz. The higher voltage avalanche diodes, erroneously called Zener diodes, exhibit too much excess noise for use as calibrated noise sources. The equivalent circuit of the Zener noise source is similar to that shown in Fig. 14-9.

A circuit diagram of a Zener diode noise generator is shown in Fig. 14-10.

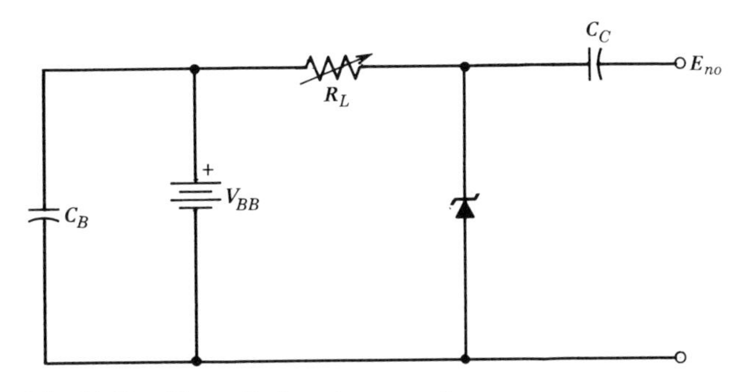

Fig. 14-10. Zener diode noise generator.

The dc supply V_{BB} must be much larger than the Zener voltage so that load resistor R_L can be large. The thermal noise current of R_L parallels the output. Capacitors C_B and C_C bypass the battery and decouple the dc output level, respectively.

A forward-biased semiconductor diode can be used as a calibrated noise voltage generator. The diode shunt resistance is 26 divided by its forward current in mA. The noise voltage E_{sh} is equal to the shot noise current times the forward resistance of the diode as discussed in Chapter 1. Often the base-emitter junction of a low-noise transistor can be used as a forward-biased noise diode. The value of the shot noise voltage of a semiconductor diode is

$$E_{sh}^2 = \frac{(0.026)^2 2q\,\Delta f}{I_{DC}} = \frac{5.46 \times 10^{-22}\,\Delta f}{I_{DC}} \tag{14-19}$$

A low-noise amplifier operating with a shorted input and a first-stage collector current of less than 10 μA may be a good calibrated noise generator. In this case the E_n noise is limited by the shot noise voltage of the first transistor so that the noise is constant if the collector current is constant. The amplifier gain can be adjusted to provide the desired output noise level. This may sound like an anomaly, using a low-noise amplifier as a calibrated noise source, but it meets the basic criteria. The amplifier noise is set by known fundamental noise mechanisms. This feature is valuable to circuit designers because they can use the same amplifier for both amplification and calibrated noise generation.

14-5 NOISE MEASUREMENT: COMMERCIAL NOISE ANALYZERS

For making electronic noise measurements, the engineer must decide whether to use general-purpose laboratory equipment or to buy a commercial noise analyzer. An advantage of assembling your own apparatus is that the same equipment is used for measuring the noise of devices and of completed systems. Often the equipment is already available in the laboratory. A disadvantage of building your own noise measuring setup is the possibility of inaccuracies. When interconnecting equipment it is possible to have a ground loop or noise pickup. Additional time and measurements are required to verify the measurement method. Another disadvantage is the time required to make noise measurements; it is necessary to make *both* noise and gain measurements at each measurement point.

Transistor or component noise analyzers are available. There are several good models on the market including types made by Hewlett-Packard and by Quan-Tech Laboratories. Commercial equipment is easy to use. Since the analyzer is designed for noise measurement, it requires a minimum amount of setup and checking time. These instruments use the sine wave method, but measure the transfer voltage gain and output noise simultaneously. The meter is calibrated to display equivalent input noise.

Naturally, commercial equipment is high in initial cost and lacks the flexibility of a home-grown system. The decision to purchase a commercial noise analyzer with its increased speed and acceptability is based on the anticipated use. It is worthwhile for incoming inspection where many units are measured under varying conditions and it is justifiable for an engineering application when there is an immediate or continuing need.

Low-frequency transistor and IC noise analyzers are made by Quan-Tech Laboratories, Inc. In addition, they manufacture noise generators, low-noise amplifiers, and wave analyzers for general-purpose noise and signal analysis. Their most versatile unit, the model 2173C-2181 transistor noise analyzer, is on the top in Fig. 14-11. The filter unit with the frequency meters is above. The lower unit contains the range selection, voltage and current controls, as well as the transistor socket. This transistor noise analyzer consists of a test amplifier for the transistor, adjustable collector current and voltage, filters, gain stabilization system, and noise meters. This noise analyzer uses the sine wave method of measurement. The gain is measured at 4000 Hz and a variable gain amplifier is servoed to maintain a transfer voltage gain of 10,000. Transfer voltage gain is measured by injecting a calibration signal across a 1-Ω emitter resistor, which is effectively in series with the source resistance. There is a special servo control to maintain the operating point at the meter setting. In operation the analyzer measures the output noise, divides by the controlled gain of 10,000, and displays the equivalent input noise simultaneously at five frequencies. This provides data over the whole frequency range, indicating the relative amount of excess noise to midband noise. The measurement procedure is simple; set the bias voltage and current, plug the transistor into the shielded socket, and read the noise meter. Noise voltage E_n and equivalent input noise E_{ni} are measured directly. Noise current I_n must be calculated from the E_{ni} measured with a large source resistance.

Specifications are

pnp- and *npn*-bipolar transistors, as well as *p*- and *n*-channel FETs
Frequencies: 10, 100 Hz; 1, 10, 100 kHz
Noise spectral density range: 3 to 3000 nV full scale
Source resistance range: 0 to 1 MΩ plus external
Collector voltage range: 0.3 to 60 V
Collector current range: 0.3 μA to 30 mA

The Hewlett-Packard Company model 4470A uses the sine wave method to measure output noise at each frequency. The gain is automatically measured

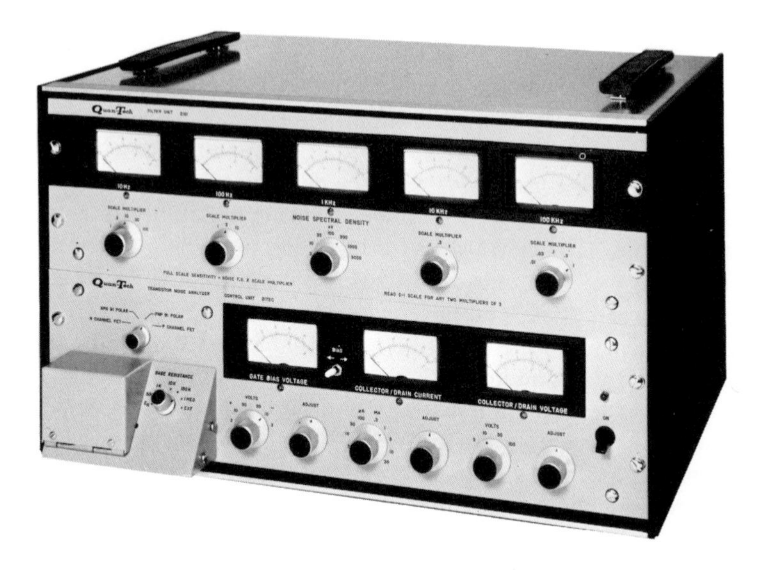

Fig. 14-11. Quan-Tech and Hewlett Packard noise analyzers.

by injecting a signal close to the measurement frequency; therefore, the gain is accurately controlled at each value of frequency and source resistance. Another feature is the direct reading of noise voltage, noise current, or NF, and not just equivalent input noise.

This instrument provides a constant 4-Hz bandwidth. This narrow bandwidth requires a long integration time at all frequencies and may increase measurement time. The noise at only one frequency can be measured at one time.

Specifications are
Bipolar transistors and FETs
Noise voltage: 3 nV
Noise current: 100 fA
Frequency range: 10 Hz to 1 MHz
Collector current range: 1 μA to 30 mA
Collector voltage range: to 15 V
Source resistance: 10 Ω to 1 MΩ plus external.

14-6 COMPARISON OF METHODS

The two methods of noise measurement, sine wave and noise generator, can be compared. The main advantage of the noise generator method is its ease of measurement. The main advantage of the sine wave method is its applicability at low frequencies and the availability of equipment.

The noise generator method is straightforward because we simply connect a noise generator to the amplifier input and adjust the generator to double the output noise power of the amplifier. In a wideband system this is done very quickly. Another advantage is the availability of low-cost calibrated noise diodes. However, these noise diodes exhibit 1/f noise below a few kHz; therefore a more expensive noise source is required for low-frequency measurements. Since the noise generator is a broadband source, the system noise bandwidth is not required. Some form of bandwidth limiting is necessary to measure the spot noise. This may be a narrowband filter, a one-third octave filter, or a wave analyzer.

When using the noise generator method, the noise source is connected to the amplifier while measuring output noise, and the probability of pickup is increased. With the sine wave method, only a small shielded resistor is connected to the input terminals. Another disdvantage of the noise generator method is the long measurement time at low frequencies. For the instrumentation and control fields, noise measurements extend down to a few Hz or less. Spot noise measurements require a bandwidth of less than 1 octave. As pointed out in Section 14-7, a noise bandwidth of 5 Hz requires an averaging time constant of 10 sec for 10% accuracy. Also, the noise generator method requires two or more measurements of output noise, the amplifier alone and the amplifier plus the noise generator.

A general rule of thumb: use the sine wave method for low and medium frequencies and the noise generator method for high frequencies and *RF*.

14-7 EFFECT OF MEASURING TIME ON ACCURACY

Although it is impossible to predict the instantaneous value of noise, the long-term average of rectified noise can be determined statistically.

Let us examine the fluctuations of the pointer of a dc meter when rectified noise is being indicated. It is desired that the meter indicate the average value of the waveform. However, if the meter circuit has a short time constant, the pointer will try to respond to the instantaneous value of the noise, instead of its average value. It is of course impossible to read a wildly fluctuating meter.

A relation between noise bandwidth, output meter time constant τ, and meter relative error σ is

$$\sigma = \frac{1}{\sqrt{2\tau \, \Delta f}} \qquad (14\text{-}20)$$

where σ is the ratio of the rms value of meter fluctuations to the average meter reading. For the same level of accuracy, narrowband measurements require a longer averaging time than do wideband measurements. For bandwidths of 1000 Hz or more, the meter time constant is usually sufficient. On the other hand, a 5-Hz bandwidth requires a 40-sec time constant for 5% accuracy, whereas 1% accuracy requires a 1000-sec time constant. This illustrates one type of difficulty encountered when making accurate noise measurements.

As a rule of thumb, use the widest possible bandwidth for shortest reading times and greatest accuracy.

Several methods of obtaining a long time constant were described in Section 14-2. The most common way is to parallel the meter with a capacitor. The capacitor must not interfere with the calibration of the meter. Use a low-leakage capacitor, such as a tantalum electrolytic, so that the capacitor does not shunt the dc signal. The capacitor also may overload the output of the amplifier by requiring larger peak currents. Test the accuracy with a sine wave by noting whether the presence of the capacitor changes the meter reading. In addition to the accuracy requirements, allow several time constants for the capacitor to charge.

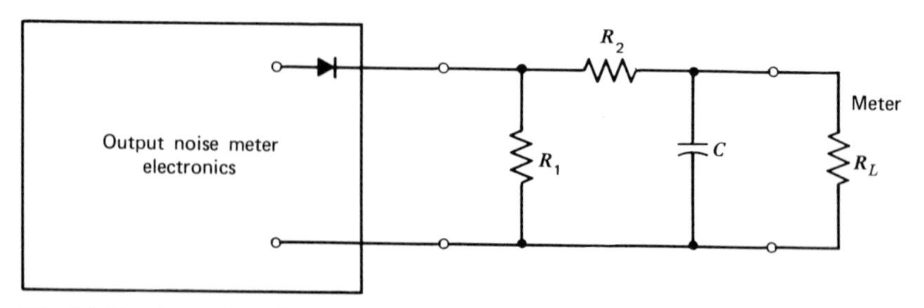

Fig. 14-12. Averaging circuit.

When the meter is connected directly to the rectifier, adding a capacitor makes it peak responding. By connecting a small shunt resistor R_1 and a large series resistor R_2, as shown in Fig. 14-12, the meter reads correctly; although the gain is reduced to $R_L/(R_2 + R_L)$. The meter resistance is R_L.

14-8 BANDWIDTH ERRORS IN SPOT NOISE MEASUREMENTS

Spot noise or noise spectral density is the noise in a 1-Hz noise bandwidth. This implies that the noise was measured with a 1-Hz bandwidth filter. In actual practice the noise is measured with a noise or wave analyzer, and the bandwidth may be significantly greater than 1 Hz. To obtain spot noise the reading is divided by the square root of the noise bandwidth.

When the noise is white, bandwidth poses no problem since the total noise voltage in a wide band is equal to the sum of identical noise contributions in narrow bands that make up the total bandwidth. Frequently, however, noise waveforms contain a significant amount of 1/f noise power. The question arises: *how much error is introduced when measuring the noise spectral density with a wide bandwidth and dividing by the square root of bandwidth?* To find an answer the spot noise is compared with the integrated noise in a bandwidth, divided by the bandwidth. Consider an amplifier with a 1/f noise power distribution. It was shown in Chapter 1 that the mean square value of 1/f noise voltage is given by

$$E_f^2\,(\Delta f) = K \ln \frac{f_h}{f_l} \tag{1-7a}$$

where f_h and f_l are the upper- and lower- cutoff frequencies in the band being considered. This relation can be rearranged to the form

$$E_f^2\,(\Delta f) = K \ln \frac{f_l + \Delta f}{f_l}$$

and written as

$$E_f^2\,(\Delta f) = K \ln \left(1 + \frac{\Delta f}{f_l}\right) \tag{14-21}$$

For purposes of analysis we define a center frequency f_o and consider that Δf is geometrically centered about f_o. Then $f_o = \sqrt{f_l f_h}$.

The bandwidth error can be calculated. Using Eq. 14-21 to integrate over a bandwidth and dividing by the square root of bandwidth, we compare this calculated value with the true value of spot noise at the center frequency f_o as shown in Table 14-2.

Table 14-2. Errors Due to Bandwidth in Spot Noise Measurements of a 1/f Noise Dominated System

Δf (as % of f_0)	Value of Wideband Noise Divided by Δf	Value of Spot Noise at f_0	% Error
0.1	1	1	0
1	1	1	0
10	0.9998	1	−0.02
30	0.9978	1	−0.22
100	0.9809	1	−1.91

Even with a 1/f noise power distribution, a bandwidth of one-third the frequency gives a negligible error. If part of the noise is white, the error is even smaller. This is convenient because it allows us to use the readily available one-third octave filters. One-third of an octave is approximately equal to one-third of the frequency. In general, it is possible to use a bandwidth equal to the frequency, geometrically centered around the frequency, and still have negligible error. In some cases, such as with non-Gaussian "popcorn" noise, the noise *power* spectral density is proportional to $1/f^2$. The mean-square noise voltage is

$$E_p^2 = \frac{K_2}{f^2} \tag{14-22}$$

and the total noise in the bandwidth from f_l to f_h is

$$E_p^2\,(\Delta f) = K_2 \int_{f_l}^{f_h} \frac{df}{f^2} \tag{14-23a}$$

Therefore

$$E_p^2\,(\Delta f) = K_2\left(\frac{f_h - f_l}{f_l \times f_h}\right) \tag{14-23b}$$

Substituting in the noise expression of Eq. 14-23b gives

$$E_p^2\,(\Delta f) = K_2\left(\frac{\Delta f}{f_o^2}\right) \tag{14-24}$$

Equation 14-24 shows that the total noise voltage of a $1/f^2$ power distribution is proportional to the square root of bandwidth Δf just as for white noise. The measurement accuracy is independent of the measurement bandwidth.

14-9 OUTPUT POWER METERS

Variations in noise readings can frequently be traced to the use of different types of meters to measure the output noise power. To measure noise accurately requires that the meter respond properly to the noise power, and have an adequate crest factor and bandwidth.

The meter must provide a response proportional to power or voltage squared. A typical meter movement is designed to measure constant amplitude, repetitive waveforms. It is calibrated to indicate the rms amplitude of a sine wave, but does not usually respond to the rms value. When such a voltmeter measures noise, whose waveform is neither sinusoidal nor of constant amplitude, we are concerned about the indication.

Random noise has a well-defined average value when rectified, but its instantaneous value cannot be defined except in terms of probability (see Chapter 1). The percentage of time that Gaussian noise exceeds a certain peak level is shown in Table 14-3. *Crest or peak factor is defined as the ratio of the peak value to the rms value of a waveform.* As shown in Table 14-3, noise has a crest factor of 3 almost 1% of the time, but it reaches 4 less than 0.01% of the time.

Table 14-3. Crest Factors for Gaussian Noise [1]

% of Time Peak is Exceeded	$\frac{\text{Peak}}{\text{Rms}}$	Peak factor in $\text{dB} = 20 \log_{10} \frac{\text{Peak}}{\text{Rms}}$
10.0	1.645	4.32
1.0	2.576	8.22
0.1	3.291	10.35
0.01	3.890	11.80
0.001	4.417	12.90
0.0001	4.892	13.79

The meter must respond to the total signal. Its bandwidth must be much greater than the noise bandwidth. As an example, when measuring broadband white noise, if the meter bandwidth is equal to the noise bandwidth of the system, the noise reading will be 29% low (see Table 14-4). If the meter 3-dB bandwidth is ten times the noise bandwidth, the error will only be −4.4%. It is clear that the output power meter must have adequate frequency response.

Table 14-4. Measurement Error due to Output Meter Bandwidth

Meter Bandwidth (BW)	Relative Reading	% Error
BW = Δf	0.707	−29.3
BW = $2\,\Delta f$	0.818	−18.2
BW = $5\,\Delta f$	0.915	− 8.5
BW = $10\,\Delta f$	0.956	− 4.4
BW = $20\,\Delta f$	0.979	− 2.1
BW = $50\,\Delta f$	0.993	− 0.7
BW = $100\,\Delta f$	0.998	− 0.2
BW = $1000\,\Delta f$	1.000	0.0

Three general requirements of the output noise meter, then, are that it must respond to the noise power, have a crest factor greater than 4, and a bandwidth greater than ten times the noise bandwidth. Consider, now, the three common types of meters; true rms, average responding, and peak responding.

True Rms Instruments

An rms responding voltmeter gives true rms indication of the noise if the voltmeter bandwidth is greater than the noise bandwidth of the system and the noise peaks do not exceed the maximum crest factor of the voltmeter. The crest factor of 4 to 5 on most new rms voltmeters is adequate except for a system with unusually high-noise spikes such as popcorn noise. To check for clipping, monitor the noise meter signal with a scope or change the range switch and observe whether the reading is identical.

Two common types of rms responding meters are the quadratic devices and those that respond to heat. Most new rms meters use a form of squaring circuit so that the meter averages the square of the instantaneous value of the signal. The meter scale is calibrated in terms of the square root of the quantity, thus giving an indication of the rms value. Squaring uses approximation methods or one of several types of quadratic responding devices. A diode *V-I* curve approximates a quadratic response when operated with less than a few tenths of a V of signal superimposed on a forward biasing voltage.

Thermal responding meters use the signal power to heat a resistor linearly and detect the temperature rise with a thermistor or thermocouple. This follows the basic definition that the rms value of an ac current has the same heating power as an equal dc current. Thermal type meters are usually slower in response and tend to be subject to burn-out on overload.

Average Responding Instruments

To measure the rms value of a signal use an rms responding meter. This may seem obvious, yet it is frequently ignored. Most ac voltmeters and wave analyzers are average-responding rectifier-type instruments. They use a diode rectifier and a dc meter that responds to the average value of one or both halves of the rectified waveform. The average value of a full-wave rectified sine wave is 0.636 of the peak amplitude; the rms is 0.707 of the peak. To indicate the rms value of a sine wave the meter has a multiplying *scale correction factor* of $0.707/0.636 = 1.11$. In other words, 1.11 times the average value gives rms of a sine wave.

A problem arises when the input signal is noise, and therefore not a sine wave. Gaussian noise has an average value of 0.798 of rms. Since an average responding voltmeter indicates 1.11 times higher, it reads $1.11 \times 0.798 = 0.885$ of the true rms value. The averaging meter reads too low. *When reading noise on an average responding meter multiply the reading by 1.13 or add 1 dB to obtain the correct value.* If you do not apply this correction, you may find that the noise of a resistor is less than thermal noise, a phenomenon that is difficult to explain.

Average responding meters may be suitable for noise measurements with proper precautions, but it is desirable to verify the reading with a calibrated noise source. The accuracy of the output meter is particularly important for the sine wave noise measurement method since there is no noise reference signal. To measure noise with an average responding meter, two more characteristics must be considered: bandwidth and crest factor of the meter. As pointed out in Table 14-2, the meter bandwidth must be much greater than the noise bandwidth of the system. The crest factor limitation is a special problem. To protect the indicator many average responding voltmeters are designed to saturate on greater than full-scale readings corresponding to crest factors of only 1.4 to 2. As shown in Table 14-3, a crest factor of 3 to 5 is desirable for Gaussian noise. There are two possible solutions: use a meter with a high-crest factor, or simply read at less than half scale. For half-scale reading, place a 6-dB attenuator ahead of the meter and multiply all readings by 2. This increases the crest factor from 2 to 4. The attenuator must have a flat frequency response.

The addition of a shunt capacitor to increase the averaging time can convert the meter to a peak detecting instrument. This problem arises when the meter is connected directly to the rectifier. The averaging circuit should have equal charge and discharge time constants.

Peak Responding Instruments

A peak reading voltmeter responds to the peak or peak-to-peak value of a signal. It can be calibrated to indicate the peak or the rms value of a sine wave. This works well when the peak value is constant, but as shown in Table 14-3, the peak value of noise is a variable.

The reading of noise on a peak responding meter depends on the charge versus discharge time constants. Peak voltmeters of different types produce different results. In general, *noise should not be measured with a peak responding meter* unless the peak amplitude of the noise has some significance such as for an application that involves level detection.

14-10 NOISE BANDWIDTH

As discussed in Chapter 1, noise bandwidth is defined differently from signal bandwidth. A commonly used definition of noise bandwidth was given:

$$\Delta f = \frac{1}{A_{vo}^2} \int_0^{\infty} A_v^2(f)\, df \tag{1-5}$$

Here we add a few comments to the discussion given earlier.

The noise bandwidth Δf of a narrowband system is illustrated graphically in Fig. 14-13. Curve *EFG* is the squared voltage transfer function of the system. Below line *EG* the gain measurement is dominated by noise, so the area below that line should be excluded from the calculation. To determine the noise bandwidth, select reference frequency f_o, usually the frequency of maximum gain. Next, the rectangle *ABCD* passing through *F* is constructed so that the area of *ABCD* is equal to the area under the power curve, *EFG*. The span *AD* is the noise bandwidth Δf.

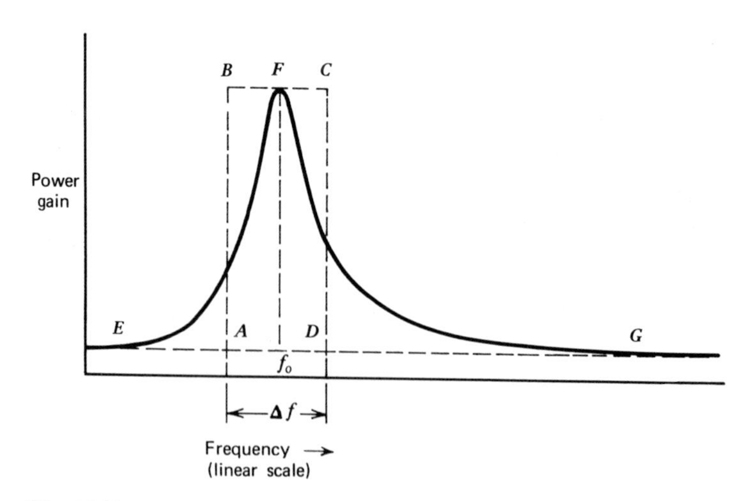

Fig. 14-13. Definition of noise bandwidth.

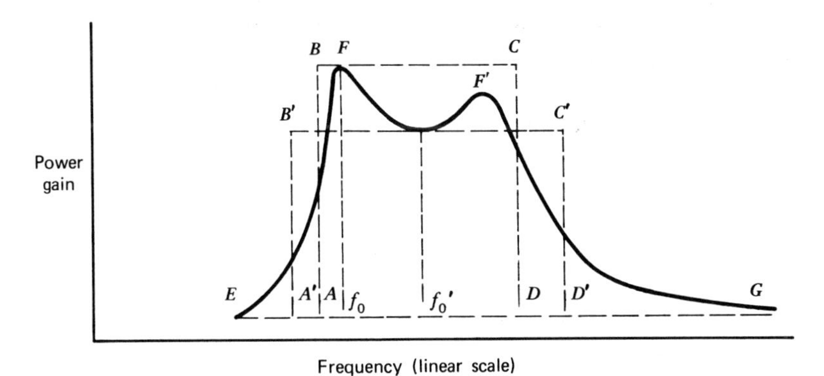

Fig. 14-14. Noise bandwidth; nonsymmetrical transfer function.

If the power gain curve is not symmetrical about the center frequency as shown in Fig. 14-14, the selection of f_o is arbitrary. Consider the noise bandwidth of curve $EFF'G$ which has two peaks at F and F'. Two reference frequencies f_o and f_0' are defined in the figure. These produce rectangular bandwidths of $ABCD$ and $A'B'C'D'$, respectively. Each of these rectangles has the same gain bandwidth product, but the peak gains differ; we see that noise bandwidth AD is smaller than $A'D'$. Either determination of noise bandwidth can be correct, although most engineers prefer to use AD.

Noise bandwidth can be measured with a calibrated noise generator. The noise generator must be the only significant noise in the system. Other noise sources should be weak in comparison. Measure the transfer voltage gain K_t at f_o. Then, with the noise generator connected to the input terminals, measure the total output noise E_{no}. The output is

$$E_{no} = K_t E_i \,\Delta f M \tag{14-25}$$

where K_t = transfer voltage gain
E_i = input noise signal in V/Hz$^{1/2}$
Δf = noise bandwidth
M = meter correction factor (if not true rms)

The unknown Δf is calculated from

$$\Delta f = \frac{E_{no}}{K_t E_i M}$$

A final word of caution is in order regarding noise bandwidth measurements and calculations. Refer again to Fig. 14-13. Notice that point G represents a frequency much larger than the noise bandwidth Δf. Suppose G = 15 MHz, Δf = 1 MHz, and we are measuring with a voltmeter having a 3-dB bandwidth of 10 MHz. This would seem acceptable since the voltmeter response is much wider than Δf. But note that the voltmeter attenuates much of the tail of the gain curve, thereby reducing the noise bandwidth below the calculated 1-MHz value. This meter effect on Δf must be considered in noise measurements.

SUMMARY

a. Two widely used techniques for noise measurement are the sine wave and the noise generator methods.

b. The sine wave method requires measurement of total output noise E_{no} and transfer gain K_t. Then

$$E_{ni} = \frac{E_{no}}{K_t}$$

c. For $R_s \to 0$, $E_{ni} = E_n$. For $R_s \to \infty$, $E_{ni} = I_n R_s$.

d. To eliminate thermal noise from R_s a capacitor can sometimes be substituted for source resistance.

e. A wave analyzer is convenient for making spectral density measurements.

f. The noise voltage of a noise generator necessary to double output noise power is equivalent to the E_{ni} of the system under test. Thus the generator must be calibrated.

g. The shot noise of a calibrated noise source is used for noise measurements.

h. Commercial noise analyzers, where economically justifiable, are very convenient.

i. Requirements for an output noise meter: (a) must respond to noise power; (b) must have a crest factor greater than 4; and (c) must have a bandwidth ten or more times the system noise bandwidth. A true rms reading meter is best; an average responding meter must be corrected by a 1.13 multiplier; and a peak responding meter is valueless.

PROBLEMS

1. Design an attenuator of the Fig. 14-4 type to match a 600-Ω signal generator. Consider that $R_2 = 10\ \Omega$. The output V_s should be 1% of V_g. Specify values for R_1 and R_3.

2. Spot noise figure is to be determined, but we do not have a measuring instrument calibrated in dB; we just have a true rms reading voltmeter. The transfer voltage gain K_t is measured to be 10,000. We adjust the level of input signal so

that at the output port $V_{so} = 100\ E_{no}$. We measure the input signal to be 1 μV and the source resistance is 1000 Ω. Find NF.

3. Design a test setup to measure excess noise of resistors. Develop a block diagram and specify the gain necessary and any special requirements for the equipment.

REFERENCE

1. Bennett, W. R., *Electrical Noise*, McGraw-Hill, New York, 1960, p. 44.

Appendix I

NOISE DATA ON BIPOLAR TRANSISTORS

The following pages contain original data on eleven types of bipolar transistors:

2N930	*npn*	Si	Fairchild
2N3801	*pnp*	Si	Motorola
2N3964	*pnp*	Si	Fairchild
2N4045	*npn*	Si	Intersil
2N4104	*npn*	Si	Texas Instruments
2N4124	*npn*	Si	Motorola
2N4125	*pnp*	Si	Motorola
2N4250	*pnp*	Si	Fairchild
2N4403	*pnp*	Si	Motorola
2N5138	*pnp*	Si	Fairchild
MPS-A18	*npn*	Si	Motorola

A manufacturer is listed in order to identify the types more easily. Because the noise mechanisms are process dependent, tests on more recently fabricated samples may not be in complete agreement with the data given here. Additional information pertaining to these transistors appears in Chapter 4.

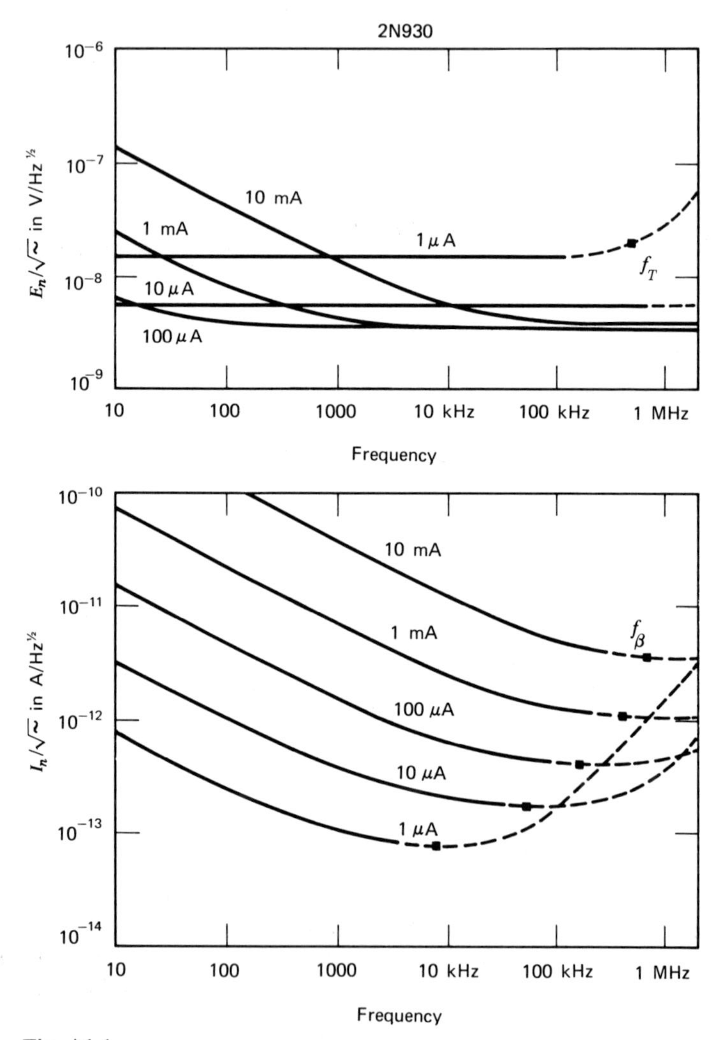
2N930
10^{-6}
10^{-7}
10^{-8}
10^{-9}
$E_n/\sqrt{\sim}$ in V/Hz$^{1/2}$
10 mA
1 mA
1 μA
10 μA
100 μA
f_T
10
100
1000
10 kHz
100 kHz
1 MHz
Frequency
10^{-10}
10^{-11}
10^{-12}
10^{-13}
10^{-14}
$I_n/\sqrt{\sim}$ in A/Hz$^{1/2}$
10 mA
1 mA
100 μA
10 μA
1 μA
f_β
10
100
1000
10 kHz
100 kHz
1 MHz
Frequency

Fig. A1-1

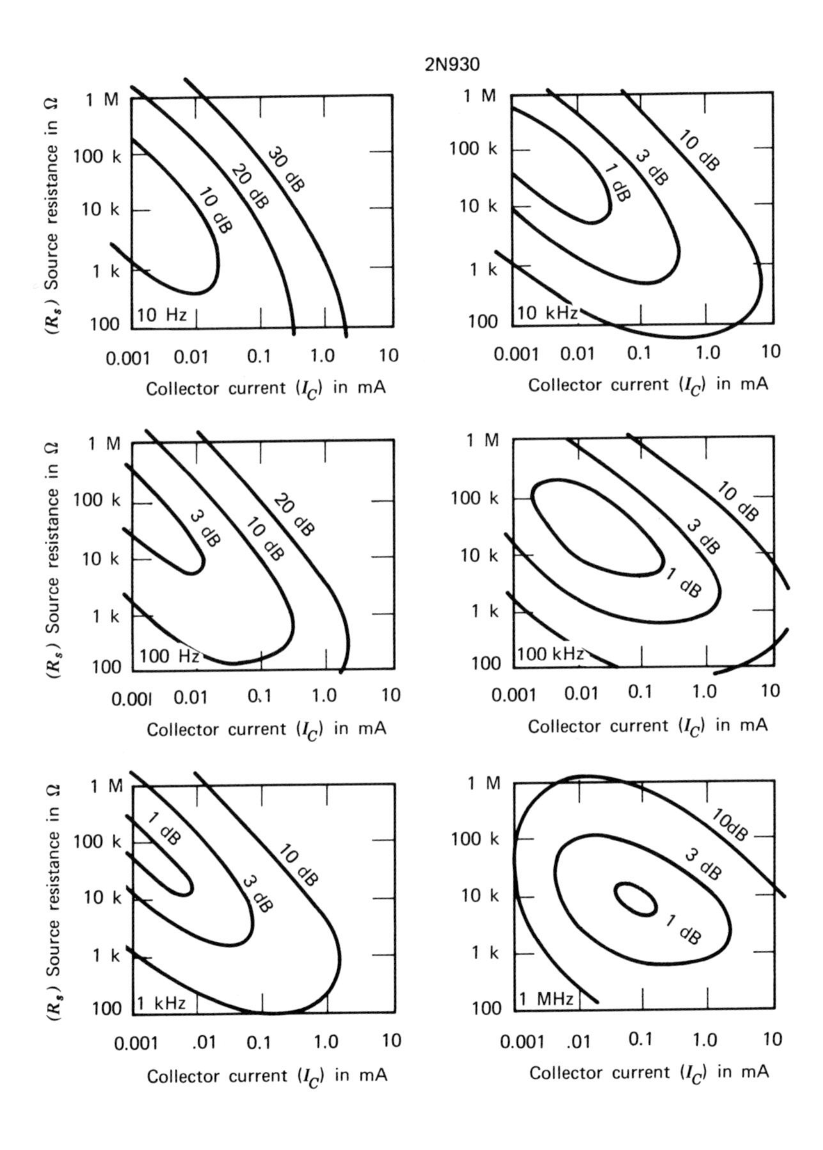
2N930
1 M
100 k
10 k
1 k
100
0.001 0.01 0.1 1.0 10
Collector current (I_C) in mA
(R_s) Source resistance in Ω
10 Hz
10 dB
20 dB
30 dB
10 kHz
1 dB
3 dB
10 dB
100 Hz
3 dB
10 dB
20 dB
100 kHz
1 dB
3 dB
10 dB
1 kHz
1 dB
3 dB
10 dB
1 MHz
1 dB
3 dB
10dB

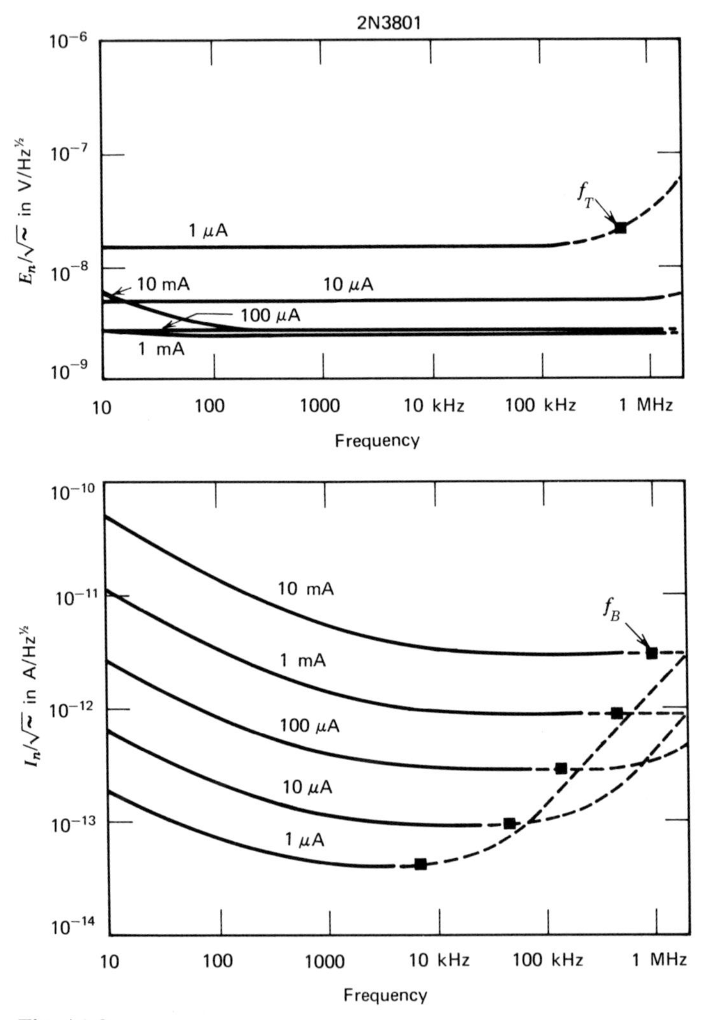
2N3801
$E_n/\sqrt{\sim}$ in V/Hz$^{1/2}$
10^{-6}
10^{-7}
10^{-8}
10^{-9}
f_T
1 μA
10 mA
10 μA
100 μA
1 mA
10
100
1000
10 kHz
100 kHz
1 MHz
Frequency
$I_n/\sqrt{\sim}$ in A/Hz$^{1/2}$
10^{-10}
10^{-11}
10^{-12}
10^{-13}
10^{-14}
f_B
10 mA
1 mA
100 μA
10 μA
1 μA
10
100
1000
10 kHz
100 kHz
1 MHz
Frequency

Fig. A1-2

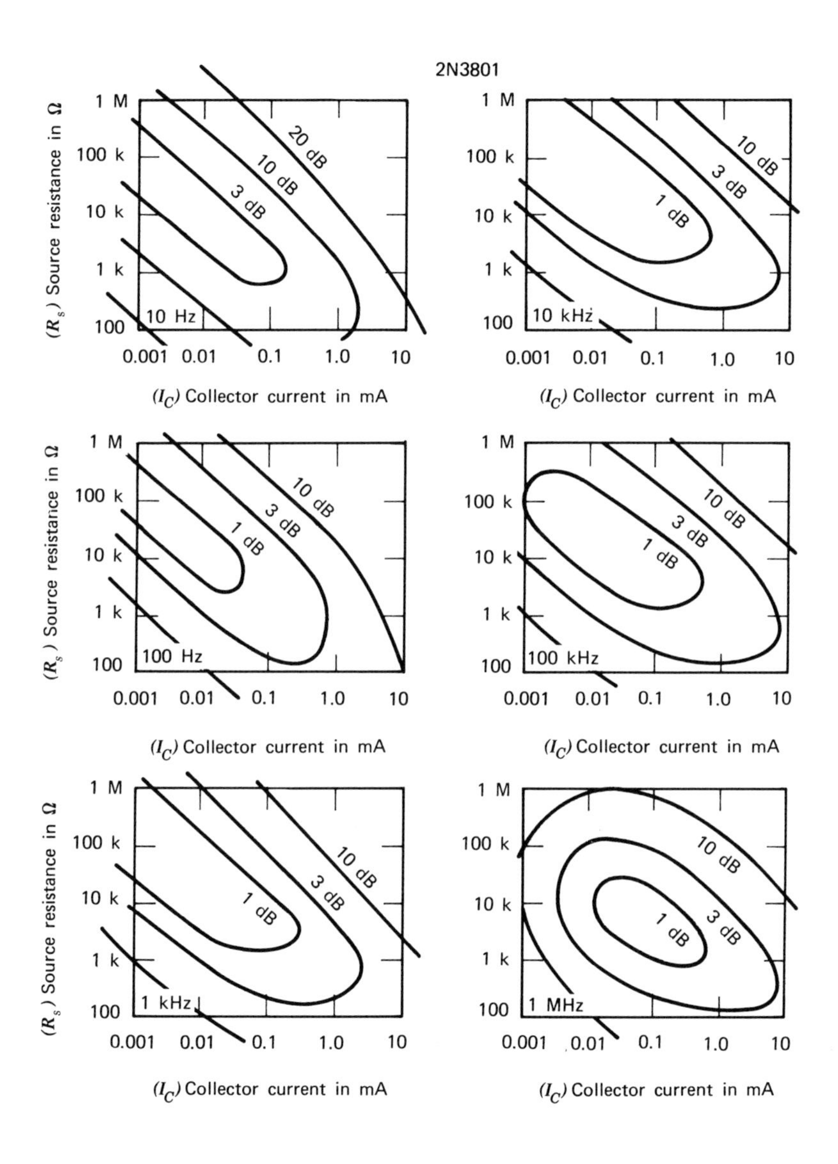
2N3801
1 M
100 k
10 k
1 k
100
(R_s) Source resistance in Ω
0.001 0.01 0.1 1.0 10
(I_C) Collector current in mA
20 dB
10 dB
3 dB
1 dB
10 Hz
10 kHz
100 Hz
100 kHz
1 kHz
1 MHz

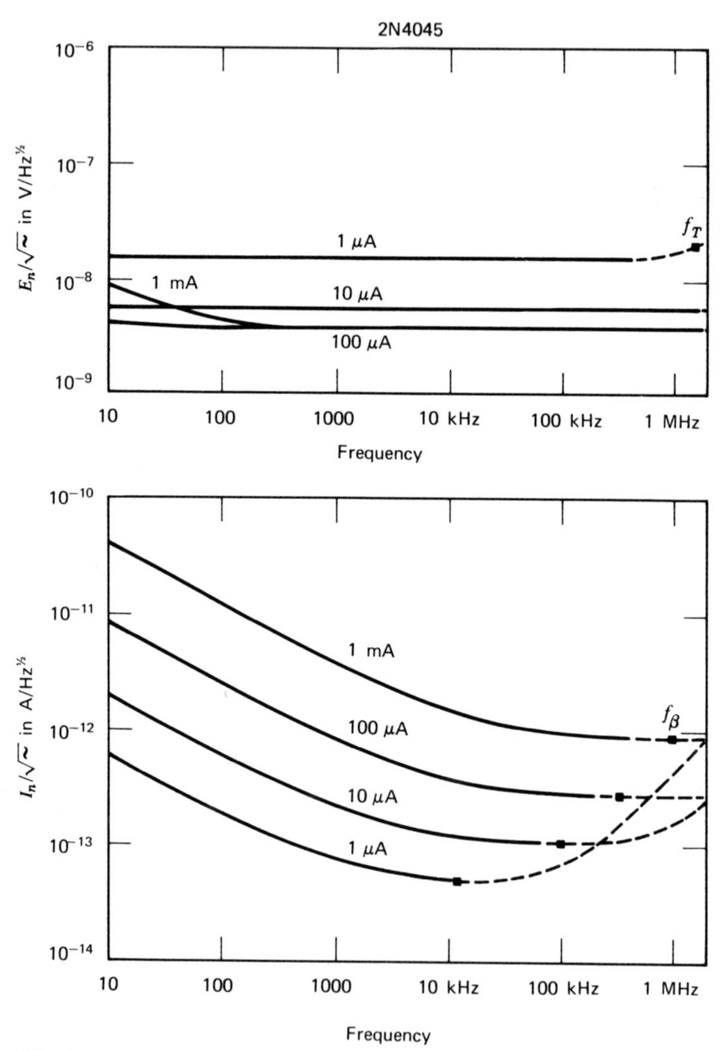
2N4045
$E_n/\sqrt{\sim}$ in V/Hz$^{1/2}$
10^{-6}
10^{-7}
10^{-8}
10^{-9}
1 μA
1 mA
10 μA
100 μA
f_T
10
100
1000
10 kHz
100 kHz
1 MHz
Frequency
$I_n/\sqrt{\sim}$ in A/Hz$^{1/2}$
10^{-10}
10^{-11}
10^{-12}
10^{-13}
10^{-14}
1 mA
100 μA
10 μA
1 μA
f_β
10
100
1000
10 kHz
100 kHz
1 MHz
Frequency

Fig. A1-3

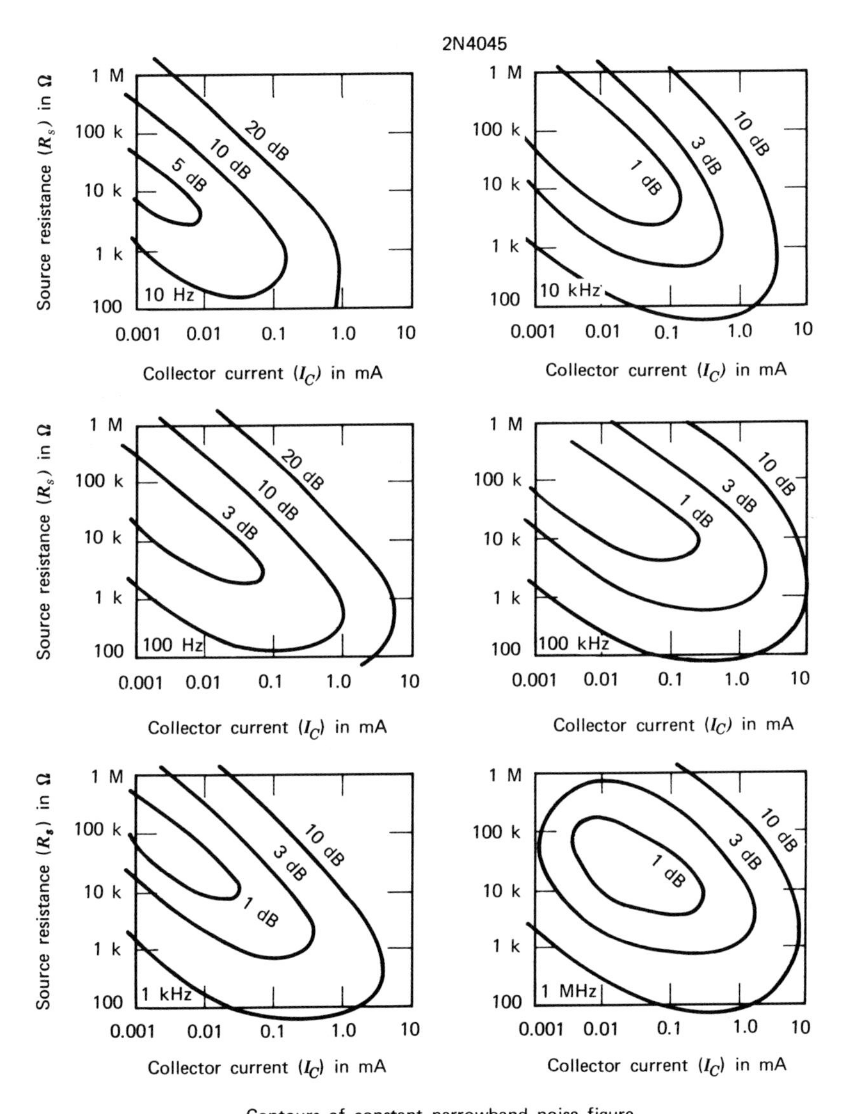

Contours of constant narrowband noise figure

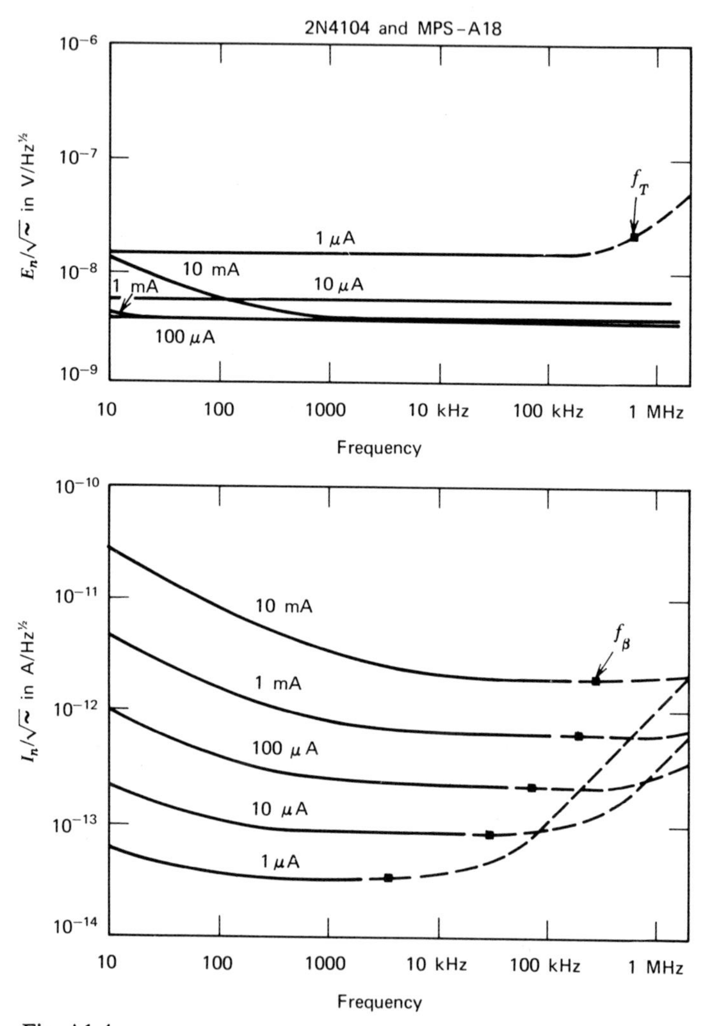
2N4104 and MPS-A18
$E_n/\sqrt{\sim}$ in V/Hz½
10^{-6}
10^{-7}
10^{-8}
10^{-9}
f_T
1 μA
10 mA
1 mA
10 μA
100 μA
10
100
1000
10 kHz
100 kHz
1 MHz
Frequency
$I_n/\sqrt{\sim}$ in A/Hz½
10^{-10}
10^{-11}
10^{-12}
10^{-13}
10^{-14}
10 mA
f_β
1 mA
100 μA
10 μA
1 μA
10
100
1000
10 kHz
100 kHz
1 MHz
Frequency

Fig. A1-4

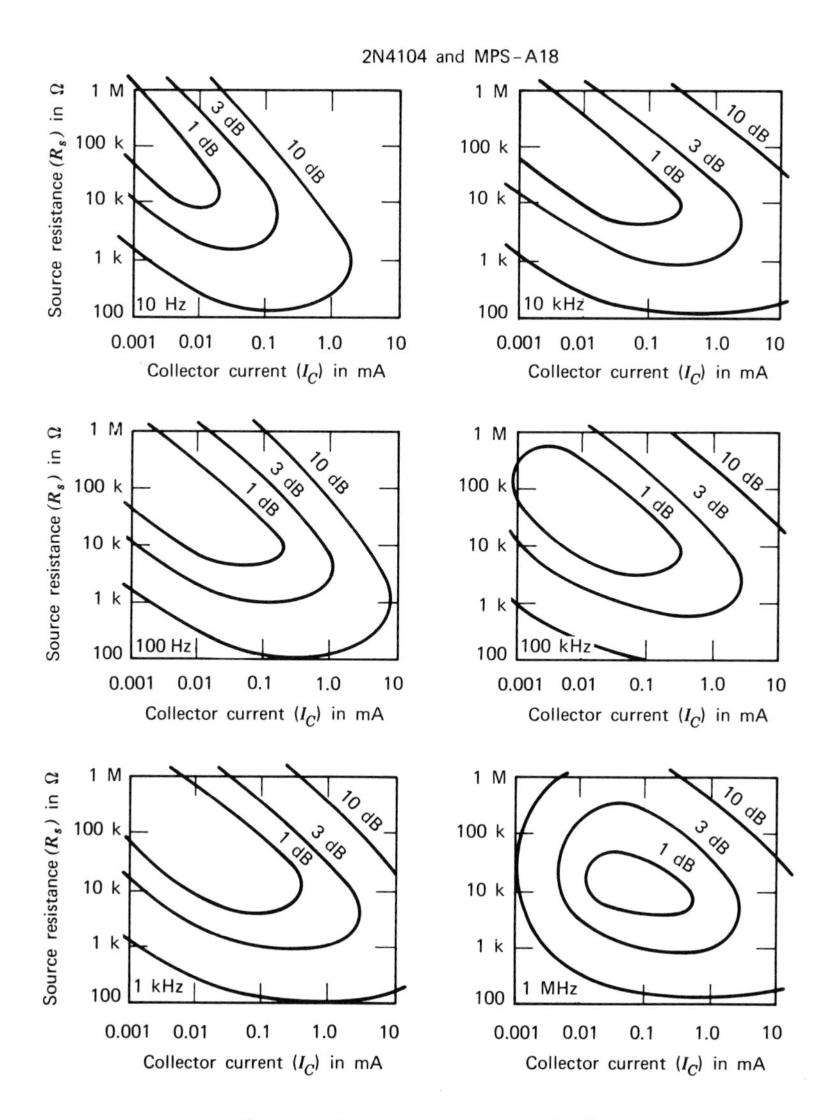

Contours of constant narrowband noise figure

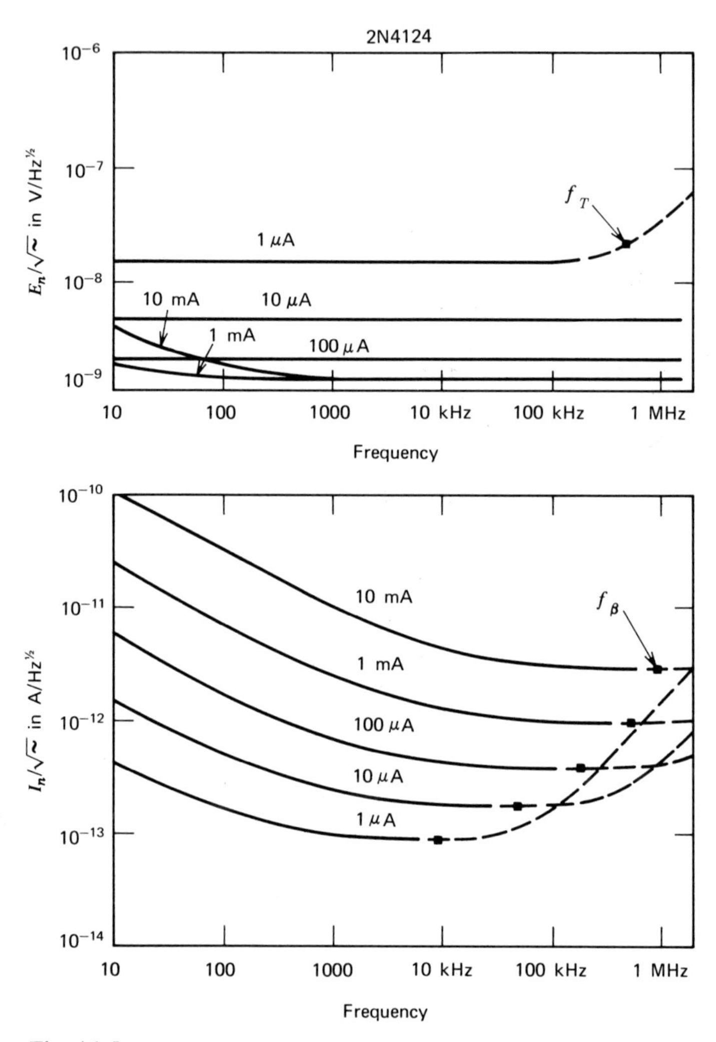
2N4124
$E_n/\sqrt{\sim}$ in V/Hz$^{1/2}$
10^{-6}
10^{-7}
10^{-8}
10^{-9}
f_T
1 μA
10 mA
10 μA
1 mA
100 μA
10
100
1000
10 kHz
100 kHz
1 MHz
Frequency
$I_n/\sqrt{\sim}$ in A/Hz$^{1/2}$
10^{-10}
10^{-11}
10^{-12}
10^{-13}
10^{-14}
10 mA
f_β
1 mA
100 μA
10 μA
1 μA
10
100
1000
10 kHz
100 kHz
1 MHz
Frequency

Fig. A1-5

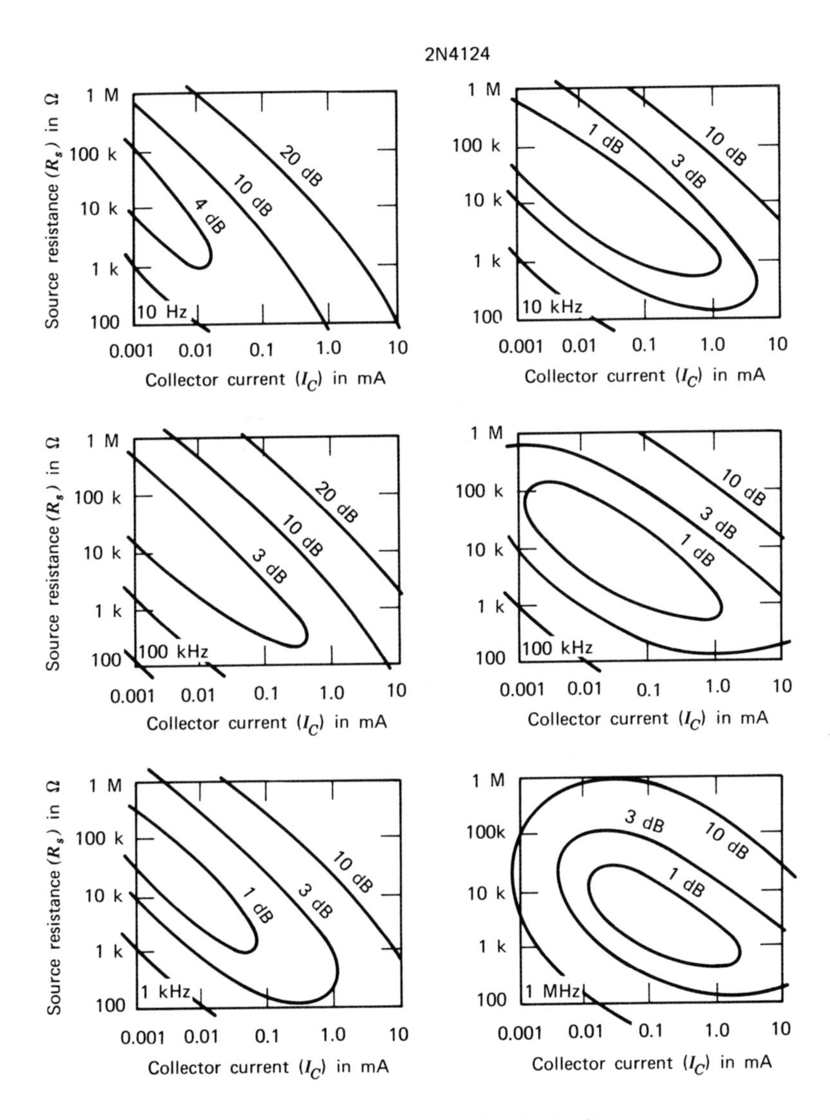

Contours of constant narrowband noise figure

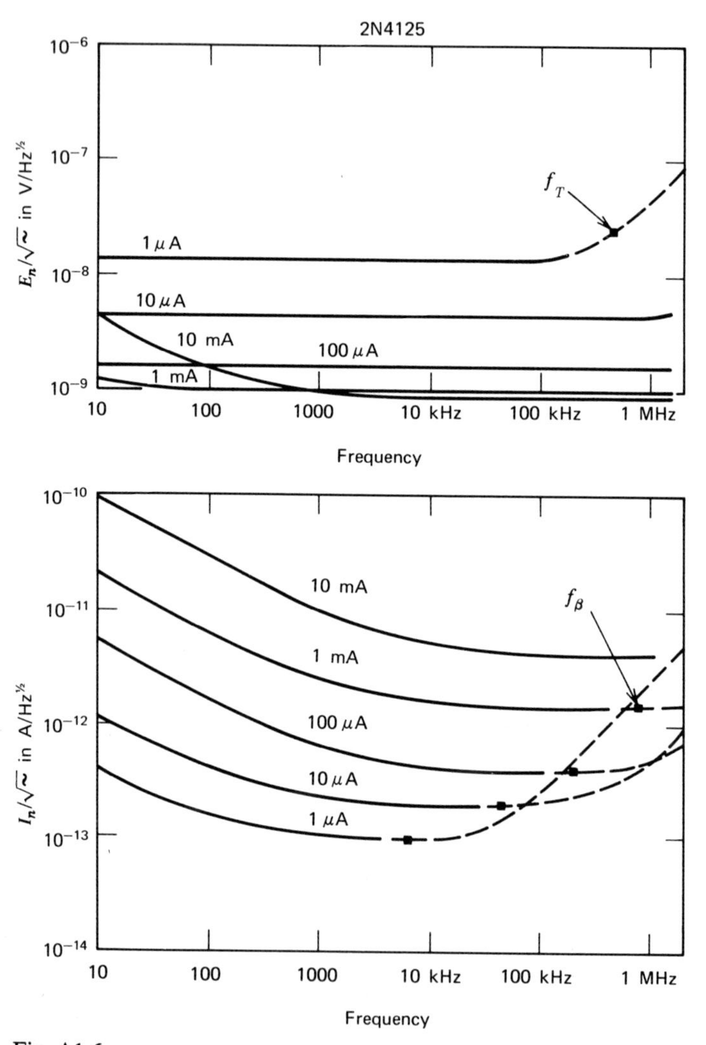
2N4125
$E_n/\sqrt{\sim}$ in V/Hz$^{1/2}$
10^{-6}
10^{-7}
10^{-8}
10^{-9}
f_T
1 μA
10 μA
10 mA
100 μA
1 mA
10
100
1000
10 kHz
100 kHz
1 MHz
Frequency
$I_n/\sqrt{\sim}$ in A/Hz$^{1/2}$
10^{-10}
10^{-11}
10^{-12}
10^{-13}
10^{-14}
10 mA
f_β
1 mA
100 μA
10 μA
1 μA
10
100
1000
10 kHz
100 kHz
1 MHz
Frequency

Fig. A1-6

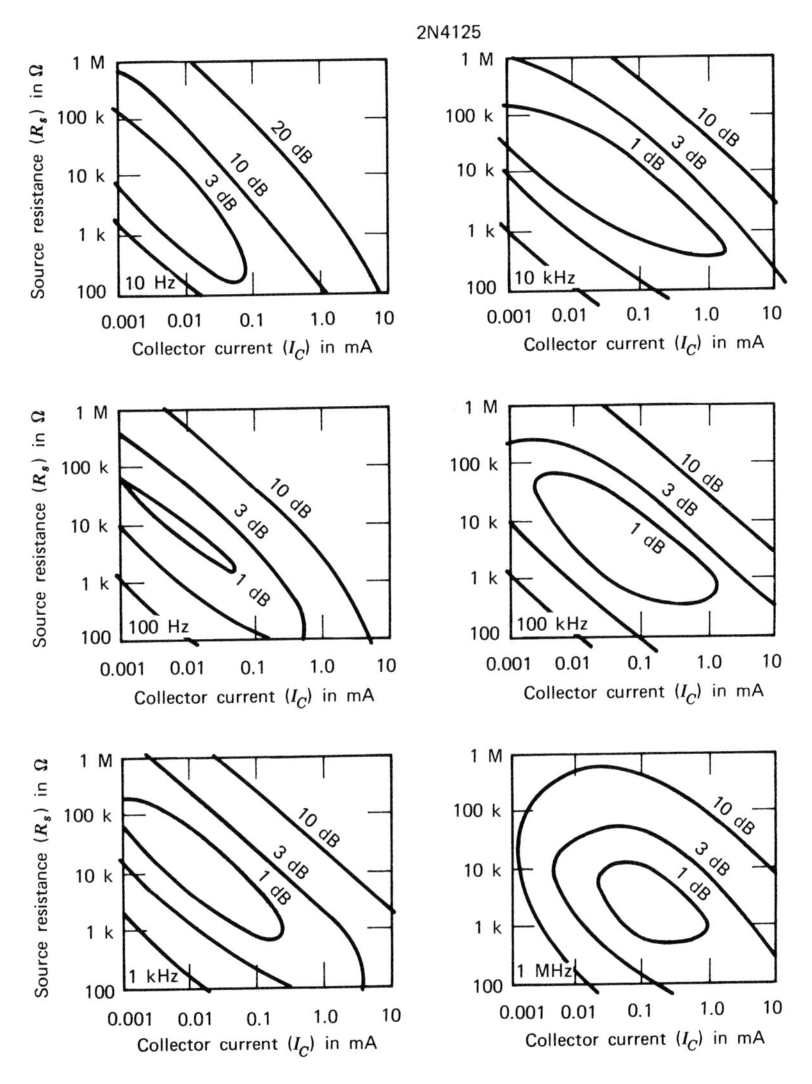

Contours of constant narrowband noise figure

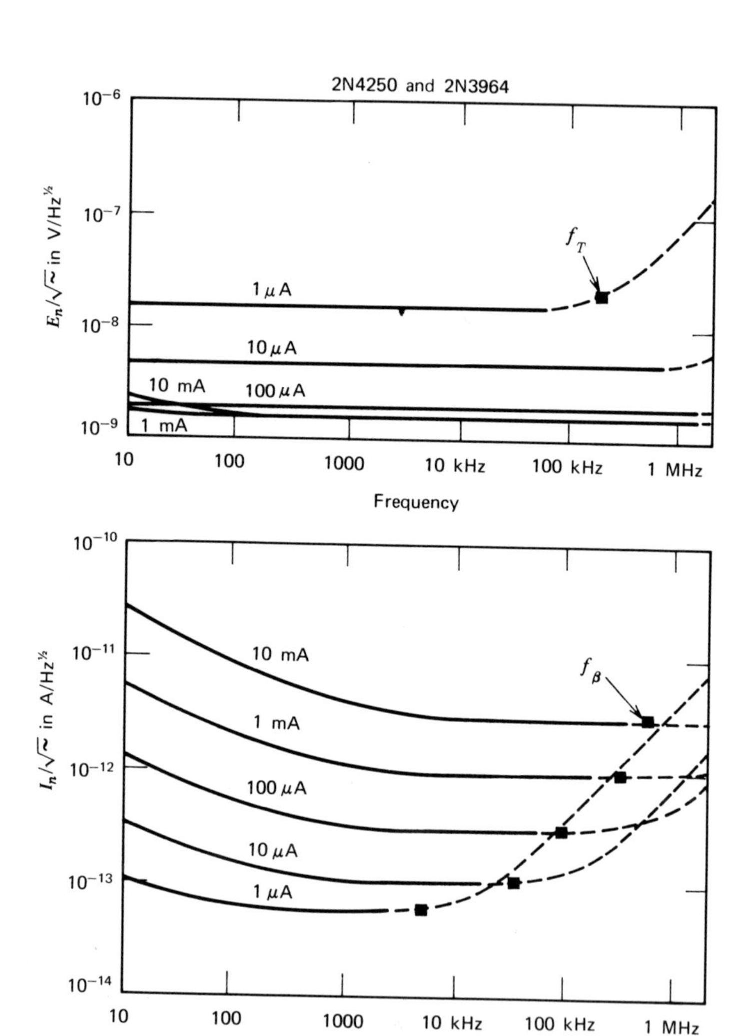
2N4250 and 2N3964
$E_n/\sqrt{\sim}$ in V/Hz$^{1/2}$
10^{-6}
10^{-7}
10^{-8}
10^{-9}
f_T
1 μA
10 μA
10 mA
100 μA
1 mA
10
100
1000
10 kHz
100 kHz
1 MHz
Frequency
$I_n/\sqrt{\sim}$ in A/Hz$^{1/2}$
10^{-10}
10^{-11}
10^{-12}
10^{-13}
10^{-14}
f_β
10 mA
1 mA
100 μA
10 μA
1 μA
10
100
1000
10 kHz
100 kHz
1 MHz
Frequency

Fig. A1-7

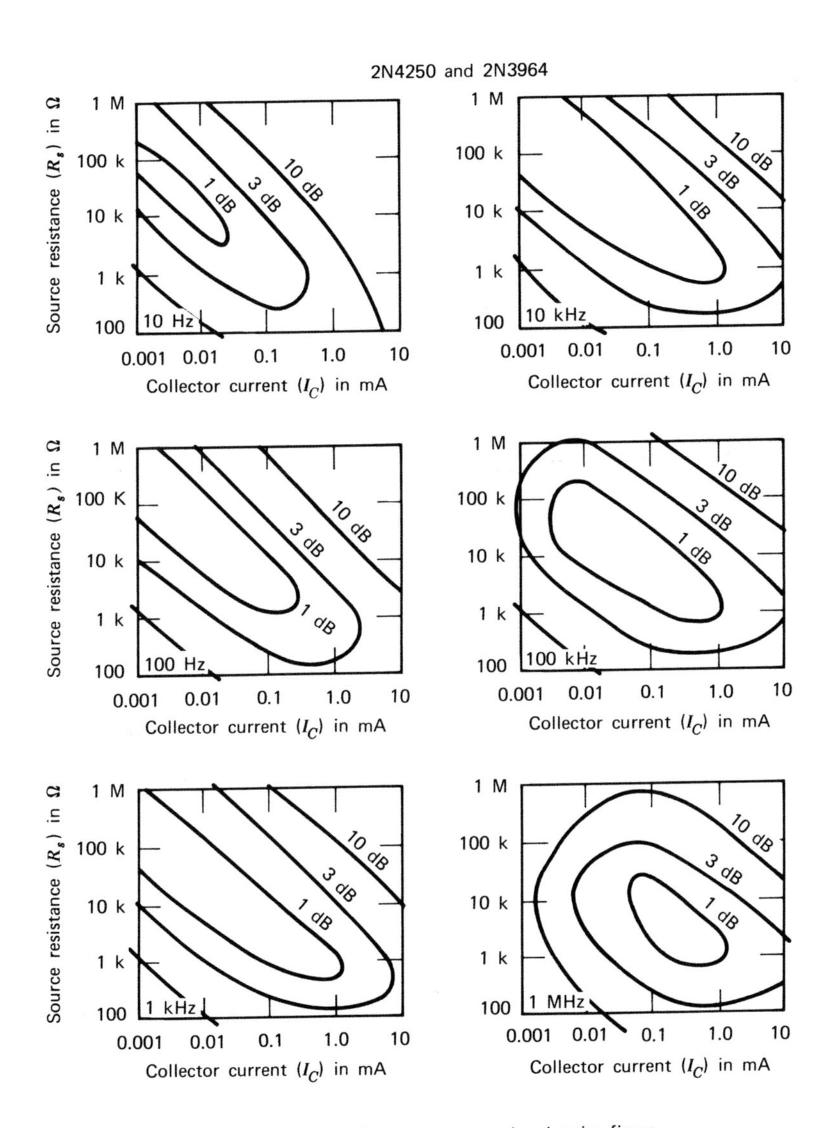

Contours of constant narrowband noise figure

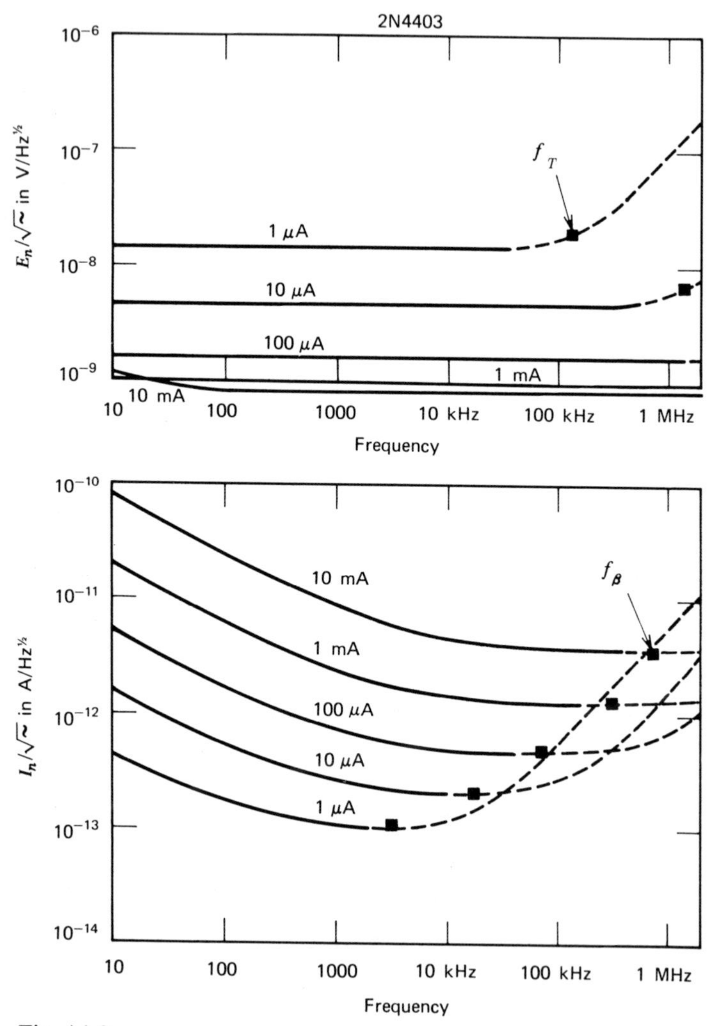
2N4403
$E_n/\sqrt{\sim}$ in V/Hz½
10^{-6}
10^{-7}
10^{-8}
10^{-9}
f_T
1 μA
10 μA
100 μA
1 mA
10 mA
10
100
1000
10 kHz
100 kHz
1 MHz
Frequency
$I_n/\sqrt{\sim}$ in A/Hz½
10^{-10}
10^{-11}
10^{-12}
10^{-13}
10^{-14}
f_β
10 mA
1 mA
100 μA
10 μA
1 μA
10
100
1000
10 kHz
100 kHz
1 MHz
Frequency

Fig. A1-8

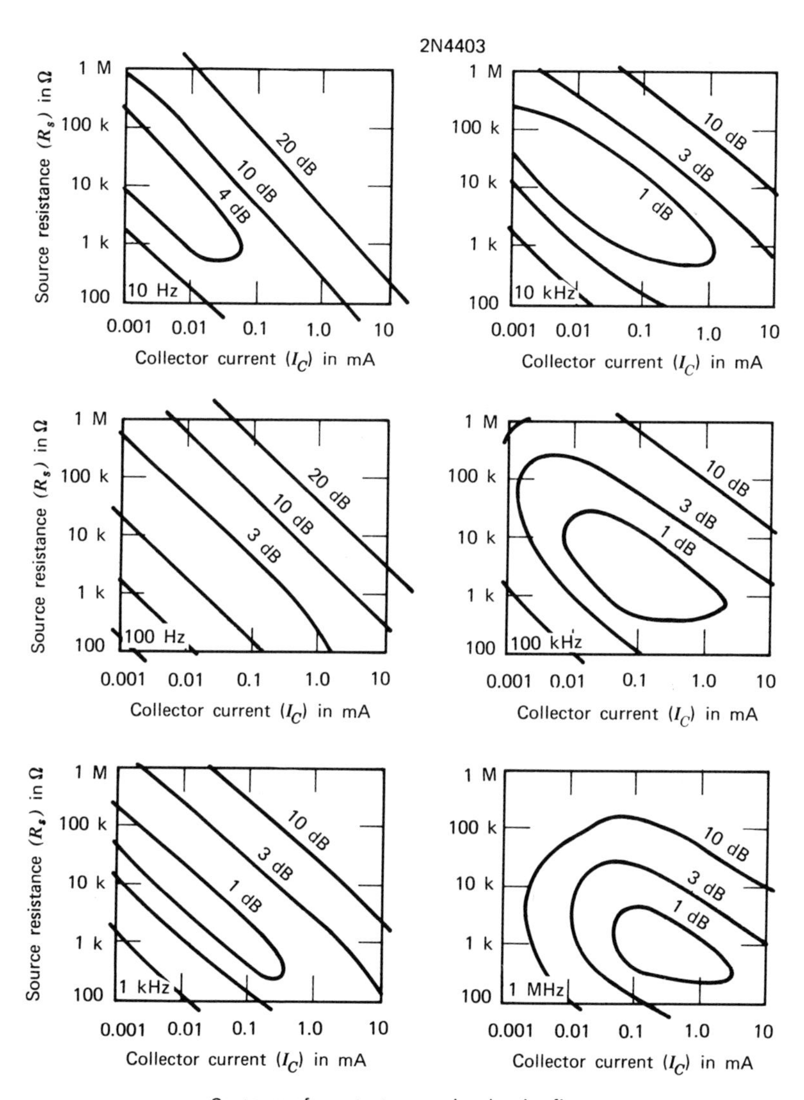

Contours of constant narrowband noise figure

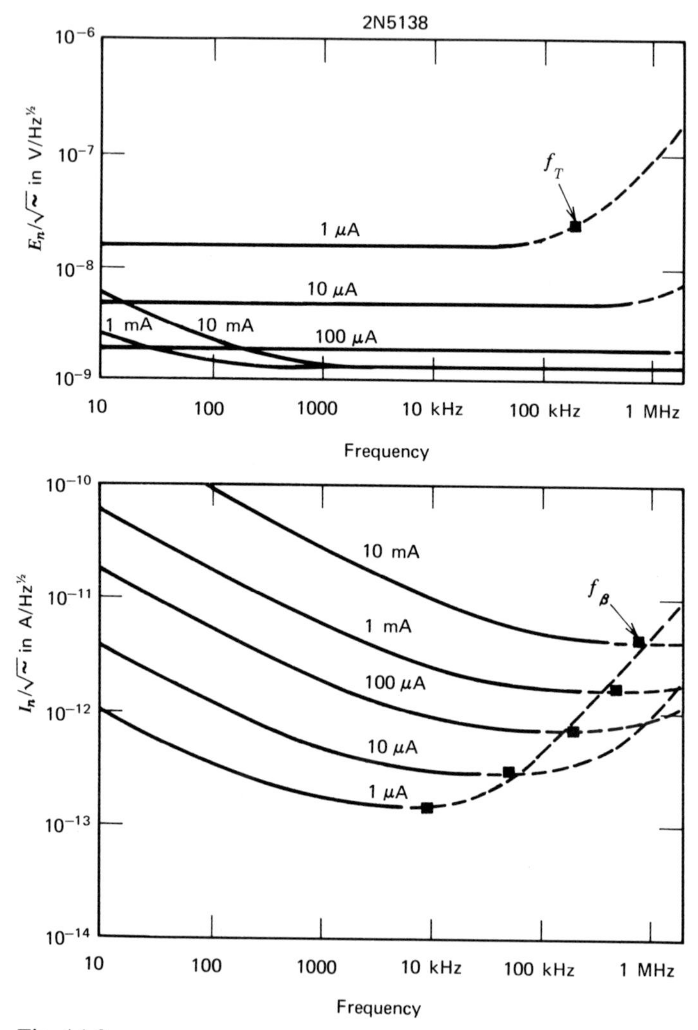
2N5138
$E_n/\sqrt{\sim}$ in V/Hz$^{1/2}$
10^{-6}
10^{-7}
10^{-8}
10^{-9}
f_T
1 μA
10 μA
1 mA
10 mA
100 μA
10
100
1000
10 kHz
100 kHz
1 MHz
Frequency
$I_n/\sqrt{\sim}$ in A/Hz$^{1/2}$
10^{-10}
10^{-11}
10^{-12}
10^{-13}
10^{-14}
10 mA
f_β
1 mA
100 μA
10 μA
1 μA
10
100
1000
10 kHz
100 kHz
1 MHz
Frequency

Fig. A1-9

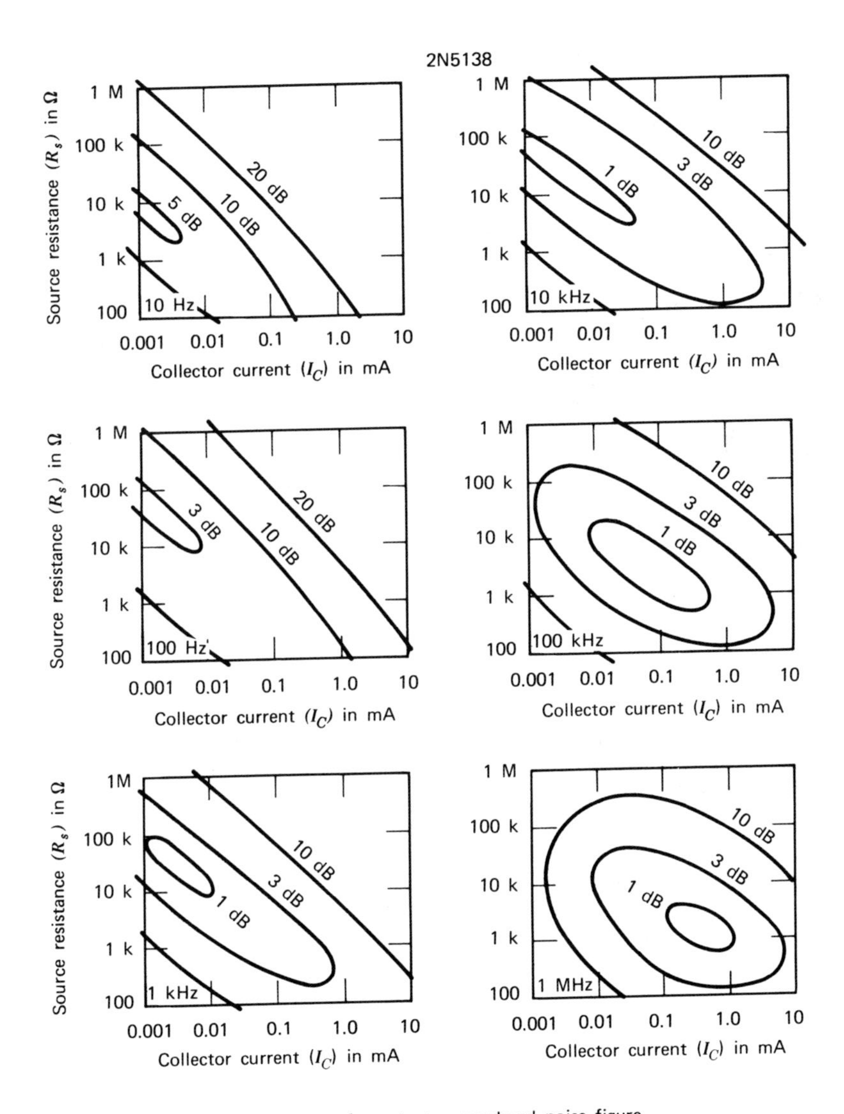

Contours of constant narrowband noise figure

Appendix II

NOISE DATA ON FIELD-EFFECT DEVICES

The following pages contain original data on seven types of JFETs and MOSFETs:

2N2609	JFET	*p*-channel	Siliconix
2N3631	MOSFET	*n*-channel	Siliconix
2N3821	JFET	*n*-channel	Texas Instruments
2N4221A	JFET	*n*-channel	Motorola
2N4869	JFET	*n*-channel	Siliconix
2N5266	JFET	*p*-channel	Motorola
2N5394	JFET	*n*-channel	Amelco

A manufacturer is listed in order to identify the types more easily. Because the noise mechanisms are process dependent, tests on more recently fabricated samples may not agree with the data given here.

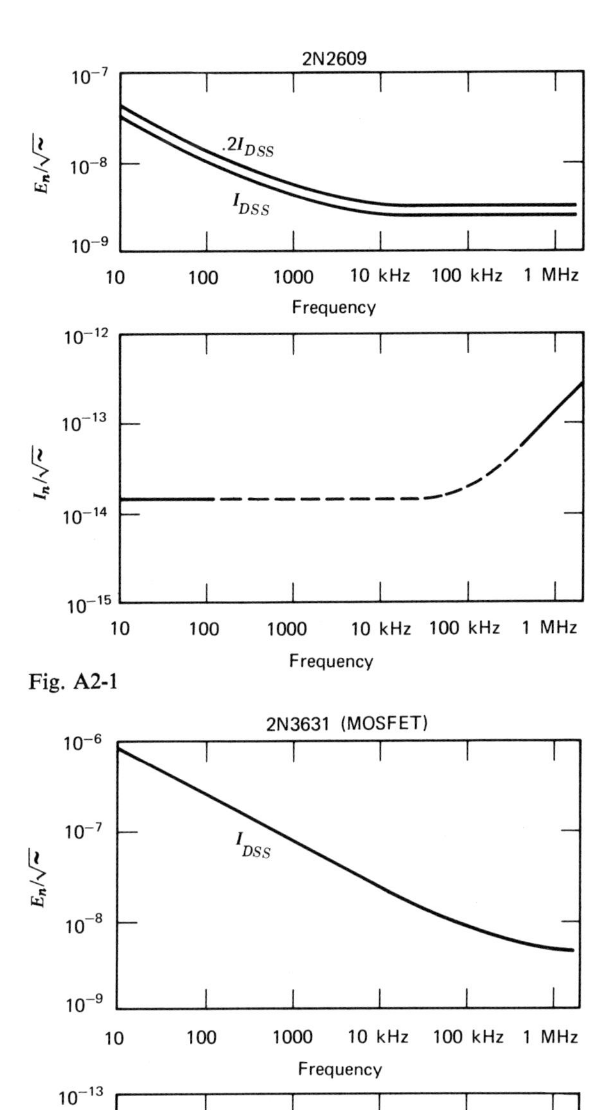

Fig. A2-1

Fig. A2-2

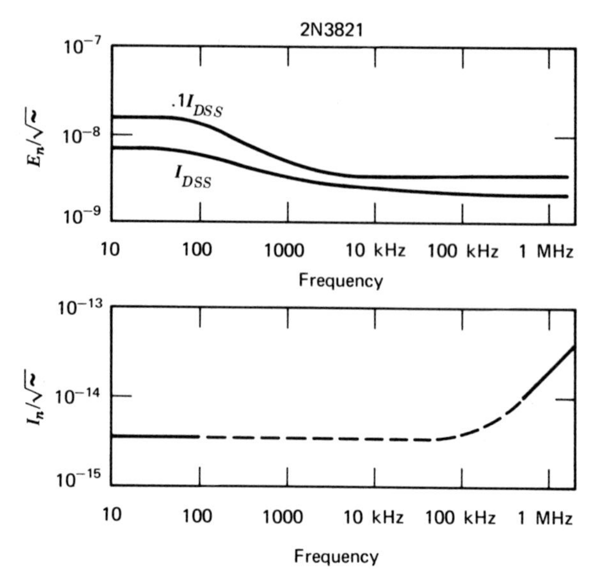

Fig. A2-3

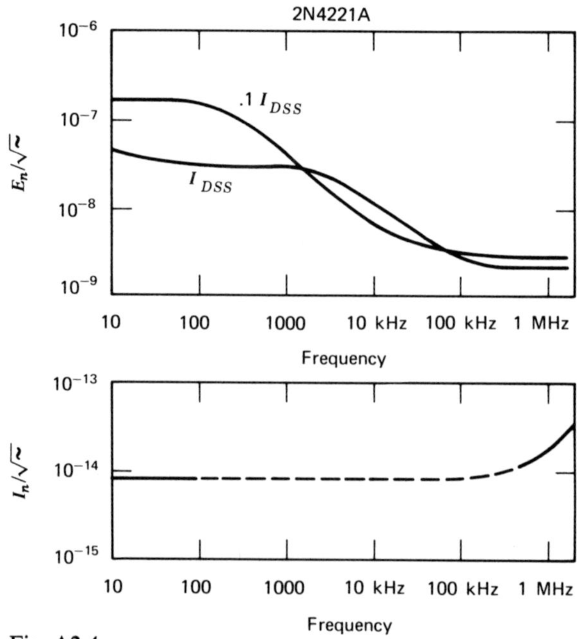

Fig. A2-4

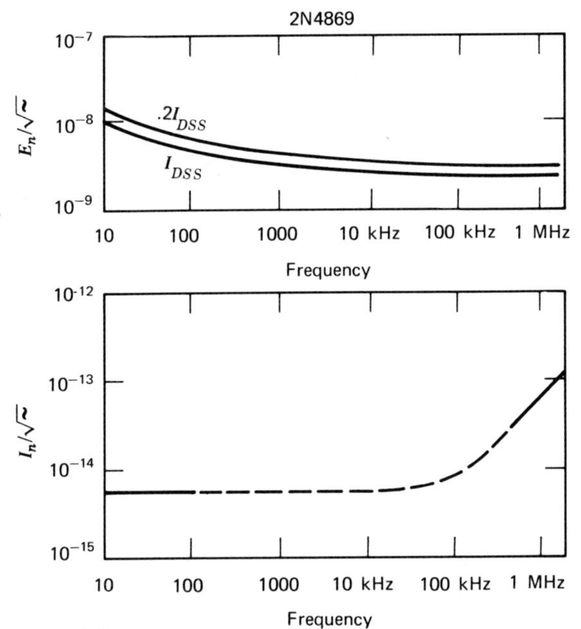

Fig. A2-5

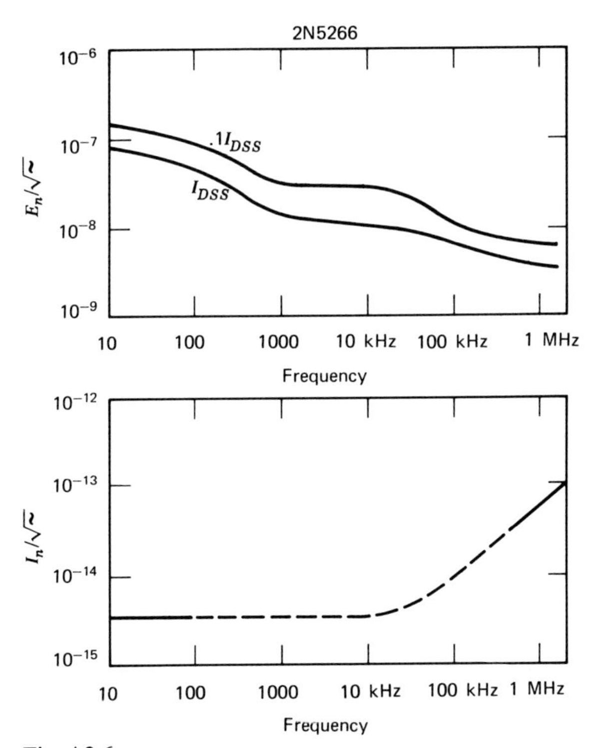

Fig. A2-6

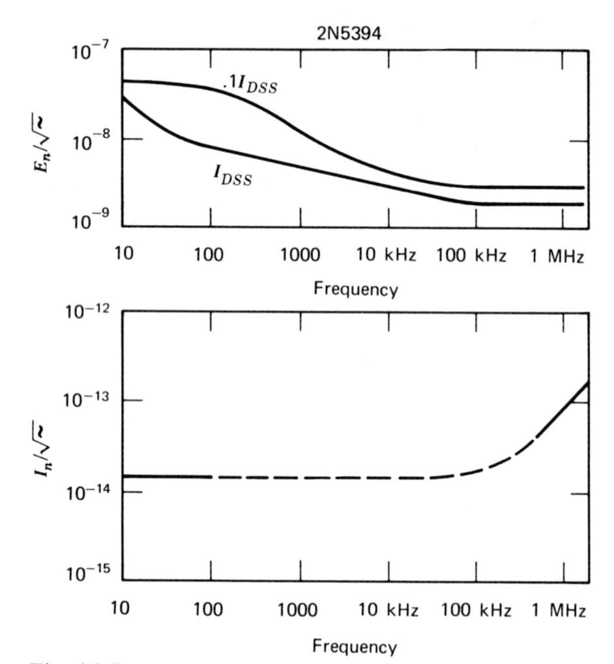

2N5394
$E_n/\sqrt{\sim}$
10^{-7}
10^{-8}
10^{-9}
$.1I_{DSS}$
I_{DSS}
10
100
1000
10 kHz
100 kHz
1 MHz
Frequency
$I_n/\sqrt{\sim}$
10^{-12}
10^{-13}
10^{-14}
10^{-15}
10
100
1000
10 kHz
100 kHz
1 MHz
Frequency

Fig. A2-7

Appendix III

NOISE DATA ON LINEAR INTEGRATED CIRCUITS

The following pages contain original data on seven types of linear ICs:

μA735	micropower op amp	Fairchild
μA739	op amp	Fairchild
μA741, 741C	op amp	Fairchild, Intersil
MC1510	video amp	Motorola
MC1556	op amp	Motorola
RM4132	micropower amp	Raytheon
LM101F	op amp	Siliconix

A manufacturer is listed in order to identify the types more easily. Because the noise mechanisms are process dependent, tests on more recently fabricated samples may not agree with the data given here.

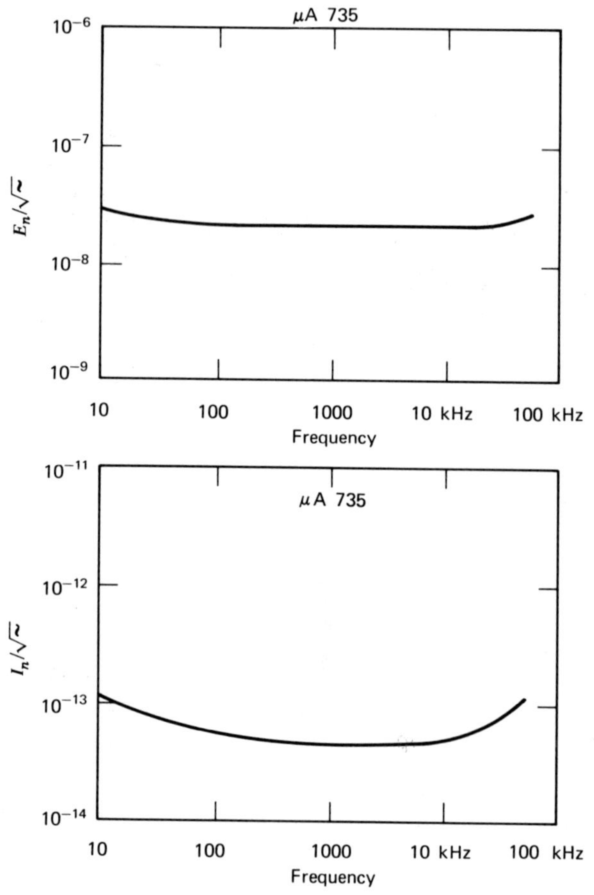
μA 735
10^{-6}
10^{-7}
$E_n/\sqrt{\sim}$
10^{-8}
10^{-9}
10
100
1000
10 kHz
100 kHz
Frequency
10^{-11}
μA 735
10^{-12}
$I_n/\sqrt{\sim}$
10^{-13}
10^{-14}
10
100
1000
10 kHz
100 kHz
Frequency

Fig. A3-1

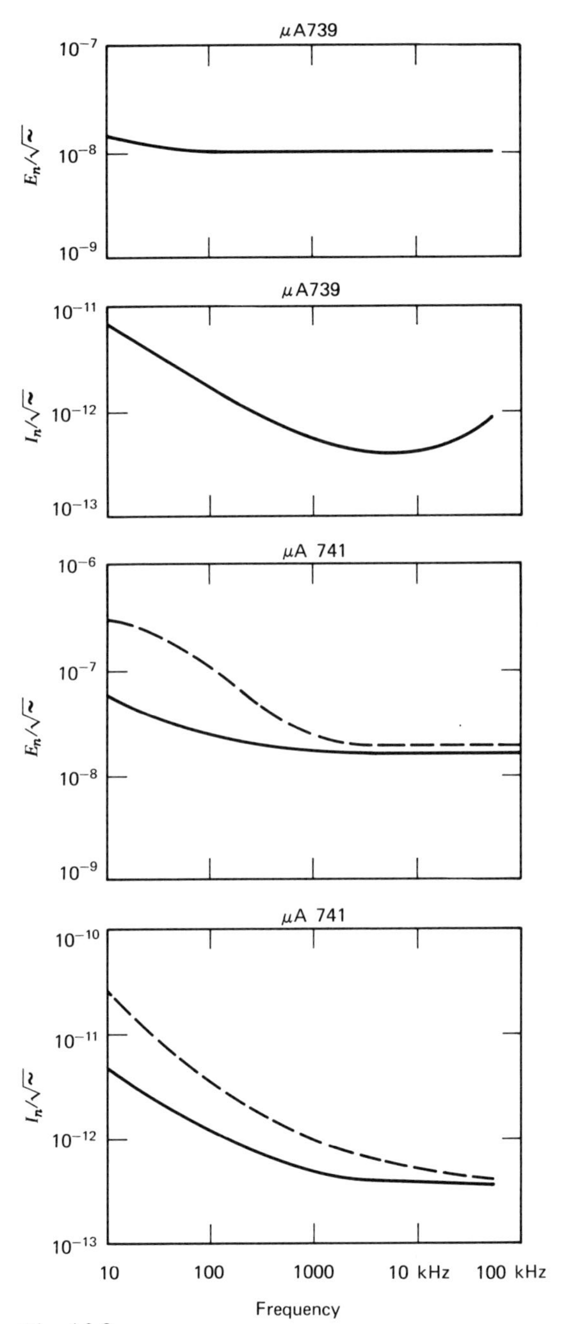
μA739
10^{-7}
$E_n/\sqrt{\sim}$
10^{-8}
10^{-9}
μA739
10^{-11}
$I_n/\sqrt{\sim}$
10^{-12}
10^{-13}
μA 741
10^{-6}
$E_n/\sqrt{\sim}$
10^{-7}
10^{-8}
10^{-9}
μA 741
10^{-10}
$I_n/\sqrt{\sim}$
10^{-11}
10^{-12}
10^{-13}
10
100
1000
10 kHz
100 kHz
Frequency

Fig. A3-2

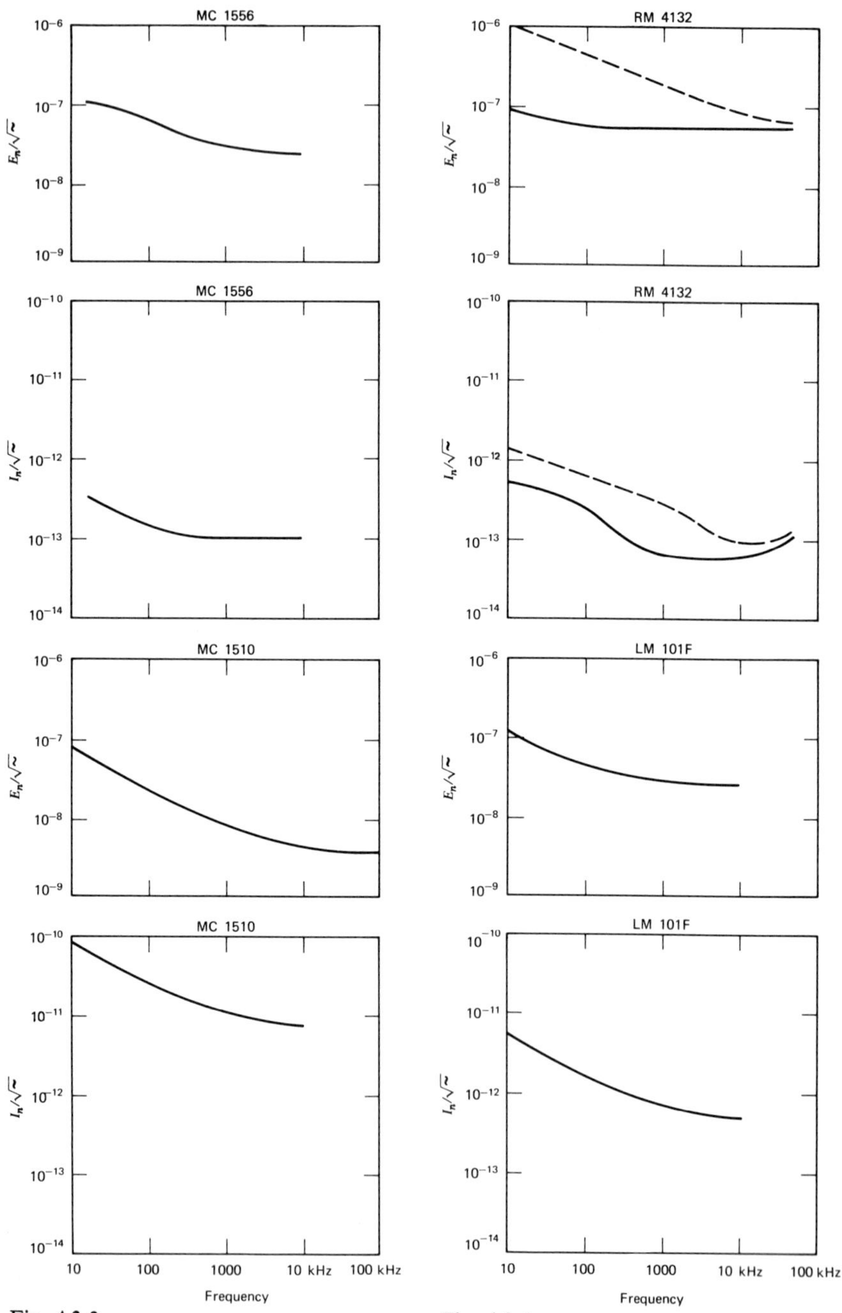

Fig. A3-3

Fig. A3-4

Appendix IV

"NOISE" COMPUTER PROGRAM

The program NOISE, discussed in Chapter 8, is included in this Appendix. This version, in the FORTRAN language, has been adapted from the original BASIC language program developed for a time-shared computer. The original is available—see reference 1, Chapter 8.

```
        DIMENSION FQ(10)
        COMMON A0,A1,A2,A3,A4,A5,A6,A7,A8,B,B0,B1,BKO,C1,C2,C3,C6,
            CKO,E,*
      1 E6,F,F0,F1,F2,F3,F4,F5,F6,F7,F8,FREQ(10),I,I0,I2,I4,I5,I6,I7,K,K0,*
      2 K1,K2,K3,K4,K5,L1,L2,N,N1,N2,N3,N4,N5,N6,N8,N9,NM,P,Q,Q0,Q1,
            Q2,R,*
      3 R1,R2,R3,R4,R6,R7,S1,S2,S3,S4,S5,SUM,T,T0,W,X1,X2,X3,X4,X5*
        REAL I,I0,I2,I4,I5,I6,I7,K,K0,K2,K4,K5,L1,L2,N,N2,N3,N5,N6,N8,
        N9*
C       READ IN VALUES FOR SENSOR MODELS
        READ(11,5) C1,C2,C3,C6,R1,R2,R3,R4,R6
        READ(11,5) I2,L1,L2,T,T0,F6
C READ IN VALUES FOR N1(CALCULATION DESIRED), N4(SOURCE USED)
        AND NM (NOISE MODEL)
        READ (11,9) N1,N4,NM
      9 FORMAT(3I2)
        IF(NM-1) 1,2,3
      1 WRITE(12,4)
      4 FORMAT(//32H NOISE MODEL IMPROPERLY SELECTED)
     10 STOP
```

```
C     READ IN VALUES FOR NOSMOD
    2 READ(11,5) E6,I6,F1,F2,F3,F4
      GO TO 60
C     READ IN VALUES FOR XISTOR
    3 READ(11,5) B,F7,F8,Q1,Q2,I4,I7,N5,N6
    5 FORMAT(9E8.1)
C     READ IN VALUES FOR AMPLGN
   60 READ (11,5) F0,F5,N8,N9,K2
      READ(11,5) S1,S2,S3,S4,S5,Q0
    6 P=3.14159
      Q=1.602E-19
      K=1.38E-23
  210 GO TO (250,250,250,280,250,250),N4
  250 GO TO (410,500,610,900,710,800),N1
  280 GO TO (400,500,600,982,700,800),N1
  400 K1=1
      CALL PRNT1
      CALL INTEG
  350 N2=SQRT(N2)
      WRITE(12,20) N2
   20 FORMAT(//23H TOTAL NOISE CURRENT = ,F6.2,9H PICOAMPS)
      STOP
  410 K1=1
      CALL PRNT2
      CALL PRNT3
      CALL INTEG
  360 N2=SQRT(N2)
      WRITE(12,21) N2
   21 FORMAT(//23H TOTAL NOISE VOLTAGE = ,F10.2,10H NANOVOLTS)
      STOP
  500 CALL READ1
      CALL PRNT5
  501 K3=2
      CALL SOURCE
  505 CALL PRNT4
  507 GO TO 501
  600 WRITE(12,27)
   27 FORMAT(//70H  SENSOR IN  LOAD IN AMPL EN/ZS  AMPL IN
      SUM NOISE PICOAMPS)
  625 READ(11,26) FQ
   26 FORMAT(10F8.0)
      IC=1
   31 F=FQ(IC)
      IF(F) 10,10,30
   30 K1=1
```

```
630 CALL KVAL
635 SUM=SQRT(N+R+E+I)
    N=SQRT(N)
    R=SQRT(R)
    E=SQRT(E)
    I=SQRT(I)
606 WRITE(12,28) N,R,E,I,SUM
    IC=IC+1
    GO TO 31
 28 FORMAT(//5F12.3)
610 CALL PRNT3
    GO TO 625
700 K1=2
    CALL PRNT1
    CALL INTEG
    GO TO 350
710 K1=2
    CALL PRNT2
    CALL PRNT3
    CALL INTEG
    GO TO 360
 38 FORMAT(I8)
800 CALL READ1
    CALL PRNT5
801 K3=2
    CALL SOURCE
820 CALL AMPLGN
    K0=K0*K5
840 CALL PRNT4
850 GO TO 801
900 K1=1
    CALL READ1
    CALL PRNT2
    CALL PRNT3
902 CALL KVAL
    JF=F
952 WRITE(12,38) JF
    SUM=SQRT(N+E+I)
    N=SQRT(N)
    R=SQRT(R)
    E=SQRT(E)
    I=SQRT(I)
    WRITE(12,28) N,R,E,I,SUM
382 IF(F-B1) 904,904,10
904 F=F*1.2589254
    GO TO 902
```

```
  982 K1=1
      CALL READ1
      CALL PRNT1
  983 CALL KVAL
      JF=F
      WRITE(12,38) JF
      SUM=SQRT(N+E+I)
      N=SQRT(N)
      E=SQRT(E)
      I=SQRT(I)
      R=SQRT(R)
      WRITE (12,28) N,R,E,I,SUM
      IF(F-B1) 984,984,10
  984 F=F*1.2589254
      GO TO 983
      END

    SUBROUTINE MODEL
    COMMON A0,A1,A2,A3,A4,A5,A6,A7,A8,B,B0,B1,BKO,C1,C2,C3,C6,CKO,E,*
  1 E6,F,F0,F1,F2,F3,F4,F5,F6,F7,F8,FREQ(10),I,I0,I2,I4,I5,I6,I7,K,K0,*
  2 K1,K2,K3,K4,K5,L1,L2,N,N1,N2,N3,N4,N5,N6,N8,N9,NM,P,Q,Q0,Q1,Q2,R,*
  3 R1,R2,R3,R4,R6,R7,S1,S2,S3,S4,S5,SUM,T,T0,W,X1,X2,X3,X4,X5*
    IF(NM-1) 1,2,3
  1 STOP
  3 CALL XISTOR
  5 RETURN
  2 CALL NOSMOD
    GO TO 5
    END

    SUBROUTINE NOSMOD
    COMMON A0,A1,A2,A3,A4,A5,A6,A7,A8,B,B0,B1,BKO,C1,C2,C3,C6,CKO,E,*
  1 E6,F,F0,F1,F2,F3,F4,F5,F6,F7,F8,FREQ(10),I,I0,I2,I4,I5,I6,I7,K,K0,*
  2 K1,K2,K3,K4,K5,L1,L2,N,N1,N2,N3,N4,N5,N6,N8,N9,NM,P,Q,Q0,Q1,Q2,R,*
  3 R1,R2,R3,R4,R6,R7,S1,S2,S3,S4,S5,SUM,T,T0,W,X1,X2,X3,X4,X5*
    SUBROUTINE FOR NOISE CONSTANTS
    REAL I,I6
    E=E6*E6*(((F1/F)+1.)+((F/F2)*(F/F2)))
    I=I6*I6*(((F3/F)+1.)+((F/F4)*(F/F4)))
    RETURN
    END

    SUBROUTINE XISTOR
    COMMON A0,A1,A2,A3,A4,A5,A6,A7,A8,B,B0,B1,BKO,C1,C2,C3,C6,CKO,E,*
  1 E6,F,F0,F1,F2,F3,F4,F5,F6,F7,F8,FREQ(10),I,I0,I2,I4,I5,I6,I7,K,K0,*
  2 K1,K2,K3,K4,K5,L1,L2,N,N1,N2,N3,N4,N5,N6,N8,N9,NM,P,Q,Q0,Q1,Q2,R,*
```

```
3 R1,R2,R3,R4,R6,R7,S1,S2,S3,S4,S5,SUM,T,T0,W,X1,X2,X3,X4,X5*
  REAL I4,N5,N6,I7,I5,I
  X1=4.0*K*T*N5+Q/ (I4*800.)
  I5=(I4/B)+I7
  X2=(2.*Q*F7*(I5**Q1)*N6*N6)/(F**Q2)
  X3=2.0*Q*I4*N5*F*F/(F8*F8)
  E=X1+X2+X3
  X4=2.0*Q*I5+(2.0*Q*F7*(I5**Q1))/(F**Q2)
  X5=2.0*Q*I4*F*F/(F8*F8)
  I=X4+X5
  RETURN
  END

        SUBROUTINE INTEG
        COMMON A0,A1,A2,A3,A4,A5,A6,A7,A8,B,B0,B1,BKO,C1,C2,C3,C6,
        CKO,E,*
      1 E6,F,F0,F1,F2,F3,F4,F5,F6,F7,F8,FREQ(10),I,I0,I2,I4,I5,I6,I7,K,K0,*
      2 K1,K2,K3,K4,K5,L1,L2,N,N1,N2,N3,N4,N5,N6,N8,N9,NM,P,Q,Q0,Q1,
           Q2,R,*
      3 R1,R2,R3,R4,R6,R7,S1,S2,S3,S4,S5,SUM,T,T0,W,X1,X2,X3,X4,X5*
C       INTEGRATION BY SIMPSONS RULE
        DIMENSION FQ(10)
        REAL N2,N3,I0,I,N
   1020 N2=0.0
   1022 READ(11,11)FQ
     11 FORMAT(10F8.0)
        M=1
        J=2
     25 IF(FQ(J))26,26,27
     27 IF(FQ(J)-100.0)12,13,13
     13 IQ=FQ(M)
        JQ=FQ(J)
   1040 WRITE(12,14)IQ,JQ
     14 FORMAT(//I8,4H TO ,I8)
        GO TO 1060
     12 WRITE(12,15)FQ(M),FQ(J)
     15 FORMAT(//F7.2,4H TO ,F7.2)
   1060 Q9=(FQ(J)-FQ(M))/24.0
        N3=0.0
        E0=0.0
        I0=0.0
        R0=0.0
        F=FQ(M)
     17 F=F+Q9
   1065 CALL KVAL
```

```
1070 E0=E0+E*2.0
     I0=I0+I*2.0
     N3=N3+N*2.0
     R0=R0+R*2.0
1085 IF(F-FQ(J)+2.0*Q9)17,17,18
  18 F=FQ(M)
1090 F=F+Q9/2.
  66 CALL KVAL
     E0=E0+E*4.0
     I0=I0+I*4.0
     N3=N3+N*4.0
     R0=R0+R*4.0
     IF(F-FQ(J)+Q9)19,19,20
  19 F=F+Q9
     GO TO 66
  20 F=FQ(M)
     QX=FQ(J)-FQ(M)
  23 CALL KVAL
     E0=E0+E
     I0=I0+I
     N3=N3+N
     R0=R0+R
     IF(F-FQ(J)+2.0*Q9)21,22,22
  21 F=F+QX
     GO TO 23
  22 N2=N2+(E0+I0+R0+N3)*Q9/6.
     XQ=Q9/6.0
     SRN=SQRT(N3*XQ)
     SRR=SQRT(R0*XQ)
     SRE=SQRT(E0*XQ)
     SRI=SQRT(I0*XQ)
     SRSM=SQRT((E0+R0+I0+N3)*XQ)
1160 WRITE(12,24)SRN,SRR,SRE,SRI,SRSM
  24 FORMAT(1H0,5F12.3)
1025 M=M+1
     J=J+1
     IF(J-10)25,25,26
  26 RETURN
     END

     SUBROUTINE AMPLGN
     COMMON A0,A1,A2,A3,A4,A5,A6,A7,A8,B,B0,B1,BKO,C1,C2,C3,C6,CKO,E,*
   1 E6,F,F0,F1,F2,F3,F4,F5,F6,F7,F8,FREQ(10),I,I0,I2,I4,I5,I6,I7,K,KO,*
   2 K1,K2,K3,K4,K5,L1,L2,N,N1,N2,N3,N4,N5,N6,N8,N9,NM,P,Q,Q0,Q1,Q2,R,*
```

```
3 R1,R2,R3,R4,R6,R7,S1,S2,S3,S4,S5,SUM,T,T0,W,X1,X2,X3,X4,X5*
  REAL N8,N9,K2,K5
  A3=F/S2
  A4=F/S1+F/S2
  A5=(A3*A3+1.0)/(A4*A4+1.0)
  A3=F/S3+F/S4
  A4=F/S4
  A6=(A3*A3+1.0)/(A4*A4+1.0)
  W=2.0*P*F
  A7=((F/F0)/SQRT(1.0+(F*F)/(F0*F0)))**N8
  A8=(1.0/SQRT((F*F)/(F5*F5)+1.0))**N9
  A0=F/S5
  A1=1.0-A0*A0
  A2=2.0*A0*(SQRT(1.0+1.0/Q0)-1.0)
  Q0=1.0+(A0*A0/(A1*A1+A2*A2))
  K5=K2*K2*A5*A6*A7*A7*A8*A8*Q0
  RETURN
  END

     SUBROUTINE KVAL
     COMMON A0,A1,A2,A3,A4,A5,A6,A7,A8,B,B0,B1,BKO,C1,C2,C3,C6,
          CKO,E,*
   1 E6,F,F0,F1,F2,F3,F4,F5,F6,F7,F8,FREQ(10),I,I0,I2,I4,I5,I6,I7,K,K0,*
   2 K1,K2,K3,K4,K5,L1,L2,N,N1,N2,N3,N4,N5,N6,N8,N9,NM,P,Q,Q0,Q1,Q2,
          R,*
   3 R1,R2,R3,R4,R6,R7,S1,S2,S3,S4,S5,SUM,T,T0,W,X1,X2,X3,X4,X5*
1280 IF(K1-1)10,1300,1600
1300 CALL MODEL
     K3=1
     CALL SOURCE
     RETURN
  10 STOP
1600 CALL KVAL2
     RETURN
     END
     SUBROUTINE KVAL2
     COMMON A0,A1,A2,A3,A4,A5,A6,A7,A8,B,B0,B1,BKO,C1,C2,C3,C6,CKO,
          E,*
   1 E6,F,F0,F1,F2,F3,F4,F5,F6,F7,F8,FREQ(10),I,I0,I2,I4,I5,I6,I7,K,K0,*
   2 K1,K2,K3,K4,K5,L1,L2,N,N1,N2,N3,N4,N5,N6,N8,N9,NM,P,Q,Q0,Q1,
          Q2,R,*
   3 R1,R2,R3,R4,R6,R7,S1,S2,S3,S4,S5,SUM,T,T0,W,X1,X2,X3,X4,X5*
     REAL I,N,K4,K0,K5
     CALL MODEL
     K3=1
```

```
      CALL SOURCE
      K3=2
      CALL SOURCE
      CALL AMPLGN
 1630 K4=K0*K5
      E=E*K4
      I=I*K4
      N=N*K4
 1670 R=R*K4
      RETURN
      END

      SUBROUTINE READ1
      COMMON A0,A1,A2,A3,A4,A5,A6,A7,A8,B,B0,B1,BKO,C1,C2,C3,C6,CKO,
         E,*
    1 E6,F,F0,F1,F2,F3,F4,F5,F6,F7,F8,FREQ(10),I,I0,I2,I4,I5,I6,I7,K,K0,*
    2 K1,K2,K3,K4,K5,L1,L2,N,N1,N2,N3,N4,N5,N6,N8,N9,NM,P,Q,Q0,Q1,Q2,R,*
    3 R1,R2,R3,R4,R6,R7,S1,S2,S3,S4,S5,SUM,T,T0,W,X1,X2,X3,X4,X5*
      READ(11,1)B0,B1
    1 FORMAT(2F8.0)
      F=B0
      RETURN
      END

      SUBROUTINE SOURCE
      COMMON A0,A1,A2,A3,A4,A5,A6,A7,A8,B,B0,B1,BKO,C1,C2,C3,C6,CKO,
         E,*
    1 E6,F,F0,F1,F2,F3,F4,F5,F6,F7,F8,FREQ(10),I,I0,I2,I4,I5,I6,I7,K,K0,*
    2 K1,K2,K3,K4,K5,L1,L2,N,N1,N2,N3,N4,N5,N6,N8,N9,NM,P,Q,Q0,Q1,Q2,R,*
    3 R1,R2,R3,R4,R6,R7,S1,S2,S3,S4,S5,SUM,T,T0,W,X1,X2,X3,X4,X5*
      GO TO (1,2,3,4,5,6),N4
    1 CALL SORCE1
      GO TO 7
    2 CALL SORCE2
      GO TO 7
    3 CALL SORCE3
      GO TO 7
    4 CALL SORCE4
      GO TO 7
    5 CALL SORCE5
      GO TO 7
    6 CALL SORCE6
    7 RETURN
      END
```

```
      SUBROUTINE SORCE1
      COMMON A0,A1,A2,A3,A4,A5,A6,A7,A8,B,B0,B1,BKO,C1,C2,C3,C6,
           CKO,E,*
    1 E6,F,F0,F1,F2,F3,F4,F5,F6,F7,F8,FREQ(10),I,I0,I2,I4,I5,I6,I7,K,K0,*
    2 K1,K2,K3,K4,K5,L1,L2,N,N1,N2,N3,N4,N5,N6,N8,N9,NM,P,Q,Q0,Q1,
           Q2,R,*
    3 R1,R2,R3,R4,R6,R7,S1,S2,S3,S4,S5,SUM,T,T0,W,X1,X2,X3,X4,X5*
      REAL I,N,K,K0
C     RESISTIVE SOURCE
      W=2.*P*F
      GO TO (20,120,170),K3
   20 A0=1.+(R4/R3)*(1.+(C2/C6))
      A1=W*R4*C2-(1./(W*R3*C6))
      A2=W*R4*(C2+C6)
      A3=(1.+A2*A2)/(W*W*C6*C6)
      E=E*(A0*A0+A1*A1)*1.E18
      I=I*A3*1.E18
      N=4.*K*T*R4*1.E18
      R=(4.*K*T*A3/R3)*1.E18
      RETURN
C     GAIN
  120 R7=R1*R3/(R1+R3)
      A1=R4*(C2+C6)+R7*(C6+C1)
      A2=W*R7*R4*(C1*(C2+C6)+C2*C6)-(1./W)
      K0=(R7*R7*C6*C6)/(A1*A1+A2*A2)
      RETURN
  170 N4=1
      RETURN
      END

      SUBROUTINE SORCE2
      COMMON A0,A1,A2,A3,A4,A5,A6,A7,A8,B,B0,B1,BKO,C1,C2,C3,
           C6,CKO,E,*
    1 E6,F,F0,F1,F2,F3,F4,F5,F6,F7,F8,FREQ(10),I,I0,I2,I4,I5,I6,I7,K,K0,*
    2 K1,K2,K3,K4,K5,L1,L2,N,N1,N2,N3,N4,N5,N6,N8,N9,NM,P,Q,Q0,Q1,
           Q2,R,*
    3 R1,R2,R3,R4,R6,R7,S1,S2,S3,S4,S5,SUM,T,T0,W,X1,X2,X3,X4,X5*
      REAL N,K,I,K0
C     BIASED RESISTANCE SOURCE
      W=2.*P*F
      GO TO (20,120,170),K3
   20 A1=C6*R3*(R2+R4)+R2*R4*(C2+C6)
      A2=W*C2*C6*R2*R3*R4-(R2+R4)/W
      A3=(R4*(C2+C6))/C6
      A4=(R2+R4)/(W*R2*C6)
```

```
      N=4.*K*T*R4*(F6/F+1.)*1.E18
      A5=((R4*R4)+(R4*R4)/(W*W*C6*C6*R3*R3))
      A6=(A1*A1+A2*A2)/(R2*R2*R3*R3*C6*C6)
      I=I*(A3*A3*A4*A4)*1.E18
      R=((4.*K*T/R3)*A3*A3+A4*A4)+(4.*K*T*A5/R2) *1.E18
      E=E*A6*1.E18
      RETURN
C     GAIN
  120 R7=R3*R1/(R3+R1)
      A1=(R2+R4)*(R7*C6+R7*C1)+R2*R4*(C2+C6)
      A2=W*C6*C2*R2*R4*R7-((R2+R4)/W)+W*C1*R2*R4*R7*(C2+C6)
      K0=R2*R2*R7*R7*C6*C6/(A1*A1+A2*A2)
      RETURN
  170 N4=2
      RETURN
      END


      SUBROUTINE SORCE3
      COMMON A0,A1,A2,A3,A4,A5,A6,A7,A8,B,B0,B1,BKO,C1,C2,C3,C6,
         CKO,E,*
    1 E6,F,F0,F1,F2,F3,F4,F5,F6,F7,F8,FREQ(10),I,I0,I2,I4,I5,I6,I7,K,K0,*
    2 K1,K2,K3,K4,K5,L1,L2,N,N1,N2,N3,N4,N5,N6,N8,N9,NM,P,Q,Q0,Q1,
         Q2,R,*
    3 R1,R2,R3,R4,R6,R7,S1,S2,S3,S4,S5,SUM,T,T0,W,X1,X2,X3,X4,X5*
      REAL L2,I,N,K,K
C     COIL  -  RLC SOURCE
      W=2.*P*F
      GO TO (20,115,170),K3
   20 A1=1.-W*W*L2*(C2+C6)-W*W*C6*C2*R4*R3
      A2=W*(C6*(R3+R4)+C2*R4-W*W*C2*C6*L2*R3)
      A3=1.-W*W*L2*(C2+C6)
      A4=W*R4*(C2+C6)
      E=E*1.E18*((A1*A1+A2*A2)/(R3*R3*W*W*C6*C6))
      I=I*1.E18*((A3*A3+A4*A4)/(W*W*C6*C6))
      R=(4.*K*T/R3)*((A3*A3+A4*A4)/(W*W*C6*C6))*1.E18
      N=4.*K*T*R4*1.E18
      RETURN
  115 R7=R1*R3/(R1+R3)
      A1=R7*W*C6
      A2=1.-W*W*L2*(C2+C6)-W*W*R4*R7*(C1*C2+C2*C6+C1*C6)
      A3=W*(R7*(C1+C6)+R4*(C2+C6)-W*C2*C6*L2*R7-W*W*L2*C1*
         R7*(C2+C6))*
      K0=A1*A1/(A2*A2+A3*A3)
      RETURN
```

```
170 N4=3
    RETURN
    END

    SUBROUTINE SORCE4
    COMMON A0,A1,A2,A3,A4,A5,A6,A7,A8,B,B0,B1,BKO,C1,C2,C3,C6,
        CKO,E,*
  1 E6,F,F0,F1,F2,F3,F4,F5,F6,F7,F8,FREQ(10),I,I0,I2,I4,I5,I6,I7,K,K0,*
  2 K1,K2,K3,K4,K5,L1,L2,N,N1,N2,N3,N4,N5,N6,N8,N9,NM,P,Q,Q0,Q1,
    Q2,R,*
  3 R1,R2,R3,R4,R6,R7,S1,S2,S3,S4,S5,SUM,T,T0,W,X1,X2,X3,X4,X5*
    REAL L1,N,I2,K,I,K0
C   BIASED DIODE SENSOR
    W=2.0*P*F
    GO TO (2015,4005,4030),K3
2015 C4=C2+C3
    R5=(R2*R3)/(R2+R3)
    A9=(1./(R5*R5))+((1.-W*W*C4*L1)/(W*L1))**2
    E=E*A9*1.E24
    N=(((4.*K*T)/R2)+(2.*Q*I2*((F6/F)+1.)))*1.E24
    R=(4.*K*T/R3)*1.E24
    I=I*1.E24
    RETURN
4005 C5=C1+C2+C3
    R7=(R1*R2*R3)/(R1*R2+R2*R3+R3*R1)
    A0=(1.-W*W*L1*C5)/(W*L1)
    K0=(1./((1./(R7*R7))+(A0*A0)))
    RETURN
4030 N4=4
    RETURN
    END

    SUBROUTINE SORCE5
    COMMON A0,A1,A2,A3,A4,A5,A6,A7,A8,B,B0,B1,BKO,C1,C2,C3,C6,
        CKO,E,*
  1 E6,F,F0,F1,F2,F3,F4,F5,F6,F7,F8,FREQ(10),I,I0,I2,I4,I5,I6,I7,K,K0,*
  2 K1,K2,K3,K4,K5,L1,L2,N,N1,N2,N3,N4,N5,N6,N8,N9,NM,P,Q,Q0,
        Q1,Q2,R,*
  3 R1,R2,R3,R4,R6,R7,S1,S2,S3,S4,S5,SUM,T,T0,W,X1,X2,X3,X4,X5*
    REAL L1,I,K,N,K
C   TRANSFORMER MODEL
    W=2.*P*F
    GO TO (2015,4105,4160),K3
2015 R9=R3/(T0*T0)
    A0=((R9+R6)*(R4+R6))/(W*L1*R9)
```

```
         A1=(R9+R6+R6+R4)/R9
         A2=(R6*(R4+R6))/(W*L1)
         A3=R6+R6+R4
         E=(E*1.E18*((A0*A0)+(A1*A1)))/(T0*T0)
         I=(I*1.E18*((A2*A2)+(A3*A3)))*T0*T0
         A4=((4.*K*T)/R9)*((A2*A2)+(A3*A3))
         A5=1.+(((R4+R6)*(R4+R6))/(W*W*L1*L1))
         R=((4.*K*T*R6)+(4.*K*T*R6)*A5+A4)*1.E18
         N=4.*K*T*R4*1.E18
         RETURN
C        GAIN
  4105   R5=R4+R6
         R7=(R1*R3)/(T0*T0*(R1+R3))
         A0=W*L1*R7
         A1=R5*(R6+R7)
         A2=W*LI*(R5+R6+R7)
         K0=(A0*A0*T0*T0)/(A1*A1+A2*A2)
         RETURN
  4160   N4=5
         RETURN
         END

         SUBROUTINE SORCE6
         COMMON A0,A1,A2,A3,A4,A5,A6,A7,A8,B,B0,B1,BKO,C1,C2,C3,C6,
            CKO,E,*
       1 E6,F,F0,F1,F2,F3,F4,F5,F6,F7,F8,FREQ(10)I,I0,I2,I4,I5,I6,I7,K,K0,*
       2 K1,K2,K3,K4,K5,L1,L2,N,N1,N2,N3,N4,N5,N6,N8,N9,NM,P,Q,Q0,Q1,
            Q2,R,*
       3 R1,R2,R3,R4,R6,R7,S1,S2,S3,S4,S5,SUM,T,T0,W,X1,X2,X3,X4,X5*
         REAL L1,L2,K,I,K0,N
C        PIEZOELECTRIC SENSOR
         GO TO (20,20,290),K3
      20 A0=W*L2-(1./(W*C2))
         A1=W*L1*R3
         A2=R3*(1.-W*W*L1*C3)
         Z0=SQRT(R4*R4+A0*A0)
         B5=ATAN(A0/R4)
         Z1=SQRT(A2*A2+W*W*L1*L1)
         B6=ATAN((W*L1)/A2)
         Z1=A1/Z1
         B6=90.-B6
         Z2=SQRT(1.+W*W*C1*C1*R1*R1)
         B7=ATAN(W*C1*R1)
         Z2=R1/Z2
         B7=B7*(   -1.)
         GO TO (160,250),K3
```

```
 160 N=4.*K*T*R4*1.E18
     R=(4.*K*T*Z1*Z1/R3)*1.E18
     I=I*Z1*Z1*1.E18
     Z3=((R4+Z1*COS(B6))**2)+((A0+Z1*SIN(B6))**2)
     E=(E*Z3/(Z1*Z1))*1.E18
     RETURN
 250 Z3=Z0*Z1*COS(B5+B6)+Z1*Z2*COS(B6+B7)+Z0*Z2*COS(B5+B7)
     Z4=Z3*Z
     Z4=Z0*Z1*SIN(B5+B6)+Z1*Z2*SIN(B6+B7)+Z0*Z2*SIN(B5+B7)
     K0=Z2*Z2*Z1*Z1/(Z3*Z3+Z4*Z4)
     RETURN
 290 N4=6
     RETURN
     END

  SUBROUTINE PRNT1
  WRITE(12,1)
1 FORMAT(//15H    FREQUENCY//70H    SENSOR IN    LOAD IN
    AMPL EN/ZS    AMPL IN    SUM NOISE  PICOAMPS)
  RETURN
  END

  SUBROUTINE PRNT2
  WRITE(12,2)
2 FORMAT(//21H VOLTAGE SOURCE MODEL//14H    FREQUENCY)
  RETURN
  END

  SUBROUTINE PRNT3
  WRITE(12,3)
3 FORMAT(//71H  SENSOR EN  LOAD EN  AMPL EN AMPL IN*ZS  SUM
  NOISE NANOVOLTS)
  RETURN
  END

   SUBROUTINE PRNT4
   COMMON A0,A1,A2,A3,A4,A5,A6,A7,A8,B,B0,B1,BKO,C1,C2,C3,C6,CKO,
      E,*
 1 E6,F,F0,F1,F2,F3,F4,F5,F6,F7,F8,FREQ(10),I,I0,I2,I4,I5,I6,I7,K,K0,*
 2 K1,K2,K3,K4,K5,L1,L2,N,N1,N2,N3,N4,N5,N6,N8,N9,NM,P,Q,Q0,Q1,Q2,
      R,*
 3 R1,R2,R3,R4,R6,R7,S1,S2,S3,S4,S5,SUM,T,T0,W,X1,X2,X3,X4,X5*
   REAL K0
   BKO=SQRT(K0)
   CKO=4.3429*ALOG(K0)
```

```
      IF=F
      WRITE(12,2)IF,BKO,CKO
    2 FORMAT(//I8,2F8.3)
  382 IF(F-B1+.1*F)3,3,4
    3 F=F*1.2589254
      RETURN
    4 STOP
      END

    SUBROUTINE PRNT5
    WRITE(12,2)
  2 FORMAT(//25H FREQUENCY GAIN DB)
    RETURN
    END
```

ANSWERS TO PROBLEMS

Chapter 1

1. 0.9, 4.0, 18.0 mV
   6.3, 28, 127 mV
   28.5, 127, 567 mV
2. 4 nV
3. (a) 0.77×10^6 maxima/sec
   (b) 1.15×10^7 crossings/sec
4. 49.5×10^3 Hz
5. 56.8 pA
   16.3 μV
6. 3.5 Ω

Chapter 2

2. (a) 1, 0 dB
   (b) 2, 3 dB
   (c) 3, 4.8 dB
3. 3 dB
5. Yes

Chapter 3

1. $E_n^2[1 + R_s(1 - \omega^2 LC)/j\omega L]^2$
3. E_{sh}^2/E_t^2

Chapter 4

3. $r_{b'e} \gg r_{bb'}$
4. $F_{opt} = 2.02$
5. 1.1, 1.12, 1.22, 1.64

Chapter 5

1. 8.5×10^3, 2.8, 9×10^{-3}, 33×10^{-8}

Chapter 6

1. (b) 1
2. 0.6×10^{-15}, 0.6×10^{-25}
3. 1.4×10^{-6}, No
4. 22 mV
5. 36 mV

Chapter 7

1. 1:5, 10 mV
2. $f_L = R_s R_{L'}/2\pi L_p(R_s + R_{L'})$

Chapter 9

2. 6.0, 1.9, 0.6, 1.2, 1.34 μV
3. -23dB, 0.004 μV/V
5. (a) 0.173
   (b) 87 Hz
   (c) $8600\angle 45^\circ$
   (d) 1:60, core loss

Chapter 10

4. 20 nV
6. (a) 3 μF; (b) 150 μF

Chapter 11

4. 0.357, 2.3×10^{-3}, 0.29×10^{-3}

Chapter 13

1. 2 nV, 2 nV, 6 pA, 1.5 pA

Chapter 14

2. 8 dB

INDEX